ICSA Book Series in Statistics

More information about this series at http://www.springer.com/series/13402

Hua He • Pan Wu • Ding-Geng (Din) Chen
Editors

Statistical Causal Inferences and Their Applications in Public Health Research

Springer

Editors
Hua He
Department of Epidemiology
School of Public Health
and Tropical Medicine
Tulane University
New Orleans, LA, USA

Pan Wu
Christiana Care Health System
Value Institute
Newark, DE, USA

Ding-Geng (Din) Chen
School of Social Work and Department
of Biostatistics
University of North Carolina
Chapel Hill, NC, USA

ISSN 2199-0980 ISSN 2199-0999 (electronic)
ICSA Book Series in Statistics
ISBN 978-3-319-41257-3 ISBN 978-3-319-41259-7 (eBook)
DOI 10.1007/978-3-319-41259-7

Library of Congress Control Number: 2016952546

Printed on acid-free paper

This Springer imprint is published by Springer Nature
The registered company is Springer International Publishing AG Switzerland

To my parents, my husband Wan Tang, and my children Yi, Wenwen, Susan, and Jacob, for their eternal love and my eternal gratitude

Hua He, Ph.D.

To my parents, my sister Bei, and my wife Liang, for their love, support, and encouragement

Pan Wu, Ph.D.

To my parents and parents-in-law, who value higher education and hard work, and to my wife, Ke; my son, John D. Chen; and my daughter, Jenny K. Chen, for their love and support

Ding-Geng (Din) Chen, Ph.D.

Preface

This book originated from a series of discussions among the editors when we were all at the University of Rochester, NY, before 2015. At that time, we had a research discussion group under the leadership of Professor Xin M. Tu that met biweekly to discuss the methodological development on statistical causal inferences and their applications to public health data. In this group, we got a closer overview of the principles and methods behind the statistical causal inferences which are needed to be disseminated to aid the further development in the area of public health research. We were convinced that this can be accomplished better through the compilation of a book in this area.

This book compiles and presents new developments in statistical causal inference. Data and computer programs will be publicly available in order for readers to replicate model development and data analysis presented in each chapter so that these new methods can be readily applied by interested readers in their research.

The book strives to bring together experts engaged in causal inference research to present and discuss recent issues in causal inference methodological development as well as applications. The book is timely and has high potential to impact model development and data analyses of causal inference across a wide spectrum of analysts, as well as fostering more research in this direction.

The book consists of four parts which are presented in 15 chapters. Part I includes Chap. 1 with an overview on statistical causal inferences. This chapter introduces the concept of potential outcomes and its application to causal inference as well as the basic concepts, models, and assumptions in causal inference.

Part II discusses propensity score method for causal inference which includes six chapters from Chaps. 2 to 7. Chapter 2 gives an overview of propensity score methods with underlying assumptions for using propensity score, and Chap. 3 addresses causal inference within Dawid's decision-theoretic framework, where studies of "sufficient covariates" and their properties are essential. In addition, this chapter investigates the augmented inverse probability weighted (AIPW) estimator, which is a combination of a response model and a propensity model. It is found that, in the linear regression with homoscedasticity, propensity variable analysis provides exactly the same estimated causal effect as that from multivariate linear regression, for both population and sample. The AIPW estimator has the property of "double robustness," and it is possible to improve the precision given that the propensity model is correctly specified.

As a critical component of propensity score analysis to reduce selection bias, propensity score estimation can only account for observed covariates, and this estimation to unobserved covariates has not been fully understood. Chapter 4 is then designed to introduce a new technique to assess the robustness of propensity score estimation methods to unobserved covariates. A real dataset on substance abuse prevention for high-risk youth is used to illustrate this technique.

Chapter 5 discusses the missing confounder data in propensity score methods for causal inference. It is well known that the propensity score methods, including weighting, matching, or stratification, have been used to control potential confounding effects in observational studies and non-randomized trials to obtain causal effects of treatment or intervention. However, there are few studies to investigate the missing confounder data problem in propensity score estimation which is unique and different from most missing covariate data problem where the goal is parameter estimation. This chapter is then to review and compare existing methods to deal with missing confounder data in propensity score methods and suggest diagnostic checking tools to select a suitable method in practice. In Chap. 6, the focus is turned to the models of propensity scores for different kinds of treatment variables. This chapter gives a thorough discussion of all methods with a comparison between parametric and nonparametric approaches illustrated by a public health dataset. Chapter 7 is to discuss the computational barrier in propensity score in the era of big data with example in optimal pair matching and consequently offer a novel solution by constructing a stratification tree based on exact matching and propensity scores.

Part III is designed for causal inference in randomized clinical studies which includes five chapters from Chaps. 8 to 12. Chapter 8 reviews important aspects of semiparametric theory and empirical processes that arise in causal inference problems with discussions on empirical process theory, which provides powerful tools for understanding the asymptotic behavior of semiparametric estimators that depend on flexible nonparametric estimators of nuisance functions. This chapter concludes by examining related extensions and future directions for work in semiparametric causal inference.

Chapter 9 discusses the structural nested models for cluster-randomized trials for clinical trials and epidemiologic studies. It is known that in clinical trials and epidemiologic studies, adherence to the assigned components is not always perfect. In this chapter, the estimation of causal effect of cluster-level adherence on an individual-level outcome is provided with two different methodologies based on ordinary and weighted structural nested models (SNMs) which are validated by simulation studies. The methods are then applied to a school-based water, sanitation, and hygiene study to estimate the causal effect of increased adherence to intervention components on student absenteeism. In Chap. 10, the causal models for randomized trials with two active treatments and continuous compliance are addressed by first proposing a structural model for the principal effects and then specifying compliance models within each arm of the study. The proposed methodology is illustrated with an analysis of data from a smoking cessation trial.

In Chap. 11, the causal ensembles for evaluating the effect of delayed switch to second-line antiretroviral regimens are proposed to deal with the challenge in randomized clinical trials of delayed switch. The method is applied for cohort studies where decisions to switch to subsequent antiretroviral regimens were left to study participants and their providers as seen from ACTG 5095. Chapter 12 is to introduce a new class of structural functional response models (SFRMs) in causal inference, especially focusing on estimating causal treatment effect in complex intervention design. SFRM is an extended version of existing structural mean models (SMMs) that is widely used in the area of randomized controlled trials to provide optimal solution in estimation of exposure-effect relationship when treatment exposure is imperfect and inconsistent to every individual subject. With a flexible model structure, SFRM is ready to address the limitations of existing approaches in causal inference when the study design contains multiple intervention layers or dynamic intervention layers and capable to offer robust inference with a simple and straightforward algorithm.

Part IV is devoted to the structural equation modeling for mediation analysis which includes three chapters from Chaps. 13 to 15. In Chap. 13, the identification of causal mediation models with an unobserved pretreatment confounder is explored on identifiability of mediation, direct, and indirect effects of treatment on outcome. The mediation effects are represented by a causal mediation model which includes an unobserved confounder, and the direct and indirect effects are represented by the mediation effects. Simulation studies demonstrate satisfactory estimation performance compared to the standard mediation approach. In Chap. 14, the causal mediation analysis with multilevel data and interference is studied since this type of data is a challenge for causal inference using the potential outcomes framework because the number of potential outcomes becomes unmanageable. Then the goal of this chapter is to extend recent developments in causal inference research with multilevel data and violations of the interference assumption to the context of mediation. This book concludes with Chap. 15 to compressively examine the causal mediation analysis using structure equation modeling by taking advantage of its flexibility as a powerful technique for causal mediation analysis.

As a general note, the references for each chapter are at the end of the chapter so that the readers can readily refer to the chapter under discussion. Thus each chapter is self-contained.

We would like to express our gratitude to many individuals. First, thanks go to Professors Xin M. Tu and Wan Tang for leading and organizing the research discussion which led the production of this book. Thanks go to Hannah Bracken, the associate editor in statistics from Springer; to Jeffrey Taub, project coordinator from Springer (http://link.springer.com); and to Professor Jiahua Chen, the coeditor of Springer/ICSA Book Series in Statistics (http://www.springer.com/series/13402), for their professional support of the book. Special thanks are due to the authors of the chapters.

We welcome any comments and suggestions on typos, errors, and future improvements about this book. Please contact Professor Hua He (hhe2@tulane.edu), Pan Wu (PWu@ChristianaCare.org), or Ding-Geng (Din) Chen (DrDG.Chen@gmail.com or dinchen@email.unc.edu).

New Orleans, LA, USA — Hua He, Ph.D.
Newark, DE, USA — Pan Wu, Ph.D.
Chapel Hill, NC, USA — Ding-Geng (Din) Chen, Ph.D.
March 2016

Contents

Contributors

Haiyan Bai Department of Educational & Human Sciences, University of Central Florida, Orlando, FL, USA

Giovanni Berzuini Department of Brain and Behavioural Sciences, University of Pavia, Pavia, Italy

Babette A. Brumback Department of Biostatistics, University of Florida, Gainesville, FL, USA

Ding-Geng Chen School of Social Work & Department of Biostatistics, Gilling School of Global Public Health, University of North Carolina, Chapel Hill, NC, USA

Tian Chen Department of Mathematics and Statistics, University of Toledo, Toledo, OH, USA

Donna L. Coffman The Methodology Center, Pennsylvania State University, University Park, PA, USA

Philip Dawid Statistical Laboratory, University of Cambridge, Cambridge, UK

Matthew C. Freeman Departments of Environmental Health, Epidemiology, and Global Health, Rollins School of Public Health, Emory University, Atlanta, GA, USA

Bo Fu Administrative Data Research Centre for England & Institute of Child Health, University College London, London, UK

Zhi Geng School of Mathematical Sciences, Peking University, Beijing, China

Debashis Ghosh Department of Biostatistics and Informatics, University of Colorado, Aurora, CO, USA

Douglas Gunzler Center for Health Care Research & Policy, MetroHealth Medical Center, Case Western Reserve University, Cleveland, OH, USA

Hui Guo Centre for Biostatistics, School of Health Sciences, The University of Manchester, Manchester, UK

Hua He Department of Epidemiology, School of Public Health & Tropical Medicine, Tulane University, New Orleans, LA, USA

Jiang He Department of Epidemiology, School of Public Health & Tropical Medicine, Tulane University, New Orleans, LA, USA

Ping He School of Mathematical Sciences, Peking University, Beijing, China

Shanjun Helian Department of Biostatistics, University of Florida, Gainesville, FL, USA

Jun Hu College of Basic Science and Information Engineering, Yunnan Agricultural University, Yunnan, China

Brent A. Johnson Department of Biostatistics and Computational Biology, University of Rochester, Rochester, NY, USA

Edward H. Kennedy University of Pennsylvania, Philadelphia, PA, USA

Li Li Eli Lilly and Company, Lilly Corporate Center, Indianapolis, IN, USA

Lin (Laura) Lin Department of Statistics & Actuarial Science, University of Waterloo, Waterloo, ON, Canada

Yan Ma Department of Epidemiology and Biostatistics, The George Washington University, Washington, DC, USA

David P. MacKinnon Department of Psychology, Arizona State University, Tempe, AZ, USA

Nathan Morris Department of Epidemiology and Biostatistics, Case Western Reserve University, Cleveland, OH, USA

Wei Pan Duke University School of Nursing, Durham, NC, USA

Richard Rheingans Chair, Department of Sustainable Development, Appalachian State University, Boone, NC, USA

Jason Roy Center for Clinical Epidemiology and Biostatistics, University of Pennsylvania, Philadelphia, PA, USA

Li Su MRC Biostatistics Unit, University of Cambridge, Cambridge, UK

Wan Tang Department of Biostatistics, School of Public Health & Tropical Medicine, Tulane University, New Orleans, LA, USA

Xin M. Tu Department of Biostatistics and Computational Biology, University of Rochester, Rochester, NY, USA

Pan Wu Value Institute, Christiana Care Health System, Newark, DE, USA

Zhenguo Wu School of Mathematical Sciences, Peking University, Beijing, China

Kai Zhang Department of Statistics and Operations Research, University of North Carolina, Chapel Hill, NC, USA

Xiaohua Douglas Zhang Faculty of Health Sciences, University of Macau, Macau, China

Yeying Zhu Department of Statistics and Actuarial Science, University of Waterloo, Waterloo, ON, Canada

Part I
Overview

Chapter 1
Causal Inference: A Statistical Paradigm for Inferring Causality

Pan Wu, Wan Tang, Tian Chen, Hua He, Douglas Gunzler, and Xin M. Tu

Abstract Inferring causation is one important aim of many research studies across a wide range of disciplines. In this chapter, we will introduce the concept of potential outcomes for its application to causal inference as well as the basic concepts, models, and assumptions in causal inference. An overview of statistical methods for causal inference will be discussed.

1 Introduction

Assessing causal effect is one important aim of many research studies across a wide range of disciplines. Although many statistical models, including the popular regression, strive to provide causal relationships among variables of interest, few

P. Wu (✉)
Value Institute, Christiana Care Health System, Newark, DE 19718, USA
e-mail: PWu@Christianacare.org

W. Tang
Department of Biostatistics, School of Public Health & Tropical Medicine, Tulane University, New Orleans, LA 70112, USA
e-mail: wtang1@tulane.edu

T. Chen
Department of Mathematics and Statistics, University of Toledo, Toledo, OH 43606, USA
e-mail: tian.chen@utoledo.edu

H. He
Department of Epidemiology, School of Public Health & Tropical Medicine, Tulane University, New Orleans, LA 70112, USA
e-mail: Hhe2@tulane.edu

D. Gunzler
Center for Health Care & Policy, MetroHealth Medical Center, Case Western Reserve University, Cleveland, OH 44109, USA
e-mail: dgunzler@metrohealth.org

X.M. Tu (✉)
Department of Biostatistics and Computational Biology, University of Rochester, Rochester, NY 14642, USA
e-mail: Xin_Tu@urmc.rochester.edu

H. He et al. (eds.), *Statistical Causal Inferences and Their Applications in Public Health Research*, ICSA Book Series in Statistics, DOI 10.1007/978-3-319-41259-7_1

can really offer estimates with a causal connotation. A primary reason for such difficulties is confounding, observed or otherwise. Unless such factors, which constitute the source of bias, are all identified and/or controlled for, the observed association cannot be attributed to causation.

For example, if patients in one treatment have a higher rate of recovery from a disease of interest than those in another treatment, we cannot generally conclude that the first treatment is more effective, since the difference could simply be due to different makeups of the groups such as differential disease severity and comorbid conditions. Alternatively, if those in the first treatment group are in better healthcare facilities and/or have easier access to some efficacious adjunctive therapy, we could also see a difference in recovery between the two groups.

An approach widely used to address such bias in epidemiology and clinical trials research is to control for covariates in the analysis. Ideally, if one can find all confounders for the relationship of interest, differences found between treatment and control groups by correctly adjusting for such covariates do represent causal effects. However, as variables collected and our understanding of covariates for relationships of interest in most studies are generally limited, it is inevitable that some residual bias remains due to exclusions of some important confounding variables in the analysis. Without being able to assess the effect of such *hidden* bias, it would still be difficult to interpret findings from such conventional methods. A well-defined concept of causation is needed to assess hidden bias.

Although observational studies are most prone to bias, or *selection* bias as in statistical lingo, randomized controlled trials (RCTs) are not completely immune to confounders. The primary sources of confounders for RCTs are treatment noncompliance and missing follow-ups. Although modern longitudinal models can effectively address the latter issue, the traditional intention-to-treat (ITT) approach based on the treatment assigned rather than eventually received generally fails to deal with the former problem, especially when treatment compliance occurs in multilayered intervention studies, an emerging paradigm for designing research studies that integrate multi-level social support to increase and sustain treatment effects [34].

Another problem of great interest in both experimental and observational studies is the causal mechanism of treatment effect. The ITT and other methods only provide a wholesome view of treatment effect, since they fail to tell us how and why such effects occur. One mechanism of particular interest is *mediation*, a process that describes the pathway from the intervention to the outcome of interest. Causal mediation analysis allows one to ascertain causation for changes of implicated outcomes along such a pathway. Mediation analysis is not only of significant theoretical interest to further our understanding of causal interplays among various outcomes of interest, but also of great practical utility to help develop alternative and potentially more efficient and cost-effective treatment modalities.

In this chapter, we give an overview of the concept of potential outcome and popular methods developed under this paradigm.

2 The Counterfactual Outcome Based Causal Paradigm

Although conceptually straightforward, a formal statistical definition of causation is actually not. This is because one often relies on randomization for the notion of causation. How would one define causation in the absence of randomization? Since randomization is only the means by which to control for confounding, we cannot use it to define causal effect. Rather, we need a more fundamental concept to help explain why randomization can address confounding to achieve causation. This is the role of *potential outcome*.

2.1 *Potential Outcomes*

The concept of *potential outcome*, the underpinnings of modern causal inference paradigm, addresses the fundamental question of causal treatment effect [27]. Under this framework, associated with every patient is an outcome for each treatment condition received, and the treatment effect is the difference between the outcomes in response to the respective treatments from the same subject. Thus, treatment effect is defined for each subject based on a subject's differential responses to different treatments, thereby free of any confounding effect and providing a conceptual basis for causal effect without relying on the notion of randomization.

Under this paradigm, causal effect is defined for each subject by the differences between the *potential* outcomes. With the concept of potential outcome, we can define causal effect without invoking the notion of randomization. For example, consider a study with two treatment conditions, say intervention and control, and let y_{i1} (y_{i0}) denote the outcome of interest from a subject in response to the intervention and control. Then the difference between the two, $\Delta_i = y_{i1} - y_{i0}$, is the causal treatment effect for the subject, since this difference is calculated from the same subject and thus is free of any confounding effect. The potential outcomes are *counterfactual*, since each subject is assigned only one treatment and thus only the one associated with the assigned treatment is observed. The statistical framework of causal effects via the potential outcome is often termed the Rubin's causal model (RCM) [9].

The concept of potential outcome allows us to see why treatment differences observed in randomized control trials (RCT) represent causal effect. Consider again a study with two treatments. Let z_i denote a random binary indicator for treatment assignment and y_{i1} (y_{i0}) denote the potential outcome corresponding to $z_i = 1$ (0). The causal effect for each subject is $\Delta_i = y_{i1} - y_{i0}$, which, unfortunately, is not observable, since only the potential outcome corresponding to the treatment actually received is observed. Thus, the causal treatment, or population-level, effect, $\Delta = E(\Delta_i)$, cannot be estimated by simply averaging the Δ_i's. For an RCT, however, we can estimate Δ by using the usual difference in the sample means between the two treatment conditions.

Let n_1 (n_0) denote the number of subjects assigned to the intervention (control) group and let $n = n_0 + n_1$. If y_{ik} denotes the potential outcome of the ith subject for the kth treatment for the n subjects, we observe y_{ik} if the subject is assigned to the kth treatment condition ($k = 0, 1$). If $y_{i_1 1}$ $\left(y_{j_0 0}\right)$ represents the observed outcome for the i_1th (j_0th) subject in the n_1 (n_0) subjects in the intervention (control) group, we can express the observed potential outcomes for the n subjects as: $y_{i1} = y_{i_1 1}$ with $i = i_1$ for $1 \leq i_1 \leq n_1$ ($y_{i0} = y_{j_0 0}$ with $i = j_0 + n_1$ for $1 \leq j_0 \leq n_0$).

The sample means for the two groups and the difference between the sample means are given by

$$\widehat{\Delta} = \overline{y}_{\cdot 1} - \overline{y}_{\cdot 0}, \quad \overline{y}_{\cdot k} = \frac{1}{n_k} \sum_{i_k=1}^{n_k} y_{i_k k}, \quad k = 0, 1. \tag{1.1}$$

For an RCT, treatment assignment is independent of potential outcome, i.e., $y_{ik} \perp z_i$, where $\perp$ denotes stochastic independence. By applying the law of iterated conditional expectation (Kowalski and Tu 2007), it follows from the independent assignment that

$$E\left(y_{ik}\right) = E\left(y_{ik} \mid z_i = k\right) = E\left(y_{i_k, k}\right), \qquad k = 0, 1. \tag{1.2}$$

It then follows from (1.1) and (1.2) that

$$\begin{aligned} E\left(\widehat{\Delta}\right) &= \frac{1}{n_1} \sum_{i_1=1}^{n_1} E\left(y_{i_1 1}\right) - \frac{1}{n_0} \sum_{j_0=1}^{n_0} E\left(y_{j_0 0}\right) \\ &= E\left(y_{i_1 1}\right) - E\left(y_{j_0 0}\right) \\ &= E\left(y_{i1} \mid z_i = 1\right) - E\left(y_{i0} \mid z_i = 0\right) \\ &= E\left(y_{i1}\right) - E\left(y_{i0}\right) \\ &= \Delta. \end{aligned} \tag{1.3}$$

Thus, the difference between the sample means does estimate the causal treatment effect in the RCT.

The above shows that standard statistical approaches such as the two sample t-test and regression models can be applied to RCTs to infer causal treatment effects. Randomization is key to the transition from the incomputable individual level difference, $y_{i1} - y_{i0}$, to the computable sample means in (1.1) in estimating the average treatment effect. For non-randomized trials such as most epidemiological studies, exposure to treatments or agents may depend on the values of the outcome variable, in which case the difference between the sample means in (1.1) generally does not estimate the average causal effect $\Delta = E\left(y_{i1} - y_{i0}\right)$. Thus, associations found in observational studies generally do not imply causation.

2.2 Selection Bias in Observational Studies

Selection bias is one of the most important confounders in observational studies. Since it is often caused by imbalance in baseline covariates before treatment assignment, it is also called *pre-treatment* confounders. The potential-outcome-based paradigm provides a framework for explicating the effect of selection bias.

Consider an observational study with two treatment conditions and let z_i continue to denote the indicator of treatment assignment. Note that in observational studies, treatment conditions are often called exposure to agents, or exposure conditions. For convenience, we continue to use treatment conditions in the discussion below unless stated otherwise. If treatment assignment is not random, z_i may not be independent of the potential outcome. Thus the condition $y_{ik} \perp z_i$ may not hold true and the identity $E\left(\widehat{\Delta}\right) = \Delta$ in (1.3) may fail, in which case $\widehat{\Delta}$ no longer estimates the causal treatment effect Δ. By considering treatment difference from the perspective of potential outcome, not only can we develop models to address selection bias, but also methods to provide degree of confidence for the causal relationship ascertained.

Note that an approach widely used to address selection bias in epidemiologic research is to include covariates as additional explanatory variables in regression analysis. However, as in the case of explaining causation using randomization, such an approach does not have a theoretical justification, since without the potential-outcome-based framework it is not possible to analytically define selection bias. Another undesirable aspect of the approach is its model dependence, i.e., relying on specific regression models to control for the effect of confounding. For example, a covariate responsible for selection bias may turn out to be statistically insignificant simply because of the use of a wrong statistical model or poor model fit. Most important, despite such adjustments, some residual bias may remain due to our limited understanding of covariates for the relationship of interest and/or the limited covariates collected in most studies. Without being able to assess the effect of such *hidden bias*, it is difficult to interpret findings from such an ad-hoc approach.

2.3 Post-treatment Confounders in Randomized Controlled Studies

In RCTs assignment of treatment is independent of potential outcomes, so standard statistical models such as regression can be applied to provide causal inference. However, this does not mean that such studies are immune to selection bias. In addition to pre-treatment selection bias discussed above, selection bias of another kind, treatment noncompliance and/or informative dropout post randomization, is also quite common in RCTs. For example, if the intervention in an RCT has so many side effects that a large proportion of patients cannot tolerate it long enough to receive the benefit, the ITT analysis is likely to show no treatment effect, even

Fig. 1.1 Causal medication diagram

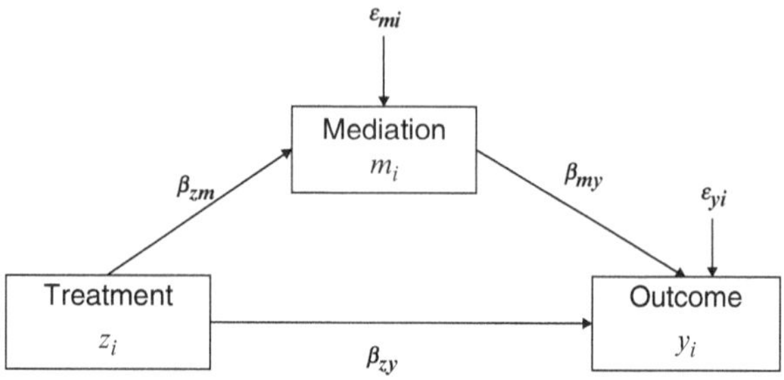

though those who continue with the intervention do benefit. Thus, we must address such downward bias in ITT estimates, if we want to estimate treatment effects for those who are either not affected by or able to tolerate the side effects.

2.4 *Mediation for Treatment Effect*

In many studies, especially those focusing on treatment research, we are also interested in how an intervention achieves its effect upon establishing the efficacy of the intervention. Mediation analysis helps answer such mechanistic questions. For example, a tobacco prevention program may teach participants how to stop taking smoking breaks at work, thereby changing the social norms for tobacco use. The change in social norms in turn reduces cigarette smoking. This mediational process is depicted in Fig. 1.1, where z_i is the indicator of treatment assignment, m_i is the mediator representing social norms, and y_i is the outcome representing tobacco use. By investigating such a mediational process through which the treatment affects study outcomes, not only can we further our understanding of the pathology of the disease and treatment, but we may also develop alternative and better intervention strategies for the disease.

Structural equation models (SEM) are generally used to model mediation effects [2, 3, 15, 17]. The mediation model in Fig. 1.1 illustrates how the treatment achieves its effect on the outcome y_i by first changing the value of the mediator m_i. For a continuous m_i and y_i, the mediation effect is modeled by the following SEM:

$$\begin{aligned} m_i &= \beta_0 + \beta_{zm} z_i + \epsilon_{mi}, \\ y_i &= \beta_1 + \beta_{zy} z_i + \beta_{my} m_i + \epsilon_{yi}, \quad \epsilon_{mi} \perp \epsilon_{yi}. \end{aligned} \tag{1.4}$$

Under the SEM framework, the parameter β_{zy} is interpreted as the *direct effect* of treatment on the outcome y_i, while $\beta_{zm}\beta_{my}$ is interpreted as the *indirect*, or *mediated*, effect of the treatment z_i on the outcome y_i through m_i. Thus, the *total effect* of treatment is viewed as the combination of the direct and indirect effects, $\beta_{zy} + \beta_{zm}\beta_{my}$.

The SEM overcomes the limitations of standard regression models to accommodate variables that serve both as a dependent and independent variable such as the mediator m_i [6, 16]. However, since it is still premised upon the classic modeling paradigm, it falls short of fulfilling the goal of providing causal effects. Causal inference for mediation analysis can also be performed under the paradigm of potential outcomes (see Sect. 3.3.1). Note that the error terms ϵ_{mi} and ϵ_{yi} in (1.4) are assumed independent. This condition, known as *pseudo-isolation* in the SEM literature and *sequential ignorability* in the causal inference literature, is critical not only for ensuring causal interpretation, but also for identifying the SEM in (1.4) as well.

3 Statistical Models for Causal Inference

Selection bias is the most important issue for observational studies. In the presence of such bias, not only models for cross-sectional data such as linear regression, but even models for longitudinal data such as mixed-effects models and structural equation models are wrongly suited for causal inference. Over the last 30 years, many methods have been proposed and a large body of literature has been accumulated to address selection bias in both observational and RCT studies. The prevailing approach is to view unobserved components of potential outcomes as missing data and employ missing data methodology to address associated technical problems within the context of causal inference. Thus, in principle, the goal of causal inference is to model or impute the missing values, or the unobserved potential outcomes, to estimate the average causal effect $\Delta = E(y_{i1} - y_{i0})$, which is not directly estimable using standard statistical methods such as the sample mean, due to the counterfactual nature of the potential outcomes (y_{i1}, y_{i0}).

In practice, these issues are further compounded by missing data, especially those that show consistent patterns such as monotone patterns resulting from study dropouts in longitudinal studies [31]. Various approaches have been developed to address the two types of confounders. These models are largely classified into one of the two broad categories: (1) parametric models and (2) semi-parametric (distribution-free) models. Since the unobserved potential outcome can be treated as missing data, the parametric and non-parametric frameworks both seek to extend standard statistical models for causal inference by treating the latent potential outcome as a missing data problem and applying missing data methods.

If treatment assignment is not random, it may depend on the observed, or missing potential outcome, or both. If the assignment mechanism is completely determined by a set of covariates such as demographic information, medical and mental health history, and indicators of behavioral problems, denoted collectively by a vector of covariates, $\mathbf{x}_i$, then the unobserved potential outcome is independent of treatment assignment once conditioned upon $\mathbf{x}_i$. This assumption, also known as the *missing at random* (MAR) mechanism in the lingo of missing data analysis [28], allows one to estimate the average causal effect $\Delta = E(y_{i1} - y_{i0})$. Thus, by identifying the unobserved potential outcome as a missing data problem, methods for missing data can be applied to develop inference procedures within the current context. For notational brevity and without the loss of generality, we continue to assume the relatively simple setting of two treatment conditions in what follows unless stated otherwise.

3.1 Causal Treatment Effects for Observational Studies

3.1.1 Case–Control Designs

Case–control studies are widely used to ascertain causal relationships in non-randomized studies. In a case–control study on the relationship between some exposure variable of interest such as smoking and disease of interest such as cancer, we first select a sample from a population of diseased subjects, or *cases*. Such a population is usually retrospectively identified by chart-reviews of patients' medical histories. We then select a sample of disease-free individuals, or *controls*, from a non-diseased population, with the same or similar socio-demographic and clinical variables, which are believed to predispose subjects to the disease of interest. Since the cases and controls are closely matched to each other in all predisposed conditions for the disease except for the exposure status, differences between the case and control groups should be attributable to the effect of exposure, or treatment.

We can justify this approach from the perspective of potential outcome. For example, if $y_{i_1 1}$ represents the outcomes from the case group, then the idea of case–control design is to find a control for each case so that the control's response $y_{j_0 0}$ would represent the case's unobserved potential outcome $y_{i_1 0}$. Thus, we may use the difference $y_{i_1 1} - y_{j_0 0}$ as an estimate of the individual-level causal effect, i.e.,

$$y_{i_1 1} - y_{j_0 0} \approx y_{i_1 1} - y_{i_1 0}.$$

Thus the computable sample average, $\overline{\Delta}_{cc} = \frac{1}{n_1} \sum_{i_1=1}^{n_1} y_{i_1 1} - \frac{1}{n_0} \sum_{j_0=1}^{n_0} y_{j_0 1}$, becomes a good approximation of the non-computable average $\overline{\Delta} = \frac{1}{n_1} \sum_{i_1=1}^{n_1} (y_{i_1 1} - y_{i_1 0})$, which is an estimate of the average causal effect Δ.

3.1.2 Matching and Propensity Score Matching

The case–control design reduces selection bias in observational studies by matching subjects in the case and control group based on pre-disposed disease conditions. For the case–control design to work well, we must be able to find good controls for the cases. If $\mathbf{x}_i$ denotes the set of covariates for matching cases and controls, we must pair each case and control with identical or similar covariates. For example, if $\mathbf{x}_i$ consists of age, gender, and patterns of smoking (e.g., frequency and years of smoking), we may try to pair each lung cancer patient with a healthy control, having same gender, same (or similar) age, and smoking patterns. As the dimension of $\mathbf{x}_i$ increases, however, matching subjects with respect to a large number of covariates can be quite difficult.

A popular approach for matching subjects is the *Propensity Score matching* (PS). This approach is premised upon the fact that treatment assignment dictated by $\mathbf{x}_i$ is characterized by the probability of receiving treatment given the covariates $\mathbf{x}_i$ [24, 25], i.e.,

$$\pi_i = \pi(\mathbf{x}_i) = \Pr(z_i = 1 \mid \mathbf{x}_i). \tag{1.5}$$

If $\mathbf{x}_i$ is a vector of covariates such that $(y_{i1}, y_{i0}) \perp z_i \mid \mathbf{x}_i$, then we can show that [25]:

$$\Pr(\mathbf{x}_i \mid z_i = 1, \pi_i) = \Pr(\mathbf{x}_i \mid z_i = 0, \pi_i).$$

The above shows that conditional on π_i, $\mathbf{x}_i$ has the same distribution between the treated ($z_i = 1$) and control ($z_i = 0$) groups. Thus, we can use the one-dimensional Propensity Score in (1.5), rather than the multi-dimensional and multi-type $\mathbf{x}_i$, to match subjects.

For example, we may model π_i using logistic regression. With an estimated $\hat{\pi}_i$, we can partition the sample by grouping together subjects with similar estimated propensity scores to create strata and compare group differences within each stratum using standard methods. We may derive causal effects for the entire sample by weighting and averaging such differences over all strata.

Although convenient to use and applicable to both parametric and semi-parametric models (e.g., the generalized estimating equations), the PS generally lacks desirable properties of formal statistical models such as estimates consistency and asymptotic normality. Another major problem is that in most studies $\mathbf{x}_i$ is only approximately balanced between the treatment groups, after matching or subclassification using the estimated propensity score, especially when the observed covariates $\mathbf{x}_i$ are not homogeneous in the treatment and control groups and/or one or more components of $\mathbf{x}_i$ are continuous. Thus, this approach does not completely remove selection bias [10], although Rosenbaum and Rubin [26] showed through simulations that creating five propensity score subclasses removes at least 90% of the bias in the estimated treatment effect. In addition, since the choice of cutpoint for creating strata using the propensity score is subjective in subclassification methods, different people may partition the sample differently, such as 5–10 for moderate and 10–20 for large sample size, yielding different estimates and even different conclusions, especially when the treatment difference straddles borderline significance. An alternative is to simply use the estimated propensity score as a covariate in standard regression analysis. This implementation is also popular, since it reduces the number of covariates to a single variable, which is especially desirable in studies with relatively small sample sizes. The approach is again ad-hoc and, like the parametric approach discussed above, its validity depends on assumed parametric forms of the covariate effects (typically linear).

3.1.3 Marginal Structural Models

A popular alternative to PS is the *marginal structural model* (MSM; [8, 21]). Like PS, MSM uses the probability of treatment assignment for addressing selection bias. But, unlike PS, it uses the propensity score as a weight, rather than a stratification variable, akin to weighting selected households sampled from a targeted region of interest in survey research [10]. By doing so, not only does the MSM completely remove selection bias, but also yields estimates with nice asymptotic properties. Another nice feature about the MSM is its readiness to address missing data, a common issue in longitudinal study data [8].

Under MSM, we model the potential outcome as

$$E(y_{ik}) = \mu_k = \beta_0 + \beta_1 k, \quad 1 \le i \le n, \quad k = 0, 1. \tag{1.6}$$

Since only one of the potential outcomes (y_{i1}, y_{i0}) is observed, the above model cannot be fit directly using standard statistical methods. If treatment assignment is random, i.e., $y_{ik} \perp z_i$, then $E(y_{ik}) = E(y_{i_k k})$ and thus

$$E(y_{i_k k}) = \beta_0 + \beta_1 k, \quad 1 \le i_k \le n_k, \quad k = 0, 1, \tag{1.7}$$

Thus for the RCT we can estimate the parameters $\boldsymbol{\beta} = (\beta_0, \beta_1)^\top$, including the average causal effect $\Delta = \beta_1$, for the model for the potential outcome in (1.6) by substituting the observed outcomes from the two treatment groups in (1.7). The above is the same argument as in Sect. 2.1, but from the perspective of a regression model.

For observational studies, z_i is generally not independent of y_{ik}. If $\mathbf{x}_i$ is a vector of covariates such that $(y_{i1}, y_{i0}) \perp z_i \mid \mathbf{x}_i$, then we can still estimate $\boldsymbol{\beta}$ by modeling the observed outcomes $y_{i_k k}$ as in (1.7), although we cannot use standard methods to estimate $\boldsymbol{\beta}$ and must construct new estimates. To this end, consider the following weighted estimating equations:

$$\sum_{i=1}^{n} \begin{pmatrix} \frac{z_i}{\pi_i} (y_{i1} - \mu_1) \\ \frac{1-z_i}{1-\pi_i} (y_{i0} - \mu_0) \end{pmatrix} = \mathbf{0}, \tag{1.8}$$

where π_i is defined in Sect. 3.1.2. Although the above involves potential outcomes, the set of equations is well defined. If the ith subject is assigned to the first (second) treatment condition, then $i = i_1$ $(i = j_0 + n_1)$ and $y_{i1} = y_{i_1 1}$ $(y_{i0} = y_{j_0 0})$ for $1 \le i_1 \le n_1$ $(1 \le j_0 \le n_0)$. It follows that

$$\begin{pmatrix} \frac{z_i}{\pi_i} (y_{i1} - \mu_1) \\ \frac{1-z_i}{1-\pi_i} (y_{i0} - \mu_0) \end{pmatrix} = \begin{cases} \begin{pmatrix} \frac{1}{\pi_i} (y_{i_1 1} - \mu_1) \\ 0 \end{pmatrix} & \text{if } z_i = 1 \\ \begin{pmatrix} 0 \\ \frac{1}{1-\pi_i} (y_{i_0 0} - \mu_0) \end{pmatrix} & \text{if } z_i = 0 \end{cases}.$$

Thus the estimating equations in (1.8) are readily computed based on the observed data. Also, the set of estimating equations is unbiased, since

$$\begin{aligned}
E\left(\frac{z_i}{\pi_i} y_{ik}\right) &= E\left[E\left(\frac{z_i}{\pi_i} y_{ik} \mid \mathbf{x}_i\right)\right] \\
&= E\left[\frac{1}{\pi_i} E\left(\frac{z_i}{\pi_i} y_{ik} \mid \mathbf{x}_i\right)\right] \\
&= E\left[\frac{1}{\pi_i} E\left(z_i \mid \mathbf{x}_i\right) E\left(y_{ik} \mid \mathbf{x}_i\right)\right] \\
&= E\left[E\left(y_{ik} \mid \mathbf{x}_i\right)\right] \\
&= \mu_k.
\end{aligned}$$

Thus, by the theory of estimating equations (Kowalski and Tu 2007), estimates of $\boldsymbol{\beta}$ obtained by solving the estimating equations in (1.8) are consistent.

3.2 *Post-treatment Confounders in Randomized Controlled Trials*

The intention-to-treat (ITT) analysis compares the outcomes of subjects by randomized groups, ignoring treatment compliance and other deviations of study protocols. As a result, the ITT yields the effect of treatment confounded by all such violations. Despite being protected from pre-treatment selection bias through randomization, ITT estimates of treatment effect are typically downwardly biased, because of the "diluted" effect by post-treatment bias due to treatment noncompliance and/or missing data.

3.2.1 Instrumental Variable Estimate

One way to address treatment noncompliance is to partition study subjects into different types based on their impacts on causal treatment effects and then characterize the causal effect for each of the types of treatment noncompliance [1, 13]. One approach that has been extensively discussed in the literature is a partition of the study sample into four types in terms of their compliance behavior:

1. Complier (CP): subjects compliant with assigned treatment (control or intervention);
2. Never-taker (NT): subjects who would take the control treatment regardless of what they are assigned;
3. Always-taker (AT): subjects who would take the intervention regardless of what they are assigned;
4. Defiers (DF): subjects who would take the opposite treatment to their assignment.

In practice, the DF generally represents a small proportion of the noncompliant group.

For the AT and NT group, $\Delta_i = y_{i1} - y_{i0} = 0$. Neither group contributes to causal effect. For DF, Δ_i is in the opposite direction of causal effect. Thus, only the CP subsample provides information for causal effect. Let $C_i = 1$ (0) if the ith subject is in the CP (otherwise). The causal effect for the CP group is

$$\Delta_{\mathrm{CP}} = E\left(y_{i1} - y_{i0} \mid C_i = 1\right). \tag{1.9}$$

The above is called the *Complier Average Causal Effect* (CACE). In contrast, the ITT effect is given by: $\Delta_{\mathrm{ITT}} = E\left(y_{i1} - y_{i0}\right)$.

If C_i is observed for each subject, then we have

$$\begin{aligned} \Delta_C &= E\left(y_{i1} \mid C_i = 1, z_i = 1\right) - E\left(y_{i0} \mid C_i = 1, z_i = 0\right) \\ &= E\left(y_{i_1 1} \mid C_{i_1} = 1\right) - E\left(y_{i_0 0} \mid C_{i_0} = 1\right), \end{aligned} \tag{1.10}$$

where C_{i_k} denotes the complier's status for the i_kth subject in the kth treatment group ($k = 0, 1$). We can then estimate $E\left(y_{i_k k} \mid C_{i_k} = 1\right)$ based on the Complier's subsample within the kth treatment condition using standard methods such as the sample mean.

In practice, we can only observe compliance status D_{i_k} for the assigned treatment condition. Although similar, D_{i_k} is generally different from C_{i_k}. For example, $D_{i_1} = 1$ includes both the CP and AT subsamples within the treated, while $D_{i_0} = 1$ includes the CP + NT subsample within the control condition. By conditioning on D_{i_k}, we can estimate

$$\Delta_D = E\left(y_{i_1 1} \mid D_{i_1} = 1\right) - E\left(y_{i_0 0} \mid D_{i_0} = 1\right).$$

However, as noted earlier, $\Delta_C \neq \Delta_D$ unless there are no AT nor NT subsample in the study population.

Let $p_1 = E\left(D_{i_1} = 1\right)$ and $p_0 = E\left(D_{i_0} = 0\right)$. Then p_1 represents the proportion of CP+AT in the intervention, while p_0 represents the proportion of AT+DF in the control condition. If we assume no DF, then p_0 becomes the proportion of AT and $p_1 - p_0$ represents the proportion of the CP group. Thus, we can express (1.9) as

$$\Delta_{\mathrm{CP}} = \frac{E\left(y_{i1} - y_{i0}\right)}{p_1 - p_0} = \frac{\Delta_{\mathrm{ITT}}}{p_1 - p_0}. \tag{1.11}$$

In other words, we can estimate the CACE by modifying the ITT estimate:

$$\widehat{\Delta}_{\mathrm{CP}} = \frac{\widehat{\Delta}_{\mathrm{ITT}}}{\widehat{p}_1 - \widehat{p}_0} = \frac{\overline{y}_{\cdot 1} - \overline{y}_{\cdot 0}}{\widehat{p}_1 - \widehat{p}_0}, \quad \widehat{p}_k = \frac{1}{n_k} \sum_{i_k=1}^{n_k} D_{i_k}, \quad k = 0, 1.$$

The identity in (1.11) depends critically on the assumption of randomization. This is because to ensure that p_k has the aforementioned interpretation, we must have

$$p_k = E(d_{ik}) = \Pr(d_{ik} = 1, z_i = k),$$

which is only guaranteed under random treatment assignment. Because of the critical role played by z_i in identifying the CP in the presence of confounding by the AT and NT subsamples, z_i is called an instrumental variable (IV) and $\widehat{\Delta}_{CP}$ is known as the IV estimate of CACE [1].

3.2.2 Principal Stratification

The IV method is limited to binary compliance variables. A notable limitation of the IV is that its estimated treatment effect only pertains to a subgroup of compliers in the study population. In most real studies, compliance varies over a range of patterns. One popular approach for allowing for graded levels of treatment compliance is the *Principal Stratification* (PST). The PST creates Principal Strata based on similar treatment compliance patterns and estimates causal effects within each Principal Stratum [4]. In the special case of IV classification of noncompliance, PST provides estimates of treatment effect for each of the four groups, albeit only CP is of primary interest. By creating graded treatment compliance categories, PST provides a more granular relationship between exposure and treatment effects.

Let s_{ik} denote a categorical outcome that indicates levels of treatment compliance for the kth treatment condition and $\mathbf{s}_i = (s_{i1}, s_{i0})^\top$. The basic principal stratification P_0 is the set of distinct $\mathbf{s}_i$, i.e., $P_0 = \{p_l; 1 \le l \le L\}$, where L is the total number of principal strata and p_l is a collection of $\mathbf{s}_i$ such that $\mathbf{s}_i = \mathbf{s}_j$ for $\mathbf{s}_i, \mathbf{s}_j \in p_l$, but $\mathbf{s}_i \neq \mathbf{s}_j$ for $\mathbf{s}_i \in p_l$, $\mathbf{s}_j \in p_m$ $(l \neq m)$. A principal stratification P is a collection of sets that are unions of sets in the basic principal stratification P_0. Thus, P is a coarser grouping of the distinct $\mathbf{s}_i$.

Consider, for example,

$$s_{ik} = \begin{cases} 1 \text{ if compliant} \\ 0 \text{ if noncompliant} \end{cases}.$$

For each subject, the potential outcome of noncompliance status $\mathbf{s}_i = (s_{i1}, s_{i0})^\top$ has four patterns, which constitutes the basic principal stratification:

$$P_0 = \{(1,1), (0,1), (1,0), (0,0)\}.$$

The four distinct patterns represent the CP $(1,1)$, the DF $(0,0)$, the AT $(1,0)$, and NT $(0,1)$ subsamples under the IV classification of treatment noncompliance. By combining some of the patterns in the basic principal stratification P_0, we can create principal stratification P to represent noncompliance patterns of interest. For example, the principal stratification $P = \{\{(1,1) \text{ or } (1,0)\}, (0,1), (0,0)\}$ no longer distinguishes between the CP and AT.

Once we establish an appropriate choice of principal stratification P, we can compare the potential outcome y_{i1} and y_{i0} within each P to define the causal effect of interest:

$$\Delta_l = E\left(y_{i1} - y_{i0} \mid \mathbf{s}_l\right), \ 1 \leq l \leq L.$$

The goal is to estimate Δ_l for each lth stratum. We may also create weighted averages to obtain overall treatment effects of interest. Inference about $\boldsymbol{\theta} = \{\Delta_l; 1 \leq l \leq L\}$ can be based on maximum likelihood or Bayesian methods [4].

In the special case of IV categorization, the PST provides more information about the relationship between noncompliance and treatment effects than the IV. In addition to the CP, PST also provides treatment effects for the AT, NT, or even the DF group.

3.2.3 Structural Mean Models

In most studies, there exists a large amount of variability in treatment noncompliance. For example, in a medication vs. placebo study, if the medication is prescribed daily for 2 weeks, exposure to medication can range from 0 to 14 days. We may group medication dosage using a graded categorical variable and apply the PST to characterize a dose–response relationship in this case. However, since this or any other grouping of the dosage variable is subjective, we may want to use the original number of days of medication use directly to more objectively characterize the dose–response relationship. Unfortunately, this will immediately increase the number of principal strata and may not provide reliable inference or the PST may simply stop working, if there is not a sufficient number of subjects within every stratum. A more sensible approach is to treat such a continuous-like treatment compliance measure as a continuous variable to study treatment effect.

In many treatment research studies, active treatments are only available to study participants. In this case, active treatment is not available to the DF and AT subsample in the control condition, in which case causal treatment effect is determined by the AT+CP subsample in the treatment group. This allows us to model treatment effect as a function of continuous dose variables.

Let s_{ik} denote a continuous compliance variable for the ith subject in the kth treatment with $k = 1$ (0) for the active treatment (control) condition. For convenience and without the loss of generality, assume that $s_{ik} \geq 0$ with 0 representing zero dose. Since the active treatment is not available to subjects in the control condition, $s_{i0} \equiv 0$ and thus s_{i0} provides no information about causal treatment effect. Thus we may model the causal effect as a function of s_{i1} only:

$$E\left(y_{i1} - y_{i0} \mid s_{i1}\right) = g\left(s_{i1}, \boldsymbol{\beta}\right). \tag{1.12}$$

where $g(s_{i1}, \boldsymbol{\beta})$ is some continuous function of s_{i1} and $\boldsymbol{\beta}$. Since $(y_{i1}, y_{i0}) \perp z_i$ for randomized studies, it follows that

$$\begin{aligned} \Delta_i(s_{i1}) &= E(y_{i1} \mid s_{i1}) - E(y_{i0} \mid s_{i1}) \\ &= E(y_{i1} \mid s_{i1}, z_i = 1) - E(y_{i0} \mid s_{i1}, z_i = 0) \\ &= E(y_{i_1 1} \mid s_{i_1 1}) - E(y_{i0} \mid s_{i1}, z_i = 0), \end{aligned} \tag{1.13}$$

where i_1 again indexes the subjects assigned to the treatment group and $y_{i_1 1}$ is the observed outcome of the subject in the treatment group. The model in (1.12) is the *Structural Mean Model* (SMM) [20].

To estimate $\Delta_i(s_{i1})$, we must evaluate $E(y_{i0} \mid s_{i1}, z_i = 0)$ so that it can be estimated with observed data. If s_{i1} is independent of y_{i0}, then we have

$$\begin{aligned} E(y_{i0} \mid s_{i1}, z_i = 0) &= E(y_{i0} \mid z_i = 0) \\ &= E(y_{i_0 0}) = \beta_0. \end{aligned} \tag{1.14}$$

This *compliance non-selective* assumption is reasonable, if, for example, s_{i1} does not correlate with disease severity. In this case, (1.13) reduces to

$$\begin{aligned} \Delta_i(s_{i1}) &= E(y_{i1} - y_{i0} \mid s_{i1}) = E(y_{i_1 1} \mid s_{i_1 1}) - E(y_{i0}) \\ &= E(y_{i_1 1} \mid s_{i_1 1}) - \beta_0. \end{aligned} \tag{1.15}$$

It then follows from (1.12)–(1.14) that

$$\begin{aligned} &E(y_{i_0 0}) = \beta_0, \ E(y_{i_1 1} \mid s_{i_1 1}) = g(s_{i_1 1}, \boldsymbol{\beta}) + \beta_0, \\ &\quad 1 \le i_k \le n_k, \ k = 0, 1, \ n = n_0 + n_1. \end{aligned} \tag{1.16}$$

Given a specific form of $g(s_{i1}, \boldsymbol{\beta})$, the SMM in (1.16) allows one to model and estimate treatment effects for continuous dose variables.

For example, if $g(s_{i1}, \boldsymbol{\beta}) = s_{i1}\beta_1$, the SMM has the form:

$$\begin{aligned} &E(y_{i_0 0}) = \beta_0, \ E(y_{i_1 1} \mid s_{i_1 1}) = \beta_0 + s_{i_1 1}\beta_1, \\ &\quad 1 \le i_k \le n_k, \ k = 0, 1, \end{aligned}$$

or equivalently,

$$\begin{aligned} E(y_{i_k k} \mid s_{i_1}) &= \beta_0 + s_{i_1} z_{i_k} \beta_1, \\ z_{i_k} &= k, \ 1 \le i_k \le n_k, \ k = 0, 1. \end{aligned}$$

Note that although s_{i_1} is missing for the control group, the above is still well defined, since $s_{i_1} z_{i_0} \equiv 0$ for all $1 \le i_0 \le n_0$.

In many studies, we may collect sufficient information, say $\mathbf{x}_i$, to explain the compliance behavior s_{i1}. In this case, we have

$$E\left(y_{i0} \mid s_{i1}, \mathbf{x}_i, z_i = 0\right) = E\left(y_{i0} \mid \mathbf{x}_i, z_i = 0\right) = E\left(y_{i_0 0} \mid \mathbf{x}_{i_0}, z_{i_0} = 0\right). \quad (1.17)$$

Under this *compliance explainable* condition [5, 32, 34], the SMM can be expressed as

$$E\left(y_{i_1 1} \mid s_{i_1 1}, \mathbf{x}_{i_1}, z_{i_1} = 1\right) = g\left(s_{i_1 1}, \mathbf{x}_{i_1}, \boldsymbol{\beta}\right) + E\left(y_{i_0 0} \mid \mathbf{x}_{i_0}, z_{i_0} = 0\right),$$
$$z_{i_k} = k, \ 1 \leq i_k \leq n_k, \ k = 0, 1.$$

In medication vs. placebo studies, if treatment compliance is also tracked for the placebo group, then it is reasonable to assume that the variable of placebo use, d_{i_0}, explains treatment compliance, if the subject is assigned to the medication group. This is because under randomization subjects cannot distinguish between medication and placebo. Thus, if we let $x_{i_0} = d_{i_0}$, then it follows that

$$E\left(y_{i_0 0} \mid \mathbf{x}_{i_0}, z_{i_0} = 0\right) = E\left(y_{i_0 0} \mid d_{i_0}, z_{i_0} = 0\right).$$

We can readily model the right-side of the above.

For example, if we model both $E\left(y_{i_0 0} \mid d_{i_0}, z_{i_0} = 0\right)$ and $g\left(s_{i_1 1}, d_{i_1}, \boldsymbol{\beta}\right)$ as a linear function, we have

$$E\left(y_{i_0 0} \mid d_{i_0}, z_{i_0} = 0\right) = \beta_0 + d_{i_0}\beta_1,$$
$$E\left(y_{i_1 1} \mid s_{i_1 1}, d_{i_1}, z_{i_1} = 1\right) = s_{i_1 1}\beta_2, \quad 1 \leq i_k \leq n_k, \ n_1 + n_0 = n.$$

We may also express the above in a compact form as

$$E\left(y_{i_k 1} \mid s_{i_1 1}, x_{i_k}, z_{i_k} = 1\right) = \beta_0 + d_{i_0}\beta_1 + z_{i_k} s_{i_1 1}\beta_2,$$
$$1 \leq i_k \leq n_k, \ n_1 + n_0 = n.$$

As before, the above is still well defined, even if s_{i_1} is missing for the control group.

In psychosocial intervention studies, the control condition offers either nothing or sessions that provide information unrelated to the intervention, such as attention or information control. In the latter case, compliance (with respect to the attention or information control) may also be tracked. However, such a dose variable, d_{i_0}, generally does not explain treatment compliance, if the subject is assigned to the intervention group, since the information disseminated through the control condition may have nothing to do with the information provided by the intervention condition. For example, in a HIV prevention intervention study for teenage girls at high risk for HIV infection, the intervention condition contains information on HIV infection, condom use and safe sex, while the control condition contains nutritional and dietary information. Thus, subjects with high compliance in the intervention may be quite different from their counterparts in the control group. This may happen if a majority of girls with high attendance in the intervention group are sexually active, while those with high attendance in the control group are more interested in the information on weight loss and healthy diets. Thus $\mathbf{x}_i$ in this study should contain variables that help explain behaviors of compliance for the intervention such as risks for unsafe sex, alcohol and drug use, and HIV knowledge.

3.3 Mechanisms of Treatment Effects

Understanding the causal pathways of treatment effect is critically important, since identification of causal mechanism not only furthers our understanding of behavioral and health issues of interest, but also allows one to develop alternative and potentially more effective and efficient intervention/prevention strategies. A popular approach for causal mechanism is mediation analysis.

3.3.1 Causal Mediation

In recent years, there has been heightened activities to develop models for causal mediation effect under the counterfactual outcome framework (e.g., [11, 12, 19, 22, 23, 33]). We give a brief review of relevant methods, focusing on the identifiability assumptions and definitions of indirect, or mediated, effect.

Let m_{ik} denote the potential outcome of a mediator, m_i, for the ith subject corresponding to the kth treatment. The potential outcome of the primary variable of interest is more complex to allow one to tease out the *direct* and *mediation* causal effects of the intervention or exposure on this variable (see the definition of direct and mediation causal effect below). Let $y_i\,(k, m_{ik'})$ denote the potential outcome of the variable of interest y_i corresponding to the kth treatment condition and mediator $m_{ik'}$ $(k, k' = 0, 1)$. Note that in practice we can only observe m_{ik} and $y_i\,(k, m_{ik})$ ($m_{ik'}$ and $y_i\,(k', m_{ik'})$), if the ith subject is assigned to the kth (k'th) treatment $(k, k' = 0, 1)$. But, in order to tease out the direct and mediation effects, we must consider $y_i\,(k, m_{ik'})$, which is not observed if $k \neq k$ [7, 19].

The *direct effect* of treatment is the effect of treatment, i.e.,

$$\zeta_i(k) = y_i\,(1, m_{ik}) - y_i\,(0, m_{ik})\,, \quad \text{for } k = 0, 1.$$

This quantity $\zeta_i(k)$ is also called the *natural direct effect* (e.g., [19]) or the *pure (total) direct effect* (e.g., [22]) corresponding to $k = 0$ (1). In addition, there is also the so-called *controlled direct effect*, $y_i(1, m) - y_i(0, m)$, which may be viewed as the treatment effect that would have been realized, had the mediator m_{ik} been controlled at level m uniformly in the population [19, 22, 23]. Note that $\zeta_i(1)$ is generally not the same as $\zeta_i(0)$ and the difference represents interaction between treatment assignment and the mediator.

The causal *mediation*, or *indirect effect, or natural indirect effect*, is the difference between the two potential outcomes, $y_i(k, m_{i1})$ and $y_i(k, m_{i0})$, of the variable of interest resulting from the two potential outcomes of the mediator, m_{i1} and m_{i0}, corresponding to the two treatment conditions $k = 1$ and $k = 0$, i.e.,

$$\delta_i(k) = y_i(k, m_{i1}) - y_i(k, m_{i0}), \quad \text{for } k = 0, 1. \tag{1.18}$$

If the treatment has no effect on the mediator, that is $m_{i1} - m_{i0} = 0$, then the causal mediation effect is zero. The quantity $\delta_i(0)$ $(\delta_i(1))$ is also referred to as the *pure indirect effect* (*total indirect effect*) [22]. As in the case of direct effect, $\delta_i(1)$ is generally different from $\delta_i(0)$.

The *total effect* of treatment is the sum of the direct and mediation effect:

$$\begin{aligned}\tau_i &= y_i(1, m_{i1}) - y_i(0, m_{i0}) = \delta_i(1) + \zeta_i(0) \\ &= \frac{1}{2}\left[\delta_i(1) + \zeta_i(1) + \delta_i(0) + \zeta_i(0)\right].\end{aligned}$$

If we assume no interaction between treatment assignment and the mediator, then $\delta_i(1) = \delta_i(0) = \delta_i$ and $\zeta_i(1) = \zeta_i(0) = \zeta_i$. The total effect of treatment in this case is simply the sum of mediation and direct effect, i.e., $\tau_i = \delta_i + \zeta_i$.

In mediation analysis, we are interested in the Average Causal Mediation Effect (ACME), $E(\delta_i(k))$, the average direct effect, $E(\zeta_i(k))$, and the average total effect, $E(\tau_i) = \frac{1}{2}\sum_{k=0,1}\left[E(\delta_i(k)) + E(\zeta_i(k))\right]$. Under no mediator by treatment assignment interaction, the average total effect reduces to $E(\tau_i) = E(\delta_i) + E(\zeta_i)$.

3.3.2 Sequential Ignorability and Model Identification

As noted in Sect. 2.4, the independence between the error terms in the SEM (1.4) plays a critical role in the causal interpretation of the mediation model. This pseudo-isolation condition plays a critical for the identifiability of the parameters of the SEM in (1.4). The issue of identifiability has also been discussed under the potential-outcome based inference paradigm [11, 22]. For example, Imai et al. [11] has shown that if $\mathbf{x}_i$ is a vector of pre-treatment covariates for the ith subject, then

$$\begin{aligned} z_i &\perp \{y_i(k', m), m_{ik}\} \mid \mathbf{x}_i = \mathbf{x}, \quad k, k' = 0, 1, \\ y_i(k', m) &\perp m_{ik} \mid z_i = k, \quad \mathbf{x}_i = \mathbf{x}, \quad k, k' = 0, 1. \end{aligned} \tag{1.19}$$

The above is called *sequential ignorability* (SI) because the first condition indicates that z_i is ignorable given the pre-treatment covariates $\mathbf{x}_i$, while the second states that the mediator m_{ik} is ignorable given $\mathbf{x}_i$ and the observed treatment assignment z_i. Although the first is satisfied by all randomized trials, the second is not. In fact, the second condition of the SI cannot be directly tested from observed data [18]. Thus, sensitivity analysis is usually carried out to examine the robustness of findings under violations of the second ignorability assumption [11].

Other assumptions have also been proposed. For example, Robins [22] proposed the following condition for the identification of *controlled direct effect*:

$$z_i \perp \{y_i(k', m), m_{ik}\} \mid \mathbf{x}_i = \mathbf{x}, \tag{1.20}$$
$$y_i(k, m) \perp m_{ik} \mid z_i = k, \quad \mathbf{x}_i = \mathbf{x}, \quad \mathbf{w}_i = \mathbf{w},$$

where $\mathbf{w}_i$ is another set of observed post-treatment variables that confound the relationship between the mediator and the outcome. Under the more stringent assumptions, the following assumption is a necessary condition for identifying the ACME [22]:

$$y_i(1, m) - y_i(0, m) = B_i,$$

where B_i is a random variable independent of m. This is the so-called non-interaction assumption, which states that the controlled direct effect of treatment does not depend on the value of the mediator.

3.3.3 Models for Causal Mediation Effect

Under the SI in (1.19), it can be shown that the ACME can be nonparametrically identified for $k = 0, 1$ [11]. Since the conditions in the SI imply $y_i(k', m) \perp z_i \mid m_{ik} = m', \mathbf{x}_i = \mathbf{x}$, it follows that for any k and k':

$$E\left(y_i\left(k, m_{ik'}\right) \mid \mathbf{x}_i\right) = \int E(y_i \mid z_i = k, m, \mathbf{x}_i) dF_{m_i \mid z_i = k', \mathbf{x}_i}(m), \tag{1.21}$$

where $F_T(\cdot)$ $\left(F_{T \mid W}(\cdot)\right)$ denotes the (conditional) cumulative distribution function (CDF) of a random variable T (T given W). We may further integrate out $\mathbf{x}_i$ to obtain the unconditional mean:

$$E\left[y_i\left(k, m_{ik'}\right)\right] = \int E\left(y_i\left(k, m_{ik'}\right) \mid \mathbf{x}_i\right) dF_{\mathbf{x}_i}(\mathbf{x}).$$

By using (1.21), we can derive direct, indirect, and total effects for the Linear SEM (LSEM) in (1.4) as well as the Generalized Linear Structural Equation Models (GLSEM), where m_i or y_i or both may be non-continuous variables. For example, by expressing the LSEM in (1.4) using the potential outcomes, we have

$$m_i(z_i) = \alpha_1 + \beta_1 z_i + \epsilon_{i1}(z_i), \tag{1.22}$$
$$y_i\left(z_i, m_i\left(z_i\right)\right) = \alpha_2 + \beta_2 m_i(z_i) + \gamma z_i + \epsilon_{i2}(z_i, m_i(z_i)),$$

Note that to indicate the dependence of the potential outcome of the mediator m_i as a function of treatment assignment, we use $m_i(z_i)$, rather than m_{ik}, in the LSEM in (1.22). The first condition in (1.19) implies

$$E\left[\epsilon_{i1}\left(z_i\right) \mid z_i = k\right] = E\left[\epsilon_{i1}\left(z_i\right)\right] = 0,$$

while the second indicates that

$$E\left[\epsilon_{i2}\left(z_i, m_i\left(z_i\right)\right) \mid m_i = m, z_i = k\right] = E\left[\epsilon_{i2}\left(k, m\right)\right] = 0.$$

It then follows that

$$\begin{aligned} E\left[y_i\left(k, m_{ik'}\right)\right] &= E_{m_i|z_i=k'}\left[E_{y_i}(y_i \mid m_i = m, z_i = k)\right] \\ &= E_{m_i|z_i=k'}\left[\alpha_2 + \beta_2 m_i + \gamma E(z_i = k)\right] \\ &= \alpha_2 + \beta_2 E\left[(\alpha_1 + \beta_1 z_i) \mid z_i = k'\right] + \gamma E(z_i = k). \end{aligned}$$

The ACME for the LSEM in (1.22) is

$$\begin{aligned} E(\delta\left(k\right)) &= E\left[y_i\left(k, m_i(1)\right)\right] - E\left[y_i\left(k, m_i(0)\right)\right] \\ &= \left[\alpha_2 + \beta_2\left(\alpha_1 + \beta_1 E\left(z_i \mid z_i = 1\right)\right) + \gamma E(z_i = k)\right] \\ &\quad - \left[\alpha_2 + \beta_2\left(\alpha_1 + \beta_1 E\left(z_i \mid z_i = 0\right)\right) + \gamma E(z_i = k)\right] \\ &= \beta_2 \beta_1. \end{aligned}$$

Thus, under no mediator by treatment assignment interaction, the mediated effect is $E\left(\delta_i(1)\right) = E\left(\delta_i\left(0\right)\right) = \beta_2 \beta_1$, which is identical to the indirect effect derived from the classic LSEM in (1.4) [2].

We can also obtain the different causal effect if there is no mediator by treatment assignment interaction. For example, if we assume an interaction of the form, $z_i m_i$, i.e.,

$$\begin{aligned} m_i(z_i) &= \alpha_1 + \beta_1 z_i + \epsilon_{i1}\left(z_i\right), \\ y_i\left(z_i, m_i\left(z_i\right)\right) &= \alpha_2 + \beta_2 m_i + \gamma z_i + \eta z_i m_i + \epsilon_{i2}(z_i, m_i(z_i)), \end{aligned} \tag{1.23}$$

then by using arguments similar to the non-interaction case above, we obtain

$$\begin{aligned} E\left(\zeta_i(k)\right) &= \gamma + \eta(\alpha_1 + \beta_1 k), \\ E\left(\delta_i(k)\right) &= \beta_1(\beta_2 + k\eta), \\ E(\tau_i) &= \beta_2\beta_1 + \gamma + \eta(\alpha_1 + \beta_1), \quad k, k' = 0, 1, \end{aligned} \tag{1.24}$$

for the indirect (mediation), direct and total causal effect. These effects are again consistent with those derived from the classic LSEM approach [15].

The identification of ACME can be extended to the GLSEM. For example, if the mediator m_i is binary, but the outcome y_i is continuous, and m_i is modeled as: $E(m_i(z_i)) = \text{logit}^{-1}(\alpha_1 + \beta_1 z_i)$, where logit^{-1} denotes the inverse of the logit link function, then under no mediator by treatment assignment interaction it follows from (1.21) that the ACME can be expressed as

$$\begin{aligned} E(\delta_i(k)) &= E[y_i(k, m_{i1}) - y_i(k, m_{i0})] \\ &= \beta_2 \left[\text{logit}^{-1}(\alpha_1 + \beta_1) - \text{logit}^{-1}(\alpha_1)\right], \end{aligned}$$

Note that others have considered mediation analyses without using the SEM paradigm. For example, Rubin [29, 30] and Jo et al. [14] considered methods to estimate the causal effect of treatment in the face of an intermediate confounding variable (mediator) based on the framework of Principal Stratification. These methods are limited in their ability to accommodate continuous mediating and outcome variables and are less popular than their SEM-based counterparts.

4 Discussion

Causal inference is widely used in biomedical, psychosocial, and related services research to investigate the causal mechanism of exposures and interventions. Not only does research on this important topic have a long history, but the body of literature in this field is quite extensive as well, containing both methodological development and applications over a wide range of disciplines. The potential outcome based causal paradigm is by far the most popular, playing a dominating role in the development of modern causal inference models and methods. For example, all popular methods, such as the propensity score, principal stratification, marginal structural and structural mean models, are developed based on this framework. Under the potential outcome based causal paradigm, these methods can be generalized for causal inferences in various different situations, as illustrated by the chapters in this book.

References

1. Angrist, J., Imbens, G.W., Rubin, D.B.: Identification of causal effects using instrumental variables (with discussion). J. Am. Stat. Assoc. **91**, 444–472 (1996)
2. Baron, R.M., Kenny, D.A.: The moderator–mediator variable distinction in social psychological research: conceptual, strategic, and statistical considerations. J. Pers. Soc. Psychol. **51**, 1173–1182 (1986)
3. Bollen, K.: Total, direct and indirect effects in structural equation models. In: Clogg, C. (ed.) Sociological Methodology, pp. 37–69. American Sociological Association, Washington, D.C (1987)

4. Frangakis, C.E., Rubin, D.B.: Principal stratification in causal inference. Biometrics **58**, 21–29 (2002)
5. Goetghebeur, E., Lapp, K.: The effect of treatment compliance in a placebo-controlled trial: regression with unpaired data. J. R. Stat. Soc. Ser. C **46**, 351–364 (1997)
6. Gunzler, D., Tang, W., Lu, N., Wu, P., Tu, X.M.: A class of distribution-free models for longitudinal mediation analysis. Psychometrika **79**(4), 543–568 (2014)
7. Hafeman, D.M., Schawrtz, S.: Opening the black box: a motivation for the assessment of mediation. Int. J. Epidemiol. **38**, 838–845 (2009)
8. Hernan, M.A., Brumback, B., Robins, J.M.: Estimating the causal effect of zidovudine on CD4 count with a marginal structural model for repeated measures. Stat. Med. **21**, 1689–1709 (2002)
9. Holland, P.: Statistics and causal inference. J. Am. Stat. Assoc. **81** 945–970 (1986)
10. Horvitz, D.G., Thompson, D.J.: A Generalization of sampling without replacement from a finite universe. J. Am. Stat. Assoc. **47**, 663–685 (1952)
11. Imai, K., Keele, L., Yamamoto, T.: Identification, inference, and sensitivity analysis for causal mediation effects. Stat. Sci. **25**, 51–71 (2010)
12. Imai, K., Keele, L., Tingley, D.: Replication data for: a general approach to causal mediation analysis. Psychol. Methods **15**(4), 309–344 (2010)
13. Imbens, G.W., Rubin, D.B.: Bayesian inference for causal effects in randomized experiments with noncompliance. Ann. Stat. **25**, 305–327 (1997)
14. Jo, B., Stuart, E.A., MacKinnon, D.P., Vinokur, A.D.: The use of propensity scores in mediation analysis. Multivar. Behav. Res., **46**(3), 425–452 (2011)
15. Judd, C., Kenny, D.: Process analysis: estimating mediation in treatment evaluations. Eval. Rev. **5**, 602–619 (1981)
16. Kowalski, J., Tu, X.M.: Modern Applied U Statistics. Wiley, New York (2007)
17. MacKinnon, D., Dwyer, J.: Estimating mediating effects in prevention studies. Eval. Rev. **17**, 144–158 (1993)
18. Manski, C.F.: Identification for Prediction and Decision. Harvard University Press, Cambridge, MA (2007)
19. Pearl, J.: Direct and indirect effects. In: Breese, J., Koller, D. (eds.) Proceedings of the 17th Conference on Uncertainty in Artificial Intelligence, pp. 411–420. Morgan Kaufmann, San Francisco, CA (2001)
20. Robins J.M.: Correcting for non-compliance in randomized trials using structural nested mean models. Commun. Stat. **23**, 2379–2412 (1994)
21. Robins, J.M.: Marginal structural models versus structural nested models as tools for causal inference. In: Halloran, M.E., Berry, D. (eds.) Statistical Models in Epidemiology: The Environment and Clinical Trials, pp. 95–134. Springer, New York (1999)
22. Robins, J.M.: Semantics of causal DAG models and the identification of direct and indirect effects. In: Green, P.J., Hjort, N.L., Richardson, S. (eds.) Highly Structured Stochastic Systems, pp. 70–81. Oxford University Press, New York, NY (2003)
23. Robins, J.M., Greenland, S.: Identifiability and exchangeability for direct and indirect effects. Epidemiology **3**(2), 143–155 (1992)
24. Rosenbaum, P.R.: The consequences of adjustment for a concomitant variable that has been affected by the treatment. J. R. Stat. Soc. Ser. A **147**, 656–666 (1984)
25. Rosenbaum, P.R., Rubin, D.B.: The central role of the propensity score in observational studies for causal effects. Biometrika **70**, 41–55 (1983)
26. Rosenbaum, P.R., Rubin, D.B.: Constructing a control group using multivariate matched sampling methods that incorporate the propensity score. Am. Stat. **39**, 33–38 (1985)
27. Rubin, D.B.: Estimating causal effects of treatments in randomized and nonrandomized studies. J. Educ. Psychol. **66**, 688–701 (1974)
28. Rubin, D.B.: Inference and missing data (with discussion). Biometrika **63**, 581–592 (1976)
29. Rubin, D.B.: Direct and indirect causal effects via potential outcomes. Scandinavian J. Stat. **31**, 161–170 (2004)
30. Rubin, D.B.: Causal inference using potential outcomes: design, modeling, decisions. J. Am. Stat. Assoc. **100**, 322–331 (2005)

31. Tang, W., He, H., Tu, X.M.: Applied Categorical Data Analysis. Chapman & Hall/CRC, Boca Raton, FL (2012)
32. Vansteelandt, S., Goetghebeur, E.: Causal inference with generalized structural mean models. J. R. Stat. Soc. Ser. B **65**, 817–835 (2003)
33. Wu, P.: A new class of structural functional response models for causal inference and mediation analysis. Ph.D. Thesis, Department of Biostatistics and Computational Biology, University of Rochester, Rochester, New York (2013)
34. Wu, P., Gunzler, D., Lu, N., Chen, T., Wyman, P., Tu, XM.: Causal inference for community based multi-layered intervention study. Stat. Med. **33**(22), 3905–3918 (2014)

Part II
Propensity Score Method for Causal Inference

Chapter 2
Overview of Propensity Score Methods

Hua He, Jun Hu, and Jiang He

Abstract The propensity score methods are widely used to adjust confounding effects in observational studies when comparing treatment effects. The propensity score is defined as the probability of treatment assignment conditioning on some observed baseline characteristics and it provides a balanced score for the treatment conditions as conditioning on the propensity score, the treatment groups are comparable in terms of the baseline covariates. In this chapter, we will first provide an overview of the propensity score and the underlying assumptions for using propensity score, we will then discuss four methods based on propensity score: matching on the propensity score, stratification on the propensity score, inverse probability of treatment weighting using the propensity score, and covariate adjustment using the propensity score, as well as the differences among the four methods.

1 Introduction

Since treatment selection is often influenced by subject characteristics, selection bias is one of the major issues when we assess the treatment effect. This is especially the case for observational studies. Most cutting-edge topics in statistical research in causal inferences attempt to address this key issue of selection bias. Variables that cause selection bias are called confounding variables, confounders, or covariates, etc. When there are confounders, treatment effects cannot be simply assessed as the observed group differences. The issue can be better illustrated under the counterfactual outcome framework for causal inference.

H. He (✉) • J. He
Department of Epidemiology, School of Public Health & Tropical Medicine, Tulane University, New Orleans, LA 70112, USA
e-mail: hhe2@tulane.edu; jhe@tulane.edu

J. Hu
College of Basic Science and Information Engineering, Yunnan Agricultural University, Yunnan, 650201, China
e-mail: hududu@ynau.edu.cn

H. He et al. (eds.), *Statistical Causal Inferences and Their Applications in Public Health Research*, ICSA Book Series in Statistics, DOI 10.1007/978-3-319-41259-7_2

Suppose we are interested in the effect of a new treatment on an outcome, say blood pressure, measured in a continuous scale. Suppose there are two groups of patients, one receives the new treatment, and the other receives control such as treatment as usual (TAU) or placebo. We are interested in assessing the treatment effects. When there are no selection bias, i.e., if the two groups are similar before the treatment, we can simply compare the observed outcomes, the blood pressures, between the two groups of patients taking the two treatments. However, bias inference may be resulted if there are selection bias, i.e., if the two groups receiving the two treatments are very different.

Under the counterfactual outcome framework, we assume that for each subject, there are two potential outcomes, one for each treatment, had the subject taken the treatment. The treatment effect is defined for each subject based on his/her differential responses to different treatments. This definition of treatment effect is free of any confounder, because all the characteristics of the same patient are the same for the two potential outcomes. However, since each subject can only take one of the two treatments, only one of them is observed and the other is missing.

More precisely, let $y_{i,j}$ denote the potential outcome for the ith subject under the jth treatment, $j = 1$ for new treatment and $j = 2$ for control. We can observe only one of the two outcomes, $y_{i,1}$ or $y_{i,2}$, depending on the treatment received by the patient. The difference between $y_{i,1}$ and $y_{i,2}$ can be attributed to the differential effect of the treatment, since there is absolutely no other confounder in this case. However, as one of $y_{i,1}$ and $y_{i,2}$ is always unobserved, standard statistical methods cannot be applied, but methods for missing data can be used to facilitate inference.

Under this paradigm of counterfactual outcomes, the mean response $E(y_{i,1} - y_{i,2})$, albeit unobserved, represents the effect of treatment for the population. Let z_i be an indicator for the first treatment, then $y_{i,1}$ ($y_{i,2}$) is observable only if $z_i = 1(2)$. Under simple randomization, the assignment of treatment is random and free of any selection bias, that is

$$E\left(y_{i,j}\right) = E\left(y_{i,j} \mid z_i = j\right), \quad 1 \leq i \leq n. \tag{2.1}$$

This shows that missing values in the counterfactual outcomes $y_{i,j}$ are missing completely at random (MCAR) and can thus be completely ignored. It follows that $E\left(y_{i,j}\right)$ can be estimated based on the observed component of each subject's counterfactual outcomes corresponding to the assigned treatment. It is for this reason that simple randomized controlled trials (RCTs) are generally considered as the gold standard approach in making causal conclusions on the treatment effects.

However, simple randomization may not always be feasible. In clinical trials, it may be preferable to adopt other randomization procedures because of cost, ethnic, and scientific reasons. For example, in some studies we often need to oversample underrepresented subjects to achieve required accuracy of estimations. In such cases, it is important to deal with the treatment selection bias and the propensity score is a very powerful tool for this task.

2 Definition of Propensity Score

To address the selection bias raised in the above more complex randomization schemes or non-randomized observational studies, assume that the treatment assignments are based on $\mathbf{x}_i$, a vector of covariates, which is always observed. In such cases, the missing mechanism for the unobserved outcome no longer follows MCAR, but rather follows missing at random (MAR) as defined by

$$(y_{i,1}, y_{i,2}) \perp z_i \mid \mathbf{x}_i. \tag{2.2}$$

Although unconditionally non-randomized , the assignment is randomized given the covariates $\mathbf{x}_i$, thus

$$E(y_{i,1} \mid \mathbf{x}_i) - E(y_{i,2} \mid \mathbf{x}_i) = E(y_{i,1} \mid z_i = 1, \mathbf{x}_i) - E(y_{i,2} \mid z_i = 2, \mathbf{x}_i).$$

So, within each pattern of the covariate $\mathbf{x}_i$, the treatment effect can be estimates simply by those subjects receiving the two treatments.

Within the context of causal inference, the MAR condition in (2.2) is known as the *strongly ignorable treatment assignment* assumption [38]. Although the treatment assignments for the whole study do not follow simple randomization, the ones within each of the strata defined by the distinct values of $\mathbf{x}_i$ do. Thus, if there is a sufficient number of subjects within each of the strata defined by the unique values of $\mathbf{x_i}$, then $E(y_{i,1} \mid \mathbf{x}_i)$ and $E(y_{i,2} \mid \mathbf{x}_i)$ can be estimated by the corresponding sample means within each strata. The overall treatment effect can then be estimated by a weighted average of these means, the weights are assigned based on the distribution of $\mathbf{x_i}$. The approach may not result in reliable estimates or simply may not work if some groups have a small or even 0 number of subjects for one or both treatment conditions. This can occur if the overall sample size is relative small, and/or the number of distinct values of $\mathbf{x}_i$ is large such as when $\mathbf{x}_i$ contains continuous components and/or $\mathbf{x}_i$ has a high dimension. However, the propensity score can help facilitate the dimension reduction.

The *propensity score (PS)* is defined as

$$e(\mathbf{x}_i) = \Pr(z_i = 1 \mid \mathbf{x}_i), \tag{2.3}$$

the probability of treatment assignment conditioning on the observed covariate $\mathbf{x}_i$ [38]. For simple randomized clinical trials, this will be a constant (and usually 0.5 if subjects are equally allotted to the two groups). However, for observational studies, subjects often make their decisions based on their own perspective of their conditions (characteristics).

Conditioning on any given propensity score, the counterfactual outcomes are independent of the treatment assignment, i.e., for any $e \in (0, 1)$,

$$E(y_{i,k} \mid z_i = 1, e_i = e) = E(y_{i,k} \mid e_i = e), \quad k = 1, 2. \tag{2.4}$$

This follows directly from (2.2), using the iterated conditional expectation argument (see [37–39]).

From (2.4), the treatment effect for subjects with a given propensity score can be estimated by the subjects actually receiving the two treatments. Thus, using the propensity score we can reduce the dimension of the covariates from dim($\mathbf{x_i}$) to 1. However, if there are continuous covariates, and hence e is also continuous, (2.4) is still not directly applicable. Methods of propensity score matching, stratification, weighting, and covariate adjustment have been developed to facilitate the causal inference using propensity scores [15, 38, 39, 43].

3 Causal Inference Based on Propensity Scores

The equation in (2.4) is fundamental to the application of propensity scores. It implies that for a given propensity score, the two treatments are directly comparable. A straightforward application would be comparing the two treatment for each given propensity score and then combining the treatment effect across all the propensity scores. First, the comparison can be performed by matching subjects in the two treatment groups by the propensity scores. This is the propensity score matching method. Instead of individual level matching, we can divide the data into subgroups according to the propensity scores, with subjects in the same subgroup having similar propensity scores, thus according to (2.4) the treatment effect for each subgroup can be estimated. This is the idea of propensity score stratification [39]. Since the propensity score is the probability of being selected for the treatment, another approach is using the inverse probability weighting method. Finally, we can treat propensity score as a covariate in regression models to control for the selection bias.

In the following we will discuss these four approaches in details, based on the assumption that the propensity score is available either by design as in some clinical trials or estimated based on some models. When the propensity scores need to be estimated, logistic regression models can be applied to model the binary treatment assignment z_i. Probit and Complementary log-log models can also be applied. The independent variables in the logistic regression models should include variables that are associated with the treatment assignment and the outcome.

3.1 Propensity Score Matching

In observational studies, it is not uncommon that there are only a limited number of subjects in the treatment group, but a much larger number of subjects in the control group. An example is that physicians have data available from hospital records for patients treated for a disease, but there is no data for subjects who don't have the disease (control). In such cases, they often seek large survey data to find controls. For example, in the study of metabolic syndrome among patients receiving clozapine by Lamberti et al. [25], they treated 93 outpatients with schizophrenia and schizoaffective disorder with clozapine. For treatment comparison purpose, they obtained a control group with more than 2700 subjects by matching the subjects in the treatment group from the National Health and Nutrition Examination Survey.

When there is a very large pool of control subjects to match, we can match each subject in the treatment group with all the key covariates. However, if the pool of control subjects is not so large and/or there are many control covariates, then the propensity score matching approach will be a useful tool because of the reduced dimensionality. The matching can be performed with 1:1 matching or more generally 1:n matching.

Different matching methods have been proposed. First, we can simply match the subjects based on the (estimated) propensity scores. When there are continuous or high dimensional covariates, we may not always be able to find subjects with the exact same propensity score to match. In this case, we can match the subject with the closest propensity score. It is recommended to select the subjects based on the logit scale (logit of the propensity score), rather than the propensity score itself. This approach is simple and easy to implement, however, it may be important to control (match) some key covariates as well. A Mahalanobis metric matching is to select the control subject with the minimum distance based on the Mahalanobis metric of some key covariates and the logit of propensity scores. For subjects with u for the key covariates and v for the logit of the propensity score, the Mahalanobis distance is defined as

$$d_{ij} = (u - v)^T C^{-1} (u - v),$$

where C is the sample covariance matrix of these variables for the full set of control subjects.

To give the propensity score a higher priority, one may combine the two matching methods. We can first select a subgroup of the control subjects based on the logit of propensity scores (caliper), and then select the control subjects from this subgroup based on the Mahalanobis metric. This approach is in general preferred over the above two methods [5, 11, 38, 40, 41].

Based on the selection criteria, the propensity score matching approach can be processed as follows. For the first subject in the treatment group, select the control subject(s). Remove them to a new data set, and repeat the process for the second subject, etc., until all the subjects in the treatment group are removed to the new data set. Ultimately, we have a new data set with matched subjects with treatment and control conditions. In these procedures, once a control is selected, it cannot be selected again to match another treated subject. This is called greedy algorithm. If the pool of control subjects is not big, one can consider reusing the matched control subjects, i.e., by putting the matched subjects back for matching again.

We may check that covariates are balanced across treatment and control groups, and then analysis can be performed based on new sample [2]. Note that the sample does not satisfy the common i.i.d assumption anymore because of the matching, hence common methods for cross-sectional data do not apply. Paired t-test may be applied for simple group comparison if the matching is 1 to 1. As for 1 to n matching, methods for dependent outcomes such as generalized estimating equations can be applied to assess the treatment effects, which has already been adjusted for covariates.

The propensity score matching approach is not only very popular in practice, but also an active methodological research topic. Applications of the propensity score matching for different scenarios, variations of the matching procedures, and new methods of inferences have been proposed, see, for example, [1–3, 5, 6, 9, 10, 12, 21, 27–29, 33, 48].

One disadvantage of the propensity matching approach is that subjects may not be able to find a matched subject in the control group. For example, if the treatment and control groups have comparable sample size, it will be very likely that there will be more subjects with high propensity scores in the treatment group than in the control group. Similarly, there will be more subjects with low propensity scores in the control group than in the treatment group. This will result in more difficulty in matching, i.e., more subjects without matched subjects. This not only suffers information loss, but also raises the question of what the matched sample represents, and hence may introduce another source of selection bias. Thus, the propensity score matching method is preferred when the control group is large so that there is no problem for every subject in the treatment group to find a matching subject.

3.2 *Propensity Score Stratification*

When the control group is much larger than the treatment group, the propensity score matching approach usually only selects a small portion of subjects in the control group, although there may be more subjects with good matching in the propensity score and key covariates available. In this case, the propensity score matching approach suffers low power. To make use of all the subjects in the control group, another common approach called stratification or subclassification can be applied. Instead of matching each individual, the propensity score stratification approach divides subjects into subgroups according to the propensity scores. More precisely, let $0 = c_0 < c_1 < c_2 < \ldots < c_m = 1$, then we can separate the sample into m groups, where the kth group consists of subjects with propensity scores falling within $I_k = (c_{k-1}, c_k]$. Under the regularity assumption that the treatments effect is a continuous function of the propensity scores, i.e., $E(y_{i,1} - y_{i,2} \mid e_i = e)$ is continuous in e, which means that subjects with comparable propensity scores should show similar treatments effect, i.e.,

$$E\left(y_{i,j} \mid e_i \in I_k\right) \approx E\left(y_{i,j} \mid z_i = j, e_i \in I_k\right), \; for\; k = 1, 2, \ldots, m, j = 1, 2$$

Hence, within each subgroup, we can estimate the treatment effects for each treatment condition by the observed outcomes for that subgroup, i.e.,

$$\widehat{E}\left(y_{i,1} \mid e_i \in I_k\right) = \frac{\sum_{i:e_i \in I_k, z_i=1} y_{i,1}}{n_{k1}}, \; \widehat{E}\left(y_{i,2} \mid e_i \in I_k\right) = \frac{\sum_{i:e_i \in I_k, z_i=2} y_{i,2}}{n_{k2}},$$

where n_{k1} and n_{k2} are the number of subjects in the kth subgroup for the treatment and control group, respectively. So the treatment effect for the kth subgroup can be estimated by

$$\widehat{E}\left(y_{i,1} \mid e_i \in I_k\right) - \widehat{E}\left(y_{i,2} \mid e_i \in I_k\right).$$

Based on the estimated treatment effect for each subgroup, we can estimate the treatment effects for the whole sample. Note that the overall treatment effects for the whole sample can be expressed as

$$\int \left[E\left(y_{i,1} \mid e_i = e\right) - E\left(y_{i,2} \mid e_i = e\right)\right] f(e) de, \tag{2.5}$$

where $f(e)$ is the density function of the propensity score e. If $E\left(y_{i,1} \mid e_i = e\right)$ is approximately a constant over $(c_{k-1}, c_k]$, then

$$\begin{aligned}\int_{c_{k-1}}^{c_k} E\left(y_{i,j} \mid e_i = e\right) f(e) de &= \left[E\left(y_{i,j} \mid e_i \in I_k\right)\right] \int_{c_{k-1}}^{c_k} f(e) de \\ &= \left[E\left(y_{i,j} \mid e_i \in I_k\right)\right] \Pr(e_i \in I_k).\end{aligned}$$

Thus, approximately, the overall treatment effect is

$$\sum_{k=1}^{m} \left[E\left(y_{i,1} \mid e_i \in I_k\right) - E\left(y_{i,2} \mid e_i \in I_k\right)\right] \Pr(e_i \in I_k),$$

which is a weighted average of the treatment effects across the subgroups. $\Pr(e \in I_k)$ can be estimated by the sample proportion

$$\widehat{\Pr}(e \in I_k) = \frac{n_{k1} + n_{k2}}{n},$$

where n is the total sample size.

This approach can be viewed as a numeric estimate of the overall treatment effect (2.5). Since the over treatment effect is an integral over the propensity score e_i, which is a scalar-valued function of $\mathbf{x}_i$ regardless of the dimensionality and density of the range of $\mathbf{x}_i$, we can estimate the integral (2.5) as a Riemann sum.

Under the propensity score stratification approach, we need to decide the cut points for the classification. In general, we can divide the subjects into comparable subgroups, i.e., based on the quantiles of the estimated propensity scores for the combined groups. In general, 5–10 groups is sufficient, and simulation studies show that such a partition seems to be sufficient to remove 90 % of the bias [39]. In the case where the treatment group is small, such a division may result in subgroups with few subject to the treatment and hence produce instable inference. In such cases, one may also choose the cut points based on the quantiles of the estimated propensity scores based on the treatments group only in order to obtain subgroups with comparable number of the subjects receiving the treatment [42, 44].

3.3 Propensity Score Weighting

Instead of comparing the treatment and control groups at each propensity score or a small interval of propensity scores, we can also correct the selection bias by the propensity score weighting approach. Note that the propensity score is the probability of a subject being assigned to a treatment group, thus, a subject in a treatment group with propensity score $e = 0.1$ would be thought of as a representative of a total $\frac{1}{e} = 10$ subjects with similar characteristics, hence in the analysis we would assign a weight of $\frac{1}{e} = 10$ to that subject when estimate the treatment effect. Similarly, since a subject in control group with propensity score $e = 0.1$ has a probability of $1 - e = 0.9$ being assigned to the control group, it also would be thought of as a representative of a total $\frac{1}{1-e} = 1.1$ subjects in the control group with similar characteristic, hence in the analysis we would assign a weight of $\frac{1}{1-e} = 1.1$ to the subject in estimating the treatment effect. This is the inverse probability weighting (IPW) approach, which has a long history in the analysis of sample survey data [22].

The mathematical justification of the propensity score weighting is the fact that

$$E\left(\frac{z_i}{e_i} y_{i,1}\right) = E\left(y_{i,2}\right) \quad \text{and} \quad E\left(\frac{1 - z_i}{1 - e_i} y_{i,2}\right) = E\left(y_{i,1}\right). \tag{2.6}$$

This weighting approach can also be applied to regression analysis. For example, suppose that there is no interaction between the treatment and the covariates, so we can assume that

$$y_{ij} = \alpha z_i + \beta \mathbf{x}_i, \; j = 1, 2. \tag{2.7}$$

The two regression models for the potential outcomes y_{ij} (2.7) can be expressed in one model of the observed outcome y_i,

$$y_i = \alpha z_i + \beta \mathbf{x}_i, \tag{2.8}$$

with weight $\frac{1}{e_i}$ for $z_i = 1$ and $\frac{1}{1-e_i}$ for $z_i = 0$. To justify this, one can easily check that the following estimating equation (EE):

$$\frac{1}{n}\sum_{i=1}^{n}\frac{z_i}{e_i}Var(y_i \mid \mathbf{x}_i)\left[y_i - (\alpha z_i + \beta \mathbf{x}_i)\right] = 0 \tag{2.9}$$

is unbiased. To account for the variation associated with estimating the propensity score, we can combine this EE in (2.9) with estimating equations for the propensity score. Note that even when e_i is known, the estimated propensity score is often preferred over the true e_i because it may fit the observed data better [20].

For the propensity score weighting approach, to provide valid inference, we need $0 < e_i < 1$, so that each subject has a positive probability to be assigned to both treatment and control groups. In other words, the subgroups must have their representatives observed in both groups. For subjects in the treatment group with extremely small e_is, the inverses of such e_i can become quite large, yielding very highly volatiled estimates. Similarly, subjects in the control group with extremely large e_is (close to 1), the weights can also become quite large and cause the estimates to be highly volatile. So, to ensure good behaviors of estimates, we need to assume

$$e_i > c > 0, \quad \text{if } z_i = 1 \quad \text{and} \quad e_i < 1 - c, \quad \text{if } z_i = 0,$$

where c is some positive constant. This assumption is similar to the bounded away from 0 assumption for regular inverse probability weight approaches for missing values.

To reduce bias and improve the stability of the propensity score weighting approach, some modified propensity score methods including the double robust estimator have been developed and discussed, see [7, 13, 16, 17, 24, 26, 27, 30, 38, 45, 47].

3.4 Propensity Score Covariate Adjustment

Propensity scores can also be used as a covariate in regression models to adjust the selection bias [11, 38, 43]. Based on (2.4), treatment effect is a function of the propensity score. Thus, without any further assumption, we can apply the non-parametric regression model

$$E(y_{ij} \mid e) = E(y_{ij} \mid z_i = j, e) = f_j(e), \tag{2.10}$$

to assess the causal effect. Without any further assumption, we can apply nonparametric curve regression methods such as local polynomial regressions to the two groups separately to estimate the two curves [8, 14]. Treatment effect may then be assessed by comparing these two estimated curves.

If we assume that the treatment effect is homogeneous across all the propensity scores, then $f_1(e) - f_2(e)$ is a constant, and $\alpha = f_1(e) - f_2(e)$ is the treatment effect. Then (2.10) can be written compactly as

$$E(y_{ij} \mid e) = \alpha z_i + f(e), \tag{2.11}$$

where α is the treatment effect. If the function $f(e)$ is further linear in e, then

$$E(y_{ij} \mid e) = \alpha z_i + \beta e. \tag{2.12}$$

Conditioning on the propensity score, since the mean of the potential outcome equals to the mean of the observed outcome, the two regression equations in (2.12) for the two groups can be written in a regular regression model

$$E(y_i) = \alpha z_i + \beta e, \tag{2.13}$$

and again the parameter α carries the information for treatment effect.

In the arguments above, the assumption of homogeneous treatment effects (2.11) is important to provide valid inference. It has been proved that under the homogeneous treatment effect, the regression model (2.13) will provide robust inference about the treatment effect, even when the parametric assumption, i.e., the function form for $f(e)$ in (2.11) is not correctly specified [11, 36]. One may check the homogeneity assumption (2.11) by testing if the interaction between the treatment and propensity score is significant. Using the propensity score stratification, we can also compare the estimated treatment effect across the groups, and test if they are the same.

Note that this propensity score covariate regression adjustment is similar to the regular covariate adjustment in regression analysis. In fact, Rosenbaum and Rubin showed the point estimate of the treatment effect is the same if the same $\mathbf{x}_i$ is used in the estimation of the propensity score and the treatment effect and the propensity score is a linear function of $\mathbf{x}_i$ (this can only be approximately true since logistic functions are not linear). The two-step procedure of propensity score covariate adjustment has the advantage that one can apply a very complicated propensity score model without worrying about the problem of over-parameterizing the model [11].

The covariate adjustment is commonly used in practice, and the methods are generalized for different scenarios [23, 46]. However, the covariance adjustment should be performed with caution [11, 19]. Standard linear regression models are based on the homoscedasticity, so it may be a problem if the variance in the treatment and control groups is very different. The above arguments are based on linear model for continuous outcomes, their application to nonlinear cases is questionable. For example, for nonlinear regression models such as logistic regression models, Austin et al. found there are considerable bias associated with treatment effect estimate if the propensity score is used as a covariate for the adjustment [4]. Even for linear models, Hade and Lu also investigated the size of the bias and recommended adjusting for the propensity score through stratification or matching followed by regression or using splines [19].

4 Example: The Genetic Epidemiology Network of Salt Sensitivity (GenSalt) Study

We use the baseline information of the Genetic Epidemiology Network of Salt Sensitivity (GenSalt) Study as an example to illustrate the methods. The objective of the GenSalt Study is to localize and identify genes related to blood pressure responses to dietary sodium and potassium intervention [18]. For each of the 3,153 participants recruited for GenSalt Study a standardized questionnaire was administered by trained staff at the baseline examination to obtain information about demographic characteristics such as age, gender, marital status, education level, employment status and baseline BMI, personal and family medical history such as history of hypertension, and lifestyle risk factors (including cigarette smoking, alcohol consumption, and physical activity level). More detailed information can be found in [18, 35]. In the example, we are interested in the effect of sport activity on blood pressure outcome at baseline.

Outcomes The primary outcome is the blood pressure (BP). In the study, there are three measures about the blood pressure, systolic BP (SBP), diastolic BP (DBP), and the mean arterial pressure (MAP) which is defined as a summation of one third of SBP and two thirds of DBP (1/3*SBP+2/3*DBP). We use MAP in this example as it involves both SBP and DBP. The baseline BP was measured every morning during the 3-day baseline observation period by trained and certified individuals using a random-zeroi sphygmomanometer according to a standard protocol adapted from procedures recommended by the American Heart Association [34]. When BP was measured, participants were in the sitting position after they had rested for 5 min. Participants were advised to avoid consumption of alcohol, coffee, or tea, cigarette smoking, and exercise for at least 30 min before their BP measurements.

Treatment Conditions The Paffenbarger Physical Activity Questionnaire was adapted for the measurement of physical activity level [31]. Data was collected on the number of hours spent in vigorous and moderate activity on a usual day during the previous 12 months for weekdays and weekends separately to account for anticipated daily variability in energy expenditure. Examples provided for vigorous activity included shoveling, digging, heavy farming, jogging, brisk walking, heavy carpentry, and bicycling on hills, and examples of moderate activity included housework, regular walking, yard work, light carpentry, and bicycling on level ground. The physical activity score was dichotomized into more activity and less activity using a cut point of 51.1 based on the 50 % sample quantile. Participants with at least 51.1 in their physical activity score were considered as receiving physical activity treatment and thus consist of the treatment group while the participants with physical activity score less than 51.1 were considered as control. We expect that participants in the treatment group would have a lower blood pressure than participants in the control group.

Covariates In addition to the demographic information such as age, gender, marital status, education level, employment status, baseline BMI, smoking and drinking status, we also considered personal medical history such as stroke, hypertension, and high cholesterol and blood chemistry results such as glucose, creatinine, total cholesterol, HDL cholesterol, LDL cholesterol, and triglycerides. All the covariates were compared between the treatment and control groups by chi-square tests for categorical variables and Wilcoxon Rank-sum tests for continuous variables. Most of the variables are significantly different between the two groups. We also compared the BP difference between the two groups, the sample difference is 4.16 mm Hg in MAP with the control group having higher MAP.

Next, we will apply propensity score methods to examine the effects of physical activity on BP.

4.1 Estimating the Propensity Score

All covariates above that were identified as potential confounder were included in the selection model to estimate the propensity scores. A forward model selection was applied to select potential interactions. The selected final model for estimating the propensity score is summarized in Table 2.1.

The Hosmer and Lemeshow goodness-of-fit test was performed to check if the model fits the data well. The p-value for the Hosmer and Lemeshow test is 0.4632, indicating that the model to estimate the propensity scores fits the data pretty well.

4.2 Propensity Score Matching

Based on the estimated propensity scores, we can match the subjects in the treatment group with subjects in the control group. In this example, we match subjects with more activity with subjects with less activity. We use the SAS macro function provided in [32] to obtain 818 pairs of matched subjects. We checked the balance of the matched groups in terms of covariates, and the propensity score matching succeeded in reducing the selection bias between the two groups. Summarized in Table 2.2 are the p-values of comparisons of covariates between the two groups mentioned above, before and after the matching.

While most of variables showed significant difference before the propensity score matching, there was no significant difference at all in the matched sample.

Paired t-test was then applied to assess the physical activity on the blood pressure based on the matched sample. After adjusting for the confounders, the treatment

Table 2.1 Parameter estimations of the propensity score model

Parameter			DF	Estimate	Standard error	Wald χ^2	Pr > χ^2
Intercept			1	−2.0373	1.6769	1.4761	0.2244
Age			1	0.0477	0.0174	7.5367	0.0060
BMI			1	−0.0684	0.0310	4.8653	0.0274
Gender	1		1	−0.5493	0.2421	5.1474	0.0233
High education	0		1	−1.9179	0.2766	48.0890	<.0001
Field center	1		1	−0.3707	0.4453	0.6929	0.4052
Field center	2		1	−0.6963	0.3666	3.6073	0.0575
Marital	0		1	2.9975	0.6468	21.4801	<.0001
Employment	1		1	1.0841	1.0879	0.9931	0.3190
Employment	2		1	−1.7077	2.1545	0.6283	0.4280
Drinking	0		1	0.9670	0.3802	6.4677	0.0110
High cholesterol	0		1	−0.4714	0.1629	8.3748	0.0038
Stroke	0		1	−0.9449	0.2406	15.4205	<.0001
Creatinine			1	0.0231	0.00641	13.0451	0.0003
GFR			1	0.0138	0.00483	8.1560	0.0043
HDL cholesterol			1	−0.0240	0.00465	26.5825	<.0001
LDL cholesterol			1	0.00677	0.00184	13.5709	0.0002
Age*gender	1		1	0.00772	0.00292	6.9849	0.0082
Age*high education	0		1	0.0289	0.00392	54.1961	<.0001
BMI*drinking	0		1	−0.0366	0.0158	5.3762	0.0204
Drinking*gender	0	1	1	−0.1678	0.0851	3.8930	0.0485
High cholesterol*gender	0	1	1	−0.3899	0.1613	5.8402	0.0157
Creatinine*field center	1		1	−0.00654	0.00406	2.5877	0.1077
Creatinine*field center	2		1	0.0112	0.00356	9.9319	0.0016
GFR*High Education	0		1	0.00471	0.00196	5.7722	0.0163
Age*marital	0		1	−0.0160	0.00469	11.7031	0.0006
BMI*marital	0		1	−0.0840	0.0290	8.3982	0.0038
Field center*marital	1	0	1	−0.2416	0.1438	2.8224	0.0930
Field center*marital	2	0	1	0.3918	0.1259	9.6792	0.0019
Age*employment	1		1	−0.0324	0.0172	3.5625	0.0591
Age*employment	2		1	0.0437	0.0338	1.6680	0.1965
Field center*employment	1	1	1	−0.1481	0.2314	0.4098	0.5221
Field center*employment	1	2	1	−0.2715	0.4233	0.4114	0.5213
Field center*employment	2	1	1	0.5910	0.1862	10.0806	0.0015
Field center*employment	2	2	1	0.0311	0.3466	0.0081	0.9285

group that has more physical activity has 1.6598 mm Hg lower in MAP than the control group with less activity. The standard error is 0.5994, and the corresponding p-value for the treatment effect is 0.0058. The adjusted effect is smaller than the unadjusted effect 4.16 mm Hg.

Table 2.2 Group comparisons pre and post propensity score matching

Variable	Before PS matching	After PS matching
Age	<.0001	0.4485
BMI	0.3598	0.9901
Gender	<.0001	0.9605
High education	<.0001	0.6923
Field center	<.0001	0.4893
Marital	<.0001	0.4355
Employment	<.0001	0.3460
Drinking	<.0001	0.9096
High cholesterol	<.0001	1.0000
Hypertension	<.0001	1.0000
Stroke	<.0001	0.7622
Creatinine	0.3870	0.9054
GFR	<.0001	0.7215
HDL cholesterol	0.0016	0.6370
LDL cholesterol	<.0001	0.3858

Table 2.3 Estimates of treatment effect for each subgroup

	Less activity			More activity			Mean
Group	Sample size	Mean	SD	Sample size	Mean	SD	Difference
1	106	88.0712788	11.2485418	503	88.857227	9.8108938	−0.7859482
2	185	91.555956	11.7847044	424	88.0452481	10.8572098	3.5107079
3	255	89.8928105	11.9811255	354	90.0043942	11.9086026	−0.1115837
4	403	91.8189505	14.1950911	206	88.9489392	13.4397874	2.8700113
5	547	96.8985036	14.8599617	62	89.9868578	12.0791987	6.9116458

4.3 Propensity Score Stratification

In the above propensity score matching approach, only a little bit more than half of the subjects were matched. Unmatched subjects were used in the estimation of the propensity score, but their information were otherwise ignored in assessing the treatment effect. To utilize all the information, we then use the propensity score stratification approach to estimate the treatment effect. We divide the whole sample into 5 subgroups according to the propensity scores. The propensity scores range from 0.0260582 to 0.2436369, 0.2437835 to 0.3666789, 0.3668133 to 0.5341626, 0.5342977 to 0.7668451 and 0.7670692 to 0.9999613 for the five subgroups, respectively. Summarized in Table 2.3 are the sample size for each subgroup for the two treatment groups, their mean/sd in blood pressures, as well as the mean difference between the two groups.

Included in the last column are the difference in the means of the blood pressure. These were the estimates of the treatment effects for the subgroups. It is clear that the treatment effects are not homogeneous across the different propensity score levels.

In groups 2, 4, and especially 5, there were benefits of physical activity, but no benefits for the physical activity were shown in groups 1 and 3.

The overall treatment effect estimated by the weighted average of the subgroup difference was 2.48. The higher activity group had 2.48 mm Hg lower than the less activity group in MAP. The p-value for testing the null hypothesis of no difference was 0.0001, indicating the difference was significant.

4.4 *Propensity Score Weighting*

We can also use the propensity score weighting approach to correct the selection bias. Using the blood pressure measures as the response and the treatment as the only predictor and weighting each subject by their inverse of the propensity scores of being assigned to the treatment group, the estimated treatment effect was −2.38 with standard error 0.45185. The more activity group had 2.38 mm Hg lower than the less activity group in MAP. The p-value was less than .0001, which indicated that the more activity group had a significant lower MAP than the less activity group. Note that there are subjects with propensity scores as small as 0.0260582 and as big as 0.9999613, so we need to be cautious about subjects with potential high influence. In fact, there are 5 subjects with weight larger than 20, with the highest weight being 47.0519.

If the subject with the highest weight is removed from the data, the estimated treatment effect would be −2.4238. In fact, this observation is not the only one with the highest impact on the estimate of treatment effect. Thus, in such situations where we have subjects with large weights, we should use the propensity score weighting approach with caution.

In the above analysis using propensity score weighting approach, the estimated propensity scores were used. For rigorous statistical inference, we should take into account the variation associated with the estimation of the propensity score. Unfortunately many inverse weighting procedures treat the weights as fixed, and do not have the capability of taking into account such variation. However, in our example, this may not be a concern since the p-value is very small.

4.5 *Propensity Score Covariate Adjustment*

Based on the analysis using the propensity score stratification approach, the treatment effects across the propensity scores did not seem to be homogeneous in this example. We can formally test this by testing the interaction between the treatment and the propensity score. The p-value for testing the interaction was <.0001, which indicated that there was significant interaction between treatment and the propensity score. We can also compare the 5 subgroups to test the null hypothesis of no treatment effect differences among the 5 subgroups. The p value for the test was 0.0005. This further confirmed that the treatment effects were significantly different across the propensity score levels.

The significant interaction between the treatment and the propensity score implies that a simple covariate adjustment is not appropriate in this case. However, for illustrative purpose, we still applied the propensity score covariate adjustment approach. We applied a linear regression model with the blood pressure measures as the response and the treatment and the propensity score as the predictor and covariate to assess the treatment effect. The estimated treatment effect was -1.86 with an SE of 0.53354, and a p-value of 0.0005. Instead of using the exact propensity score, we also used the stratified ranks as covariate. The estimated treatment effect was -2.05 with a SE of 0.5290, and a p-value of 0.0001.

So far, we have illustrated all the propensity score approaches using the Gensalt study as an example. Based on results obtained from different approaches of adjustment based on the propensity scores, the estimated treatment effects range from 1.86 to 2.37, which are smaller than the unadjusted difference of 4.16 mm Hg in MAP. All the results shows that more activity is beneficial to the blood pressure outcome.

5 Discussion

Selection bias may produce biased estimates in observational and non-randomized studies if it is not appropriately addressed. Propensity score is a powerful tool in adjusting such selection bias. In this book chapter, we discussed several common approaches based on propensity scores to correct selection bias. All these approaches depend on the validity of the propensity score model, i.e., a model for the treatment assignment to estimate the probability of treatment assignment. Among the approaches, the propensity score weighting and covariate adjustment approaches directly use the propensity scores in the analysis while propensity score matching and stratification methods do not explicitly rely on the propensity scores in subsequent analysis. They only use the propensity score to find matched subjects either at an individual or group level. Thus the propensity score matching and stratification approaches may be less sensitive to misspecification of the propensity score model.

It is important to note that all the approaches based on propensity scores can only address observed selection bias. All the arguments are based on the assumption that the propensity score, as the probabilities of being assigned to the treatment is correctly modeled and estimated. The propensity score approaches do not have any capability to account for unobserved factors.

We have discussed the use of propensity scores in the context of assessment of treatment effect. Since the methods essentially deal with the missing values in the potential outcomes, the methods can be naturally adapted to handle missing values. For example, we have successfully applied the stratification of propensity scores to verification bias problems in statistical analysis of diagnostic studies [20].

Appendix: SAS Program Codes

All the analysis for the examples in Sect. 4 were performed using SAS. The SAS program codes are included here for readers who are interested in applying the methods for their data analyses.

- Logistic regression model for estimation of the propensity scores.
- The fitted values are saved in variable prob in data set preds.

```
proc logistic data=path.comb;
class High_Cholesterol  Stroke Drinking  Gender High_Education
Field_Center  Marital Employment;
model act_b50=Age BMI Gender High_Education Field_Center
Marital Employment
Drinking High_Cholesterol Stroke Creatinine GFR
HDL_Cholesterol LDL_Cholesterol Age*Gender Age*High_Education
BMI*Drinking Drinking*Gender High_Cholesterol*Gender Creatinine
*Field_Center Creatinine*Field_Center
GFR*High_Education Age*Marital BMI*Marital Field_Center*Marital
Field_Center*Marital Age*Employment Age*Employment Field_Center
*Employment
Field_Center*Employment Field_Center*Employment Field_Center
*Employment/lackfit;
output out=preds pred=prob;
run;
```

Macro %OneToManyMTCH was used for the propensity score matching. The macro can be copied from [32]

```
%OneToManyMTCH(work, preds,act_b50,hid,pid,Matches,1);
```

- Paired t-test for matched subjects

```
* first generate paired variables
proc sort data=Matches;
by match_1 act_b50;
run;

data paired;
set Matches;
control=B_MAP;
treated=lag(B_MAP);
if mod(_n_,2)=0 then output;
run;
* paired t-test
proc t-test data=dd ;
paired treated* control;
run;
```

- Propensity score stratification

```
proc rank data=preds groups=5 out=r;
ranks rnk;
var prob;
run;
```

- Propensity score weighting

```
data preds;set preds;
w=1/prob*(1-act_b50)+act_b50*1/(1-prob);
run;

proc reg data=preds;
weight w;
model B_MAP=act_b50 ;
run;
```

- Propensity score covariate adjustment

```
proc reg data=preds;
model B_MAP=act_b50 prob;
run;
```

References

1. Austin, P.C.: A critical appraisal of propensity-score matching in the medical literature between 1996 and 2003. Stat. Med. **27**(12), 2037–2049 (2008)
2. Austin, P.C.: Some methods of propensity-score matching had superior performance to others: results of an empirical investigation and Monte Carlo simulations. Biom. J. **51**(1), 171–184 (2009)
3. Austin, P.C.: A comparison of 12 algorithms for matching on the propensity score. Stat. Med. **33**(6), 1057–1069 (2014)
4. Austin, P.C., Grootendorst, P., Anderson, G.M.: A comparison of the ability of different propensity score models to balance measured variables between treated and untreated subjects: a monte carlo study. Stat. Med. **26**(4), 734–753 (2007)
5. Austin, P.C., Small, D.S.: The use of bootstrapping when using propensity-score matching without replacement: a simulation study. Stat. Med. **33**(24), 4306–4319 (2014)
6. Baycan, I.O.: The effects of exchange rate regimes on economic growth: evidence from propensity score matching estimates. J. Appl. Stat. **43**(5), 914–924 (2016)
7. Berk, R.A., Freedman, D.A.: Weighting regressions by propensity scores. Eval. Rev. **32**, 392–400 (2008); Berk, R.A., Freedman, D.A.: Statistical Models and Causal Inference, pp.279–294. Cambridge University Press, Cambridge (2010)
8. Cleveland, W.S.: Lowess: a program for smoothing scatterplots by robust locally weighted regression. Am. Stat. **35**(1), 54 (1981)
9. Cottone, F., Efficace, F., Apolone, G., Collins, G.S.: The added value of propensity score matching when using health-related quality of life reference data. Stat. Med. **32**(29), 5119–5132 (2013)

10. Cuong, N.V.: Which covariates should be controlled in propensity score matching? Evidence from a simulation study. Stat. Neerlandica **67**(2), 169–180 (2013)
11. d'Agostino, R.B.: Tutorial in biostatistics: propensity score methods for bias reduction in the comparison of a treatment to a non-randomized control group. Stat. Med. **17**(19), 2265–2281 (1998)
12. Dehejia, R.H., Wahba, S.: Propensity score-matching methods for nonexperimental causal studies. Rev. Econ. Stat. **84**(1), 151–161 (2002)
13. Ertefaie, A., Stephens, D.A.: Comparing approaches to causal inference for longitudinal data: inverse probability weighting versus propensity scores. Int. J. Biostat. **6**(2), Art. 14, 24 (2010)
14. Fan, J., Gijbels, I.: Local Polynomial Modelling and Its Applications. Chapman and Hall, London (1996)
15. Frölich, M.: A note on the role of the propensity score for estimating average treatment effects. A note on "On the role of the propensity score in efficient semiparametric estimation of average treatment effects" [Econometria **66**(2), 315–331 (1998); mr1612242] by J. Hahn. Econ. Rev. **23**(2), 167–174 (2004)
16. Fujii, Y., Henmi, M., Fujita, T.: Evaluating the interaction between the therapy and the treatment in clinical trials by the propensity score weighting method. Stat. Med. **31**(3), 235–252 (2012)
17. Funk, M.J., Westreich, D., Wiesen, C., Stürmer, T., Brookhart, M.A., Davidian, M.: Doubly robust estimation of causal effects. Am. J. Epidemiol. **173**(7), 761–767 (2011)
18. Group, G.C.R., et al.: Genetic epidemiology network of salt sensitivity (gensalt): rationale, design, methods, and baseline characteristics of study participants. J. Hum. Hypertens. **21**, 639 (2007)
19. Hade, E.M., Lu, B.: Bias associated with using the estimated propensity score as a regression covariate. Stat. Med. **33**(1), 74–87 (2014)
20. He, H., McDermott, M.: A robust method for correcting verification bias for binary tests. Biostatistics **13**(1), 32–47 (2012)
21. Heckman, J.J., Todd, P.E.: A note on adapting propensity score matching and selection models to choice based samples. Econ. J. **12**, S1, S230–S234 (2009)
22. Horvitz, D.G., Thompson, D.J.: A generalization of sampling without replacement from a finite universe. J. Am. Stat. Assoc. **47**, 663–685 (1952)
23. Jiang, D., Zhao, P., Tang, N.: A propensity score adjustment method for regression models with nonignorable missing covariates. Comput. Stat. Data Anal. **94**, 98–119 (2016)
24. Kim, J.K., Im, J.: Propensity score adjustment with several follow-ups. Biometrika **101**(2), 439–448 (2014)
25. Lamberti, J., Olson, D., Crilly, J., Olivares, T., Williams, G., Tu, X., Tang, W., Wiener, K., Dvorin, S., Dietz, M.: Prevalence of the metabolic syndrome among patients receiving clozapine. Am. J. Psychiatry **163**(7), 1273–1276 (2006)
26. Lee, B.K., Lessler, J., Stuart, E.A.: Improving propensity score weighting using machine learning. Stat. Med. **29**(3) , 337–346 (2010)
27. Li, F., Zaslavsky, A.M., Landrum, M.B.: Propensity score weighting with multilevel data. Stat. Med. **32**(19), 3373–3387 (2013)
28. Loux, T.M.: Randomization, matching, and propensity scores in the design and analysis of experimental studies with measured baseline covariates. Stat. Med. **34**(4), 558–570 (2015)
29. Lu, B., Qian, Z., Cunningham, A., Li, C.-L.: Estimating the effect of premarital cohabitation on timing of marital disruption: using propensity score matching in event history analysis. Sociol. Methods Res. **41**(3) , 440–466 (2012)
30. Lunceford, J.K., Davidian, M.: Stratification and weighting via the propensity score in estimation of causal treatment effects: a comparative study. Stat. Med. **23**(19), 2937–2960 (2004)
31. Paffenbarger, R., Blair, S., Lee, I., et al.: Measurement of physical activity to assess health effects in free-living populations. Med. Sci. Sports Exerc. **25**(1), 60–70 (1993)
32. Parsons, L.: Performing a 1: N case-control match on propensity score. In: Proceedings of the 29th Annual SAS Users Group International Conference, pp.165–29 (2004)

33. Peikes, D.N., Moreno, L., Orzol, S.M.: Propensity score matching: a note of caution for evaluators of social programs. Am. Stat. **62**(3), 222–231 (2008)
34. Perloff, D., Grim, C., Flack, J., Frohlich, E., Hill, M., McDonald, M., et al.: Human blood pressure determination by sphygmomanometer. Circulation **88**(5), 2460–2470 (1993)
35. Rebholz, C.M., Gu, D., Chen, J., Huang, J.-F., Cao, J., Chen, J.-C., Li, J., Lu, F., Mu, J., Ma, J., Hu, D., Ji, X., Bazzano, L.A., Liu, D., He, J., Forthe GenSalt Collaborative ResearchGroup.: Physical activity reduces salt sensitivity of blood pressure. Am. J. Epidemiol. **176**(7), 106–113 (2012)
36. Robins, J.M., Mark, S.D., Newey, W.K.: Estimating exposure effects by modelling the expectation of exposure conditional on confounders. Biometrics **48**, 479–495 (1992)
37. Rosenbaum, P.R.: Observational Studies. Springer, New York (2002)
38. Rosenbaum, P.R., Rubin, D.B.: The central role of the propensity score in observational studies for causal effects. Biometrika **70**, 41–55 (1983)
39. Rosenbaum, P.R., Rubin, D.B.: Reducing bias in observational studies using subclassification on the propensity score. J. Am. Stat. Assoc. **79** (1984), 516–524.
40. Rubin, D.B.: Using multivariate matched sampling and regression adjustment to control bias in observational studies. J. Am. Stat. Assoc. **74**(366a), 318–328 (1979)
41. Rubin, D.B.: Bias reduction using mahalanobis-metric matching. Biometrics **36**(2), 293–298 (1980)
42. Senn, S., Graf, E., Caputo, A.: Stratification for the propensity score compared with linear regression techniques to assess the effect of treatment or exposure. Stat. Med. **26**(30), 5529–5544 (2007)
43. Sobel, M.E.: Causal inference in the social sciences. J. Am. Stat. Assoc. **95**(450), 647–651 (2000)
44. Stampf, S., Graf, E., Schmoor, C., Schumacher, M.: Estimators and confidence intervals for the marginal odds ratio using logistic regression and propensity score stratification. Stat. Med. **29**(7–8), 760–769 (2010)
45. Ukoumunne, O.C., Williamson, E., Forbes, A.B., Gulliford, M.C., Carlin, J.B.: Confounder-adjusted estimates of the risk difference using propensity score-based weighting. Stat. Med. **29**(30), 3126–3136 (2010)
46. Vansteelandt, S., Daniel, R.M.: On regression adjustment for the propensity score. Stat. Med. **33**(23), 4053–4072 (2014)
47. Williamson, E.J., Forbes, A., White, I.R.: Variance reduction in randomised trials by inverse probability weighting using the propensity score. Stat. Med. **33**(5), 721–737 (2014)
48. Xu, Z., Kalbfleisch, J.D.: Propensity score matching in randomized clinical trials. Biometrics **66**(3), 813–823 (2010)

Chapter 3
Sufficient Covariate, Propensity Variable and Doubly Robust Estimation

Hui Guo, Philip Dawid, and Giovanni Berzuini

Abstract Statistical causal inference from observational studies often requires adjustment for a possibly multi-dimensional variable, where dimension reduction is crucial. The propensity score, first introduced by Rosenbaum and Rubin, is a popular approach to such reduction. We address causal inference within Dawid's decision-theoretic framework, where it is essential to pay attention to sufficient covariates and their properties. We examine the role of a propensity variable in a normal linear model. We investigate both population-based and sample-based linear regressions, with adjustments for a multivariate covariate and for a propensity variable. In addition, we study the augmented inverse probability weighted estimator, involving a combination of a response model and a propensity model. In a linear regression with homoscedasticity, a propensity variable is proved to provide the same estimated causal effect as multivariate adjustment. An estimated propensity variable may, but need not, yield better precision than the true propensity variable. The augmented inverse probability weighted estimator is doubly robust and can improve precision if the propensity model is correctly specified.

1 Introduction

Causal effects can be identified from well-designed experiments, such as randomised controlled trials (RCT), because treatment assignment is entirely unrelated to subjects' characteristics, both observed and unobserved. Suppose there are two treatment arms in an RCT: treatment group and control group. Then the average

H. Guo (✉)
Centre for Biostatistics, School of Health Sciences, The University of Manchester, Jean McFarlane Building, Oxford Road, Manchester M13 9PL, UK
e-mail: hui.guo@manchester.ac.uk

P. Dawid
Statistical Laboratory, University of Cambridge, Wilberforce Road, Cambridge CB3 0WB, UK
e-mail: apd25@cam.ac.uk

G. Berzuini
Department of Brain and Behavioural Sciences, University of Pavia, Pavia, Italy
e-mail: giomanuel_b@hotmail.com

H. He et al. (eds.), *Statistical Causal Inferences and Their Applications in Public Health Research*, ICSA Book Series in Statistics, DOI 10.1007/978-3-319-41259-7_3

causal effect (ACE) can simply be estimated as the outcome difference of the two groups from the observed data. However, randomised experiments, although ideal and to be conducted whenever possible, are not always feasible. For instance, to investigate whether smoking causes lung cancer, we cannot randomly force a group of subjects to take cigarettes. Moreover, it may take years or longer for development of this disease. Instead, a retrospective case–control study may have to be considered. The task of drawing causal conclusion, however, becomes problematic since similarity of subjects from the two groups will rarely hold, e.g., lifestyles of smokers might be different from those of non-smokers. Thus, we are unable to 'compare like with like' — the classic problem of confounding in observational studies, which may require adjusting for a suitable set of variables (such as age, sex, health status, diet). Otherwise, the relationship between treatment and response will be distorted, and lead to biased inferences. In general, linear regressions, matching or subclassification are used for adjustment purpose. If there are multiple confounders, especially for matching and subclassification, identifying two individuals with very similar values of all confounders simultaneously would be cumbersome or impossible. Thus, it would be sensible to replace all the confounders by a scalar variable. The propensity score [19] is a popular dimension reduction approach in a variety of research fields.

2 Framework

The aim of statistical causal inference is to understand and estimate a 'causal effect', and to identify scientific and in principle testable conditions under which the causal effect can be identified from observational studies. The philosophical nature of 'causality' is reflected in the diversity of its statistical formalisations, as exemplified by three frameworks:

1. Rubin's potential response framework [21–23] (also known as Rubin's causal model) based on counterfactual theory;
2. Pearl's causal framework [16, 17] richly developed from graphical models;
3. Dawid's decision-theoretic framework [6, 7] based on decision theory and probabilistic conditional independence.

In Dawid's framework, causal relations are modelled entirely by conditional probability distributions. We adopt it throughout this chapter to address causal inference; the assumptions required are, at least in principle, testable.

Let X, T and Y denote, respectively, a (typically multivariate) confounder, treatment, and response (or outcome). For simplicity, Y is a scalar and X a multi-dimensional variable. We assume that T is binary: 1 (treatment arm) and 0 (control arm). Within Dawid's framework, a non-stochastic regime indicator variable F_T, taking values $\emptyset$, 0 and 1, is introduced to denote the treatment assignment mechanism operating. This divides the world into three distinct regimes, as follows:

1. $F_T = \emptyset$: the observational (idle) regime. In this regime, the value of the treatment is passively observed and treatment assignment is determined by Nature.

2. $F_T = 1$: the interventional treatment regime, i.e., treatment T is set to 1 by manipulation.
3. $F_T = 0$: the interventional control regime, i.e., treatment T is set to 0 by manipulation.

For example, in an observational study of custodial sanctions, our interest is in the effect of custodial sanction, as compared to probation (noncustodial sanction), on the probability of re-offence. Then $F_T = \emptyset$ denotes the actual observational regime under which data were collected; $F_T = 1$ is the (hypothetical) interventional regime that always imposes imprisonment; and $F_T = 0$ is the (hypothetical) interventional regime that always imposes probation. Throughout, we assume full compliance and no dropouts, i.e., each individual actually takes whichever treatment they are assigned to. Then we have a joint distribution P_f of all relevant variables in each regime $F_T = f$ $(f = 0, 1, \emptyset)$.

In the decision-theoretic framework, causal assumptions are constructed as assertions that certain marginal or conditional distributions are common to all regimes. Such assumptions can be formally expressed as properties of conditional independence, where this is extended to allow non-stochastic variables such as F_T [4, 5, 7]. For example, the 'ignorable treatment assignment' assumption in Rubin's causal model (RCM) [19] can be expressed as

$$Y \perp\!\!\!\perp F_T \mid T, \tag{3.1}$$

read as 'Y is independent of F_T given T'. However, this condition will be most likely inappropriate in observational studies where randomisation is absent.

Causal effect is defined as the response difference by manipulating treatment, which purely involves interventional regimes. In particular, the population-based average causal effect (ACE) of the treatment is defined as

$$\text{ACE} := \text{E}(Y \mid F_T = 1) - \text{E}(Y \mid F_T = 0), \tag{3.2}$$

or alternatively,

$$\text{ACE} := \text{E}_1(Y) - \text{E}_0(Y).^{1} \tag{3.3}$$

Without further assumptions, by its definition ACE is not identifiable from the observational regime.

[1] For convenience, the values of the regime indicator F_T are presented as subscripts.

3 Identification of ACE

Suppose the joint distribution of (F_T, T, Y) is known and satisfies (3.1). Is ACE identifiable from data collected in the observational regime? Note that (3.1) demonstrates that the distribution of Y given $T = t$ is the same, whether t is observed in the observational regime $F_T = \emptyset$, or in the interventional regime $F_T = t$. As discussed, this assumption would not be satisfied in observational studies, and thus, direct comparison of response from the two treatment groups cannot be interpreted as the causal effect from observational data.

Definition 1. The 'face-value average causal effect' (FACE) is defined as

$$\text{FACE} := \text{E}_\emptyset(Y|T = 1) - \text{E}_\emptyset(Y|T = 0). \tag{3.4}$$

It would be hardly true that FACE = ACE, as we would not expect the conditional distribution of Y given $T = t$ is the same in any regime. In fact, identification of ACE from observational studies requires, on one hand, adjusting for confounders, on the other hand, interplay of distributional information between different regimes. One can make no further progress unless some properties are satisfied.

3.1 Strongly Sufficient Covariate

Rigorous conditions must be investigated so as to identify ACE.

Definition 2. X is a covariate if:

Property 1.

$$X \perp\!\!\!\perp F_T.$$

That is, the distribution of X is the same in any regime, be it observational or interventional. In most cases, X are attributes determined prior to the treatment, for example, blood types and genes.

Definition 3. X is a sufficient covariate for the effect of treatment T on response Y if, in addition to Property 1, we have

Property 2.

$$Y \perp\!\!\!\perp F_T | (X, T).$$

Property 2 requires that the distribution of Y, given X and T, is the same in all regimes. It can also be described as 'strongly ignorable treatment assignment, given X' [19]. We assume that readers are familiar with the concept and properties of

Fig. 3.1 Sufficient covariate

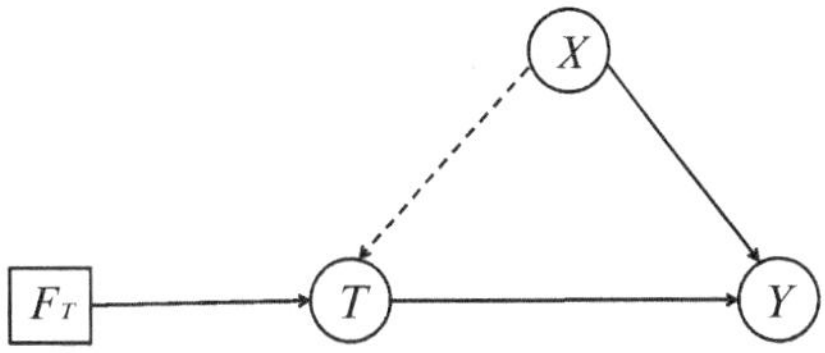

directed acyclic graphs (DAGs). Then Properties 1 and 2 can be represented by means of a DAG as shown in Fig. 3.1. The dashed arrow from X to T indicates that T is partially dependent on X, i.e., the distribution of T depends on X in the observational regime, but not in the interventional regime where $F_T = t$.

Definition 4. X is a *strongly sufficient covariate* if, in addition to Properties 1 and 2, we have

Property 3. $\mathrm{P}_\emptyset(T = t \mid X) > 0$ with probability 1, for $t = 0, 1$.

Property 3 requires that, for any $X = x$, both treatment and control groups are observed in the observational regime.

Lemma 1. *Suppose X is a strongly sufficient covariate. Then, considered as a joint distributions for (Y, X, T), P_t is absolutely continuous with respect to $\mathrm{P}_\emptyset$ (denoted by $\mathrm{P}_t \ll \mathrm{P}_\emptyset$), for $t = 0$ and $t = 1$. That is, for every event A determined by (X, T, Y),*

$$\mathrm{P}_\emptyset(A) = 0 \quad \Longrightarrow \quad \mathrm{P}_t(A) = 0. \tag{3.5}$$

Equivalently, if an event A occurs with probability 1 under the measure $\mathrm{P}_\emptyset$, then it occurs with probability 1 under the measure P_t $(t = 0, 1)$.

Proof. Property 2, expressed equivalently as $(Y, X, T) \perp\!\!\!\perp F_T \mid (X, T)$, asserts that there exists a function $w(X, T)$ such that

$$\mathrm{P}_f(A \mid X, T) = w(X, T)$$

almost surely (a.s.) in each regime $f = 0, 1, \emptyset$. Let $\mathrm{P}_\emptyset(A) = 0$. Then a.s. $[\mathrm{P}_\emptyset]$,

$$0 = \mathrm{P}_\emptyset(A \mid X) = w(X, 1)\mathrm{P}_\emptyset(T = 1 \mid X) + w(X, 0)\mathrm{P}_\emptyset(T = 0 \mid X).$$

By Property 3, for $t = 0, 1$,

$$w(X, t) = 0 \tag{3.6}$$

a.s. $[\mathrm{P}_\emptyset]$. As $w(X, t)$ is a function of X, it follows that (3.6) holds a.s. $[\mathrm{P}_t]$ by Property 1. Consequently,

$$w(X, T) = 0 \quad \text{a.s.} \quad [\mathrm{P}_t], \tag{3.7}$$

since a.s. $[\mathrm{P}_t]$, $T = t$ and $w(X, T) = w(X, t)$ for any bounded function w. Then by (3.7),

$$\mathrm{P}_t(A) = \mathrm{E}_t\{\mathrm{P}_t(A \mid X, T)\} = \mathrm{E}_t\{w(X, T)\} = 0.$$

Lemma 2. *For any integrable $Z \preceq^{2} (Y, X, T)$, and any versions of the conditional expectations,*

$$E_t(Z \mid X) = E_t(Z \mid X, T) \quad a.s.\ [\mathrm{P}_t]. \tag{3.8}$$

Proof. Let $j(X, T)$ be an arbitrary but fixed version of $\mathrm{E}_t(Z \mid X, T)$. Then $j(X, T) = j(X, t)$ a.s. $[\mathrm{P}_t]$, and $j(X, t)$ serves as a version of $\mathrm{E}_t(Z \mid X, T)$ under $[\mathrm{P}_t]$. So

$$\mathrm{E}_t(Z \mid X) = \mathrm{E}_t\{j(X, T) \mid X\} = \mathrm{E}_t\{j(X, t) \mid X\} = j(X, t) \quad \text{a.s.} \quad [\mathrm{P}_t].$$

Thus $j(X, t)$ is a version of $\mathrm{E}_t(Z \mid X)$ under $[\mathrm{P}_t]$ and (3.8) follows.

Since $\mathrm{E}_t(Z \mid X)$ is a function of X, then by Property 1, $j(X, t)$ is a version of $\mathrm{E}_t(Z \mid X)$ in any regime. Let $g(X, T)$ be some arbitrary but fixed version of $\mathrm{E}_\emptyset(Z \mid X, T)$.

Theorem 1. *Suppose that X is a strongly sufficient covariate. Then for any integrable $Z \preceq (Y, X, T)$, and with notation as above,*

$$j(X, t) \stackrel{*}{=} g(X, t) \tag{3.9}$$

almost surely in any regime.

Proof. By Property 2, there exists a function $h(X, T)$ which is a common version of $\mathrm{E}_f(Z \mid X, T)$ under $[\mathrm{P}_f]$ for $f = 0, 1, \emptyset$. Then $h(X, T)$ serves as a version of $\mathrm{E}_\emptyset(Z \mid X, T)$ under $[\mathrm{P}_\emptyset]$, and a version of $\mathrm{E}_t(Z \mid X, T)$ under $[\mathrm{P}_t]$. As $j(X, T)$ is a version of $\mathrm{E}_t(Z \mid X, T)$,

$$j(X, T) = h(X, T) \quad \text{a.s.} \quad [\mathrm{P}_t],$$

and consequently

$$j(X, t) = h(X, t) \quad \text{a.s.} \quad [\mathrm{P}_t].$$

Since $j(X, t)$ and $h(X, t)$ are functions of X, by Property 1

$$j(X, t) = h(X, t) \quad \text{a.s.} \quad [\mathrm{P}_f] \tag{3.10}$$

[2]The $\preceq$ symbol is interpreted as 'a function of'.

for $f = 0, 1, \emptyset$. We also have that $g(X, T) = h(X, T)$ a.s. $[P_\emptyset]$, and so, by Lemma 1, a.s. $[P_t]$. Then $g(X, t) = h(X, t)$ a.s. $[P_t]$, where $g(X, t)$ and $h(X, t)$ are both functions of X. By Property 1,

$$g(X, t) = h(X, t) \quad \text{a.s.} \quad [P_f] \tag{3.11}$$

for $f = 0, 1, \emptyset$. Thus (3.9) holds by (3.10) and (3.11).

3.2 *Specific Causal Effect*

Let X be a covariate.

Definition 5. The *specific causal effect* of T on Y, relative to X, is

$$\text{SCE} := \text{E}_1(Y \mid X) - \text{E}_0(Y \mid X).$$

We annotate SCE_X to express SCE as a function of X and write SCE(x) to indicate that X takes specific value x. Because it is defined in the interventional regimes, SCE has a direct causal interpretation, i.e., SCE(x) is the average causal effect in the subpopulation with $X = x$.

Although we do not assume the existence of potential responses, when this assumption is made we might proceed as follows. Take X to be the pair $\mathbf{Y} = (Y(1), Y(0))$ of potential responses—which is assumed to satisfy Property 1. Then $\text{E}_t(Y \mid X) = Y(t)$, and consequently

$$\text{SCE}_{\mathbf{Y}} = Y(1) - Y(0),$$

which is the definition of 'individual causal effect', ICE, in Rubin's causal model. Thus, although the formalisations of causality are different, SCE in Dawid's decision-theoretic framework can be regarded as a generalisation of ICE in Rubin's causal model.

We can easily prove that, for any covariate X, ACE $= \text{E}(\text{SCE}_X)$, where the expectation may be taken in any regime. Since by Property 1,

$$\text{E}_\emptyset\{\text{E}_t(Y \mid X)\} = \text{E}_t\{\text{E}_t(Y \mid X)\} = \text{E}_t(Y),$$

for $t = 0, 1$. Thus by subtraction, ACE $= \text{E}_f(\text{SCE}_X)$ for any regime $f = 0, 1, \emptyset$ and therefore the subscript f can be dropped. Hence, ACE is identifiable from observational data so long as SCE_X is identifiable from observational data. If X is a strongly sufficient covariate, by Theorem 1, $\text{E}_t(Y \mid X)$ is identifiable from the observational regime. It follows that SCE can be estimated from data purely collected in the observational regime. Then ACE expressed as

$$\text{ACE} = \text{E}_{\emptyset}(\text{SCE}_X) \tag{3.12}$$

is identifiable, from the observational joint distribution of (X, T, Y). Formula (3.12) is Pearl's 'back-door formula' [17] because by the property of modularity, P(X) is the same with or without intervention on T and thus can be taken as the distribution of X in the observational regime.

3.3 Dimension Reduction of Strongly Sufficient Covariate

Suppose X is a multi-dimensional strongly sufficient covariate. The adjustment process might be simplified if we could replace X by some reduced variable $V \preceq X$, with fewer dimensions—so long as V is itself a strongly sufficient covariate. Now since V is a function of X, Properties 1 and 3 will automatically hold for V. We thus only need to ensure that V satisfies Property 2: that is,

$$Y \perp\!\!\!\perp F_T \mid (V, T). \tag{3.13}$$

Since two arrows initiate from X in Fig. 3.1, possible reductions may be naturally considered, on the pathways from X to T, and from X to Y. Indeed, the following theorem gives two alternative sufficient conditions for (3.13) to hold. However, (3.13) can still hold without these conditions.

Theorem 2. *Suppose X is a strongly sufficient covariate and $V \preceq X$. Then V is a strongly sufficient covariate if either of the following conditions is satisfied:*

(a) ***Response-sufficient reduction:***

$$Y \perp\!\!\!\perp X \mid (V, F_T = t), \tag{3.14}$$

or

$$Y \perp\!\!\!\perp X \mid (V, T, F_T = \emptyset), \tag{3.15}$$

for $t = 0, 1$. It is indicated in (3.14) that, in each interventional regime, X contributes nothing towards predicting Y once we know V. In other words, as long as V is observed, X need not be observed to make inference on Y. While (3.15) implies that in the observational regime, knowing X is of no value of predicting Y if V and T are known.

(b) ***Treatment-sufficient reduction:***

$$T \perp\!\!\!\perp X \mid (V, F_T = \emptyset). \tag{3.16}$$

That is, in the observational regime, treatment does not depend on X conditioning on the information of V.

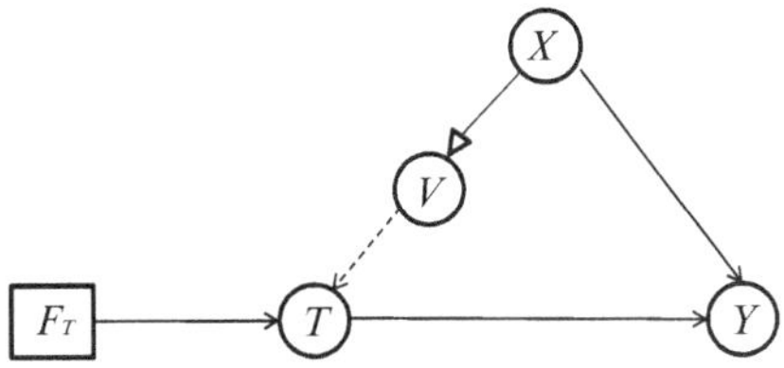

Fig. 3.2 Treatment sufficient reduction

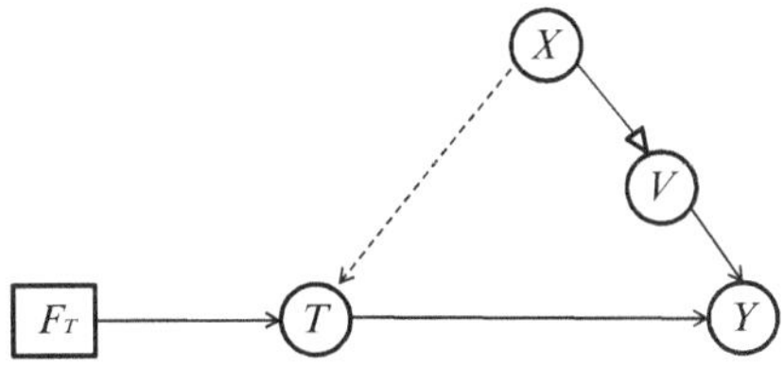

Fig. 3.3 Response sufficient reduction

Both of the two reductions in Theorem 2 were proved in [9]. An alternative proof of (b) can be implemented graphically [9], which results in a DAG as Fig. 3.2[3] off which (3.16) and (3.13) can be directly read.

A graphical approach to (a) does not work since Property 3 is required. However, while not serving as a proof, Fig. 3.3 conveniently embodies the conditional independencies Properties 1, 2 and the trivial property $V \perp\!\!\!\perp T \mid (X, F_T)$, as well as (3.13).

4 Propensity Analysis

Here we further discuss the treatment-sufficient reduction, which does not involve the response. This brings in the concept of *propensity variable*: a minimal treatment-sufficient covariate, for which we investigate the unbiasedness and precision of the estimator of ACE. Also the asymptotic precision of the estimated ACE, as well as the variation of the estimate from the actual data, will be analysed. In a simple normal linear model that applied for covariate adjustment, two cases are considered: homoscedasticity and heteroscedasticity. A non-parametric approach—subclassification will also be conducted, for different covariance matrices of X of the two treatment arms. The estimated ACE obtained by adjusting for multivariate X and by adjusting for a scalar propensity variable will then be compared theoretically and through simulations [9].

[3]The hollow arrow head, pointing from X to V, is used to emphasise that V is a function of X.

4.1 Propensity Score and Propensity Variable

The propensity score (PS), first introduced by Rosenbaum and Rubin, is a balancing score [19]. Regarded as a useful tool to reduce bias and increase precision, it is a very popular approach to causal effect estimation. PS matching (or subclassification) method, widely used in various research fields, exploits the property of *conditional (within-stratum) exchangeability*, whereby individuals with the same value of PS (or belonging to a group with similar values of PS) are taken as comparable or exchangeable. We will, however, mainly focus on the application of PS within a linear regression. The definitions of the balancing score and PS given below are borrowed from [19].

Definition 6. A *balancing score* $b(X)$ is a function of X such that, in the observational regime,[4] the conditional distribution of X given $b(X)$ is the same for both treatment groups. That is,

$$X \perp\!\!\!\perp T \mid (b(X), F_T = \emptyset).$$

It has been shown that adjusting for a balancing score rather than X results in unbiased estimate of ACE, with the assumption of strongly ignorable treatment assignment [19]. One can trivially choose $b(X) = X$, but it is more constructive to find a balancing score to be a many to one function.

Definition 7. The *propensity score*, denoted by Π, is the probability of being assigned to the treatment group given X in the observational regime:

$$\Pi := \mathrm{P}_{\emptyset}(T = 1 \mid X).$$

We shall use the symbol π to denote a particular realisation of Π. By (3.16) and Definitions 6 and 7, we assert that PS is the coarsest balancing score. For a subject i, PS is assumed to be positive, i.e., $0 < \pi_i < 1$. Those with the same value of PS are equally likely to be allocated to the treatment group (or equivalently, to the control group), which provides observational studies with the randomised-experiment-like property based on measured X. This is because the characteristics of the two groups with the same or similar PS are 'balanced'. Therefore, the scalar PS serves as a proxy of multi-dimensional variable X, and thus, it is sufficient to adjust for the former instead of the latter. In observational studies, PS is generally unknown because we do not know exactly which components of X have impact on T and how the treatment is associated with them. However, we can estimate PS from the observational data.

[4]Rosenbaum and Rubin do not define the balancing score and the PS explicitly for observational studies, although they do aim to apply the PS approach in such studies.

PS analysis for causal inference is based on a sequence of two stages:

Stage 1: PS Estimation. It is estimated by the observed T and X, and normally by a logistic regression of T on X for binary treatment. Note that the response Y is irrelevant at this stage. Because we can estimate PS without observing Y, there is no harm in finding an 'optimal' regression model of T on X by repeated trials.
Stage 2: Adjusting for PS. Various adjustment approaches have been developed, e.g., linear regression. If we are unclear about the conditional distribution of Y given T and PS, non-parametric adjustment such as matching or subclassification could be applied instead.

Although two alternatives for dimension reductions have been provided, in practice, this type of reduction may be more convenient in many cases. For example, certain values of the response may occur rarely and only after long observation periods after treatment. In addition, it may sometimes be tricky to determine a 'correct' form for a regression model of Y on X, T and F_T. Swapping the positions of X and T, Eq. (3.16) can be re-expressed as

$$X \perp\!\!\!\perp T \mid (V, F_T = \emptyset), \tag{3.17}$$

which states that the observational distribution of X given V is the same for both treatment arms. That is to say, V is a *balancing score* for X.

The treatment-sufficient condition (b) can be equivalently interpreted as follows. Consider the family $\mathscr{Q} = \{Q_0, Q_1\}$ consisting of observational distributions of X for the two groups $T = 0$ and $T = 1$. Then Eq. (3.16), re-expressed as (3.17), says that V is a *sufficient statistic* (in the usual Fisherian sense [8]) for this family. In particular, a *minimal* treatment-sufficient reduction is obtained as a minimal sufficient statistic for $\mathscr{Q}$: i.e., any variable almost surely equal to a one-one function of the likelihood ratio statistic $\Lambda := q_1(X)/q_0(X)$, where $q_i(\cdot)$ is a version of the density of Q_i.

Definition 8. A *propensity variable* is a minimal treatment-sufficient covariate, or a one–one function of the likelihood ratio statistic Λ.

The concept of a propensity variable is derived from PS which is related to Λ in the following way:

$$\Pi = \mathrm{P}_{\emptyset}(T = 1 \mid X) = \theta\, \Lambda/(1 - \theta + \theta \Lambda), \tag{3.18}$$

where $0 < \theta := \mathrm{P}_{\emptyset}(T = 1) < 1$ by Property 3.

It is entirely possible, from the above discussion, that a different propensity variable will be obtained if we start from a different strongly sufficient covariate.

4.2 *Normal Linear Model (Homoscedasticity)*

The above theory will be illustrated by a simple example under linear-normal homoscedastic parametric assumptions.

4.2.1 Model Construction

Suppose we have a scalar response variable Y, and a $(p \times 1)$ strongly sufficient covariate X that satisfies Properties 1–3. Let the conditional distribution of Y given (X, T, F_T) be specified as

$$Y \mid (X, T, F_T) \sim \mathcal{N}(d + \delta T + b'X, \phi), \tag{3.19}$$

where the symbol $\sim$ stands for 'is distributed as' and the symbol $\mathcal{N}$ stands for normal distribution, with parameters d and δ (scalar), b $(p \times 1)$ and ϕ (scalar). Note that here and in the following models, we assume no interactions between variables in X although interactions can be formally dealt with via dummy variables. Suppose X is a strongly sufficient covariate, then the coefficient δ of T in (3.19) is the average causal effect ACE, which can be easily proved as follows:

$$\begin{aligned} \text{ACE} &= \text{E}(\text{SCE}_X) = \text{E}\{\text{E}_1(Y \mid X)\} - \text{E}\{\text{E}_0(Y \mid X)\} \\ &= \text{E}(d + \delta + b'X) - \text{E}(d + b'X) = \delta \qquad \text{by} \quad (3.19). \end{aligned}$$

It is readily seen that the specific causal effect SCE_X is a constant and equals δ.

From (3.19), the *linear predictor* $\text{LP} := b'X$ satisfies the conditional independence properties in Condition (a) of Theorem 2. Thus, LP is a response-sufficient reduction of X, and $\text{E}(Y \mid \text{LP}, T) = d + \delta T + \text{LP}$, with coefficient δ of T that does not depend on the regime by virtue of the sufficiency condition.

Now assume that our model for the observational distribution of (T, X) is as follows:

$$\text{P}_\emptyset(T = 1) = \theta \tag{3.20}$$

$$X \mid (T, F_T = \emptyset) \sim \mathcal{N}(\mu_T, \Sigma) \tag{3.21}$$

with parameters $\theta \in (0, 1)$, $\mu_0(p \times 1)$, μ_1 $(p \times 1)$, and covariance matrix Σ $(p \times p$, positive definite, identical in the two treatment groups). The corresponding marginal distribution of X is a multivariate normal mixture

$$X \mid F_T = \emptyset \sim (1 - \theta)\,\mathcal{N}(\mu_0, \Sigma) + \theta\,\mathcal{N}(\mu_1, \Sigma), \tag{3.22}$$

in the observational regime, and because we have assumed Property 1 to hold, also in the interventional regime. The observational distribution of T given X is given by (3.18), with

$$\log \Lambda = \log\{\mathrm{P}_\emptyset(X \mid T = 1)\} - \log\{\mathrm{P}_\emptyset(X \mid T = 0)\}$$
$$= -\frac{1}{2}(\mu_1' \Sigma^{-1} \mu_1 - \mu_0' \Sigma^{-1} \mu_0) + \mathrm{LD}, \tag{3.23}$$

where

$$\mathrm{LD} := \gamma' X, \tag{3.24}$$

with

$$\gamma := \Sigma^{-1}(\mu_1 - \mu_0). \tag{3.25}$$

LD is Fisher's *linear discriminant* [15], best separating the pair of multivariate normal observational distributions for $X \mid T = 0$ and $X \mid T = 1$.

Suppose V is a linear sufficient covariate—a linear function of X that is itself a sufficient covariate. We have proved that the coefficient of T in the observational linear regression of Y on T and V is δ [9]. From (3.23) we see that LD is a propensity variable which is a linear strongly sufficient covariate. We deduce that under the given distributions, the coefficient of T in the observational regression of Y on T and LD is δ.

Theorem 3. *The coefficient of T in the linear regression of Y on* (T, LD) *is the same as that in the linear regression of Y on* (T, X).

Theorem 3 states that it is algebraically true that X and Fisher's linear discriminant LD generate identical coefficient of T in linear regressions, which does not have a direct link to the regimes and causality whatsoever. In our linear normal model, δ is interpreted as ACE and can be identified from the observational data simply because we have assumed that X is a strongly sufficient covariate. Applying Theorem 3 to the empirical distribution of (Y, T, X) from a sample, we deduce Corollary 1 as follows.

Corollary 1. *Suppose we have data on* (Y, T, X) *for a sample of individuals. Let* LD^* *be the sample linear discriminant for T based on X. Then the coefficient of T in the sample linear regression of Y on T and* LD^* *is the same as that in the sample linear regression of Y on T and X.*

Rosenbaum and Rubin [19, Sect. 3.4] also give this result with a brief non-causal argument: whenever the sample dispersion matrix is used in both the form of LD and regression adjustment, the estimated coefficient of T must be the same.

As discussed [9], here is a paradox: we regard adjustment for the propensity variable as an adjustment for the treatment assignment process, by regressing Y on T and the estimated propensity variable LD^*. However, from the result of Corollary 1, it appears that what we actually adjust for is the full set of covariates X, which makes the treatment assignment process completely irrelevant.

4.2.2 Precision in Propensity Analysis

One might intuitively think that the precision of the estimated ACE would be improved if we were to adjust for a scalar variable—the sample-based propensity variable LD^*, rather than p-dimensional variable X. However, Corollary 1 tells us that adjusting for LD^* does not increase the precision of our estimator. In fact, whether one adjusts for LD^* and for all the p predictors makes *absolutely no difference* to our estimate, and thus, to its precision. Similar conclusions have been drawn in [10, 28, 30]. Our intuition is that the increased precision obtained by regressing on V is offset by the overfitting error involved in selecting V.

Previous evidence [11, 18, 25] supports the claim that the estimated propensity variable outperforms the true propensity variable. That is, adjusting for the former yields higher precision of the estimated ACE than the latter. These two types of adjustment correspond to regressing Y on (T, LD) and on (T, LD^*) in our model and both provide an unbiased estimator of ACE. The claim obviously cannot be always valid by simply considering a special case: $\mathrm{LD} = \mathrm{LP}$, because by Corollary 1, regressing on LD^* is the same as adjusting for LP^*, which by the Gauss–Markov theorem will be less precise than regressing on the true linear predictor LP (or equivalently LD). Nevertheless, the claim is likely to hold when LD is not highly correlated with LP because LD is a less precise response predictor.

4.2.3 Asymptotic Variance Analysis

To gain a closer insight into the variance of the estimated ACE, by adjusting for the true propensity variable PV (if known) and the estimated propensity variable EPV, we consider a toy example in which the parameters in (3.19)–(3.21) are set as follows:

$$p = 2, \quad P_{\emptyset}(T = 1) = \theta \in (0, 1), \quad b = (b_1, b_2)',$$

the covariance matrix Σ is diagonal with identical entries τ, and

$$\mathrm{E}(X_2 \mid T = 1) = \mathrm{E}(X_2 \mid T = 0) = \mathrm{E}(X_2) \tag{3.26}$$

By the setting of Σ, we see that

$$X_1 \perp\!\!\!\perp X_2 \mid T. \tag{3.27}$$

It is also clear that the true PV is just X_1, by minimal treatment-sufficient reduction and related Eqs. (3.23)–(3.25). The conditions according to our model setting are expressed by a DAG as shown in Fig. 3.4.

In practice, all the parameters are unknown, and consequently the exact form of PV is not known. What one would normally do is adjust for the whole set

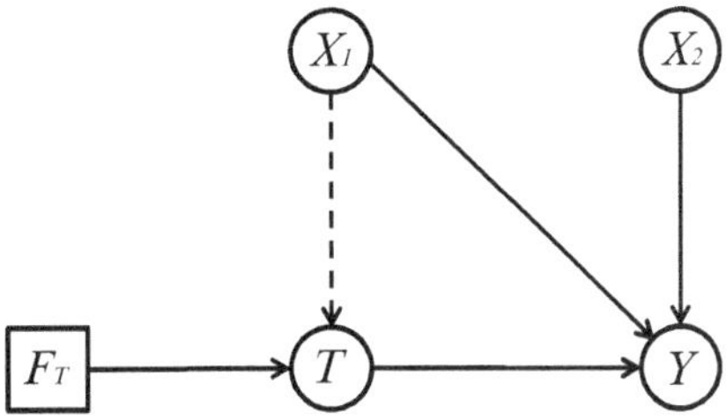

Fig. 3.4 Propensity variable X_1 and response predictor $X = (X_1, X_2)$

of the observed X, which is equivalent to adjusting for LD* (or EPV) in the linear regression approach by Corollary 1. In particular, two linear regressions are considered as follows:

$\underline{M_0}$: Y on (T, X),
$\underline{M_1}$: Y on (T, X_1).

Then the design matrix is $(1, T, X_1, X_2)'$ for M_0 and $(1, T, X_1)'$ for M_1. Let $\widehat{\beta_{M_0}}$ and $\widehat{\beta_{M_1}}$, respectively, be the least square estimators of the parameters in M_0 and M_1. The asymptotic variance of $\widehat{\beta_{M_0}}$ for sample size n is then given as

$$\mathrm{Var}_{.asy}(\widehat{\beta_{M_0}}) = \frac{A^{-1}\mathrm{Var}(Y \mid T, X)}{n} = \frac{A^{-1}\phi}{n},$$

where

$$A = \begin{pmatrix} 1 & \theta & \mathrm{E}(X_1) & \mathrm{E}(X_2) \\ \theta & \theta & \mathrm{E}(TX_1) & \mathrm{E}(TX_2) \\ \mathrm{E}(X_1) & \mathrm{E}(TX_1) & \mathrm{E}(X_1{}^2) & \mathrm{E}(X_1X_2) \\ \mathrm{E}(X_2) & \mathrm{E}(TX_2) & \mathrm{E}(X_1X_2) & \mathrm{E}(X_2{}^2) \end{pmatrix}.$$

By solving A^{-1} and extract the (2, 2)th element which is variance multiplier of the coefficient of T, we have that

$$\mathrm{Var}_{.asy}(\widehat{\delta_{M_0}}) = \frac{(W_{X_1X_1}W_{X_2X_2} - W_{X_1X_2}{}^2)\phi}{n\theta(1-\theta)(V_{X_1X_1}V_{X_2X_2} - V_{X_1X_2}{}^2)},$$

where

$$\begin{aligned} W_{X_1X_2} &= \mathrm{E}(X_1X_2) - \mathrm{E}(X_1)\mathrm{E}(X_2) = \mathrm{Cov}(X_1, X_2), \\ V_{X_1X_2} &= \mathrm{E}(X_1X_2) - \theta\mathrm{E}(X_1 \mid T=1)\mathrm{E}(X_2 \mid T=1) \\ &\quad -(1-\theta)\mathrm{E}(X_1 \mid T=0)\mathrm{E}(X_2 \mid T=0), \\ W_{X_1X_1} &= \mathrm{E}(X_1{}^2) - [\mathrm{E}(X_1)]^2 = \mathrm{Var}(X_1), \\ W_{X_2X_2} &= \mathrm{E}(X_2{}^2) - [\mathrm{E}(X_2)]^2 = \mathrm{Var}(X_2), \end{aligned}$$

and

$$V_{X_1X_1} = \mathrm{E}(X_1{}^2) - \theta[\mathrm{E}(X_1 \mid T = 1)]^2 - (1-\theta)[\mathrm{E}(X_1 \mid T = 0)]^2,$$
$$V_{X_2X_2} = \mathrm{E}(X_2{}^2) - \theta[\mathrm{E}(X_2 \mid T = 1)]^2 - (1-\theta)[\mathrm{E}(X_2 \mid T = 0)]^2.$$

By (3.26),

$$V_{X_2X_2} = \mathrm{Var}(X_2) = W_{X_2X_2}$$

and

$$V_{X_1X_2} = \mathrm{Cov}(X_1, X_2) = W_{X_1X_2},$$

where, by (3.27),

$$\mathrm{Cov}(X_1, X_2) = \mathrm{E}\{\mathrm{Cov}(X_1 \mid T, X_2 \mid T)\} + \mathrm{Cov}\{\mathrm{E}(X_1 \mid T), \mathrm{E}(X_2 \mid T)\} = 0.$$

Hence,

$$\mathrm{Var}_{.asy}(\widehat{\delta_{M_0}}) = \frac{\phi \mathrm{Var}(X_1)/[n\theta(1-\theta)]}{\mathrm{E}(X_1{}^2) - \theta[\mathrm{E}(X_1 \mid T = 1)]^2 - (1-\theta)[\mathrm{E}(X_1 \mid T = 0)]^2}. \quad (3.28)$$

For M_1, by (3.27),

$$\begin{aligned}\mathrm{Var}_{.asy}(\widehat{\delta_{M_1}}) &= \frac{W_{X_1X_1}}{n\theta(1-\theta)V_{X_1X_1}} \cdot \mathrm{Var}(Y \mid T, X_1) \\ &= \frac{W_{X_1X_1}}{n\theta(1-\theta)V_{X_1X_1}} \cdot \{\phi + b_2{}^2 \mathrm{Var}(X_2 \mid T, X_1)\} \\ &= \frac{(\phi + b_2{}^2\tau)\mathrm{Var}(X_1)/[n\theta(1-\theta)]}{\mathrm{E}(X_1{}^2) - \theta[\mathrm{E}(X_1 \mid T = 1)]^2 - (1-\theta)[\mathrm{E}(X_1 \mid T = 0)]^2}. \quad (3.29)\end{aligned}$$

Comparing (3.28) and (3.29), we have that $\mathrm{Var}_{.asy}(\widehat{\delta_{M_0}}) < \mathrm{Var}_{.asy}(\widehat{\delta_{M_1}})$ unless X_2 is random noise rather than the linear predictor, i.e., $b_2 = 0$ which equalises the two asymptotic variances.

Lemma 3. *Under the given distributional assumptions (3.19)–(3.21), suppose the propensity variable* LD *is not the same as the linear predictor* LP, *and* LD *is independent of variables that are merely response predictors. Then the asymptotic variance of the estimated ACE from the linear regression by adjusting for the estimated propensity variable* LD* *is more precise than that by adjusting for the population propensity variable* LD.

4.2.4 Simulations

Simulations are carried out for numerical illustration. Suppose we have the following true values for the parameters in (3.19)–(3.21): $p = 2, d = 0, \delta = 0.5, b = (0, 1)', \phi = 1, \theta = 0.5, \mu_1 = (1, 0)', \mu_0 = (0, 0)', \Sigma = I_2$.

Then the population linear predictor is LP $= X_2$, with

$$Y \mid (X, T, F_T) \sim \mathcal{N}(\frac{1}{2}T + X_2, 1),$$

while the population linear discriminant LD $= X_1$ which is not predictive to Y. Since for any regime $f = 0, 1, \emptyset$,

$$\mathrm{E}_f(Y \mid X_1, T) = \mathrm{E}_f\{\mathrm{E}_f(Y \mid X, T) \mid X_1, T\} = \frac{1}{2}T$$

and

$$\mathrm{Var}_f(Y \mid X_1, T) = \mathrm{E}_f\{\mathrm{Var}_f(Y \mid X, T) \mid X_1, T\} + \mathrm{Var}\{\mathrm{E}_f(Y \mid X, T) \mid X_1, T\} = 2.$$

The conditional distribution of Y given (X_1, T), for any regime, is then given by

$$Y \mid (X_1, T, F_T) \sim \mathcal{N}(\frac{1}{2}T, 2).$$

To investigate the performance of the population-based as well as sample-based LP and PV, we now consider four linear regression models:

M_0: Y on T and X ($X = (X_1, X_2)$),
M_1: Y on T and X_1,
M_2: Y on T and X_2,
M_3: Y on T and LD*,

where M_0 is the full model with all parameters unknown. In M_1, by setting $b_2 = 0$, the true linear discriminant LD $= X_1$ is fitted. While fitting the true linear predictor LP $= X_2$, equivalent to setting $b_1 = 0$, we get M_2. Note that all these models are 'true'. For M_1 the true value of b_1 is 0, and the true residual variance is 2, as against 1 for M_0 and M_2. Finally, for any dataset with no information of parameters, we construct the estimated propensity variable LD*, and then fit the model M_3.

In each model M_k, for $k = 0, 1, 2, 3$, the least-squares estimator $\widehat{\delta_k}$ is unbiased for $\delta = 0.5$. By the Gauss–Markov theorem and Corollary 1,

$$\mathrm{Var}(\widehat{\delta_0}) = \mathrm{Var}(\widehat{\delta_3}) \geq \mathrm{Var}(\widehat{\delta_2}).$$

Asymptotically, we have that $\mathrm{Var}.asy(\widehat{\delta_0}) = \mathrm{Var}.asy(\widehat{\delta_3}) = 5/n$, $\mathrm{Var}.asy(\widehat{\delta_2}) = 4/n$, and $\mathrm{Var}.asy(\widehat{\delta_1}) = 10/n$. It is indeed asymptotically less precise to adjust for PV than for its estimate in our model, which is in accordance with Lemma 3.

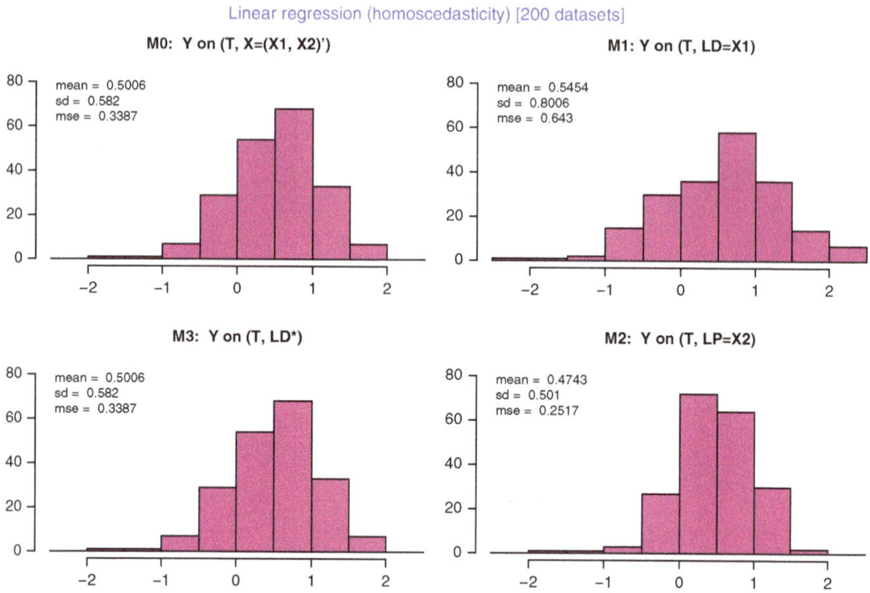

Fig. 3.5 Estimates of ACE by regression on (clockwise): *1* X_1 and X_2. *2* Population linear discriminant (propensity variable) X_1. *3* Population linear predictor X_2. *4* Estimated linear discriminant (propensity variable) LD*

For the sample analysis, 200 simulated datasets are generated, each of size $n = 20$. Shown in Fig. 3.5 are the empirical distributions of $\widehat{\delta_k}$ for all four models. Unsurprisingly, in terms of precision (from high to low), first comes the LP; next is the estimated propensity variable LD* (or the estimated linear predictor LP*), or equivalently, X (= (X_1, X_2)); and last comes the true propensity variable LD $= X_1$.

4.3 Normal Linear Model (Heteroscedasticity)

Investigation in the homoscedasticity case is simple because PV is equivalent to LD, where linearity makes analysis straightforward. If covariance matrices of the conditional distribution of X for the two treatment groups are not identical, it turns out that adjusting for PV is not appropriate.

Suppose now that, keeping all other distributional assumptions of Sect. 4.2 unchanged, (3.21) is re-specified as

$$X \mid (T, F_T = \emptyset) \sim \mathcal{N}(\mu_T, \Sigma_T)$$

with different covariance matrices Σ_0 and Σ_1 for $T = 0$ and $T = 1$. The distribution of X in all regimes then becomes

$$X \mid F_T \sim (1 - \theta)\,\mathcal{N}(\mu_0, \Sigma_0) + \theta\,\mathcal{N}(\mu_1, \Sigma_1).$$

Accordingly,

$$\log \Lambda = c + \text{QD}$$

where

$$c = -\frac{1}{2}\left\{\log(\det \Sigma_1) - \log(\det \Sigma_0) + \mu_1' \Sigma_1^{-1} \mu_1 - \mu_0' \Sigma_0^{-1} \mu_0\right\}$$

and

$$\text{QD} := \left(\Sigma_1^{-1}\mu_1 - \Sigma_0^{-1}\mu_0\right)' X - \frac{1}{2} X' \left(\Sigma_1^{-1} - \Sigma_0^{-1}\right) X. \tag{3.30}$$

QD is the *quadratic discriminant* including a linear term and a quadratic term of X, distinguishing the observational distributions of X given $T = 0, 1$. We see that QD is a minimal treatment-sufficient covariate, and thus a PV but no longer a linear function of X.

Because of the balancing property of PS (or PV), it now follows that ACE $=$ $\text{E}(\text{SCE}_{\text{QD}})$, with

$$\text{SCE}_{\text{QD}} = \text{E}_1(Y \mid \text{QD}) - \text{E}_0(Y \mid \text{QD}).$$

Since QD is quadratic in X, Y is no longer linear in QD, the coefficient of T by adjusting for PV ($=$ QD) in the linear regression does not provide exact ACE. However, as computation of the expectations in the above formula is non-trivial, one might wish to replace $\text{E}_{\emptyset}(Y \mid T, \text{QD})$ by the linear regression of Y on (T, QD), and approximate the estimated ACE. Alternatively, one can take non-parametric approaches such as matching or subclassification on QD [20]. A number of papers on various matching approaches for causal effects have been collected in [24]. More recently, statistical software becomes available for multivariate and PS matching in R [27].

Now we discuss subclassifications and linear regressions based on QD, compared to linear regressions based on LP and LD. The linear discriminant is again in the form

$$\text{LD} = (\mu_1 - \mu_0)' \Sigma^{-1} X,$$

but with $\Sigma = (1 - \theta)\Sigma_0 + \theta \Sigma_1$, the sum of the weighted dispersion matrices of the two treatment groups. From the formulae of QD and LD, we conclude that it is LD that comprises all variables with expectations depending on T. In a DAG representation of this scenario, each of such variables must have an arrow pointing to T. However, the genuine PV ($=$ QD) may depend on all the components of X, according to its quadratic term in (3.30). Only with homoscedasticity, PV is equivalent to LD and includes all variables associated with T.

Although LD is not a sufficient covariate here, Theorem 3 still applies. It enables us to identify ACE from the linear regression of Y on (T, LD), which is equivalent to the linear regression of Y on (T, X). However, other authors claim that only if LD is highly correlated with PS, adjustment for LD works well in regressions [19]. This may attribute to different scenarios considered, i.e., in our model Y is linearly related to X while non-linear in X in theirs.

4.3.1 Simulations

Simulated data is based on the above model, with the parameters: $p = 20$, $d = 0$, $\delta = 0.5$, $\theta = 0.5$, $b = (0, 1, \ldots, 0)'$, $\mu_0 = (0, \ldots, 0)'$ and $\mu_1 = (0.5, 0, 0, \ldots, 0)'$. Also, Σ_0 is set, diagonally, to 0.8 for the first ten entries and to 1.3 for the remaining entries, and Σ_1 the identity matrix.

We then have, for the population, that

$$\text{LD} = \frac{5}{9}X_1,$$

$$\text{PV} = \text{QD} = \frac{1}{2}X_1 + \frac{1}{8}\sum_{i=1}^{10} X_i^2 - \frac{3}{26}\sum_{j=11}^{20} X_j^2,$$

and $\text{LP} = X_2$. By estimating μ_0 and μ_1, Σ_0 and Σ_1 from observed data, we can compute sample-based LD* and QD*.

The results from 200 simulated datasets, each of size 500, are given in Fig. 3.6. The first three plots (clockwise) are from the linear regressions of Y on, respectively (T, X_2), (T, LD), and (T, QD). The last plot is the result of subclassification on PV (= QD). That is, 500 observations are divided into 5 subclasses with equal number of observations in each, based on the values of QD. Within each subclass, units from the two treatment groups are roughly comparable such that the average difference of the response may be interpreted as the estimated SCE. Then ACE is estimated by summing over SCEs, each weighted by $1/5$. Note that the sample size has increased, since we must have at least one observation for each treatment in each subclass.

Since LD and QD are practically unknown, they need be estimated from the observed data. Also, we do not know exactly the response predictors or the confounders, full set of the observed X may have to be used for analysis.

Figure 3.7 gives the results from the same 200 datasets as above. Again, the first three plots are the results of linear regressions of Y, but on, respectively, (T, X), (T, LD^*), and (T, QD^*), where LD* and QD* are the sample linear and quadratic discriminants. Shown in the last plot is the result of subclassification on EPV (= QD*). Unsurprisingly, by comparing the mean, standard deviation and mean squared error of the estimated ACE, regression of Y on $(T, \text{LP} = X_2)$ comes the best among all eight approaches in Figs. 3.6 and 3.7. Regressing on (T, X) is no better than regressing on (T, X_2) because all variables except X_2 in X are not predictors but noise of Y. In confirmation of the theory in Sect. 4.2.1, regressing on LD*,

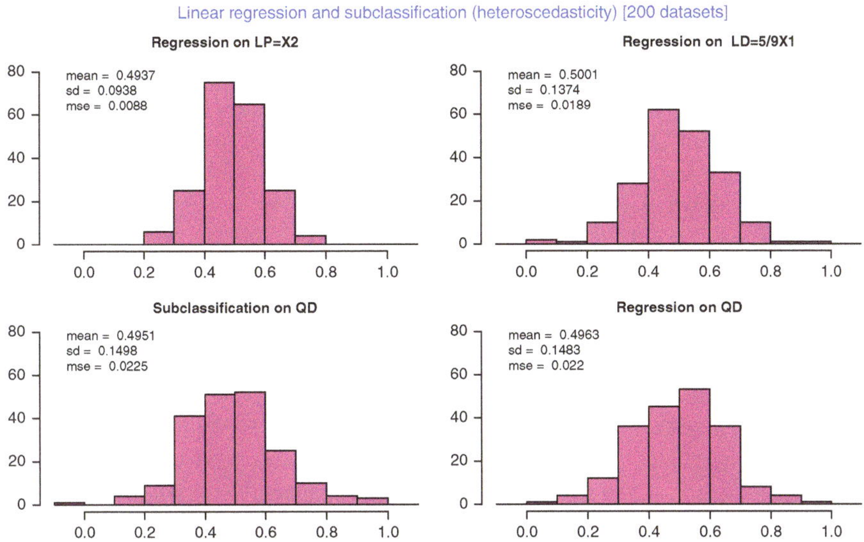

Fig. 3.6 Estimates of ACE by four different methods (clockwise): *1* Regression on population linear predictor LP $= X_2$. *2* Regression on population linear discriminant LD $= \frac{5}{9}X_1$. *3* Regression on population quadratic discriminant (propensity variable) QD. *4* Subclassification on QD

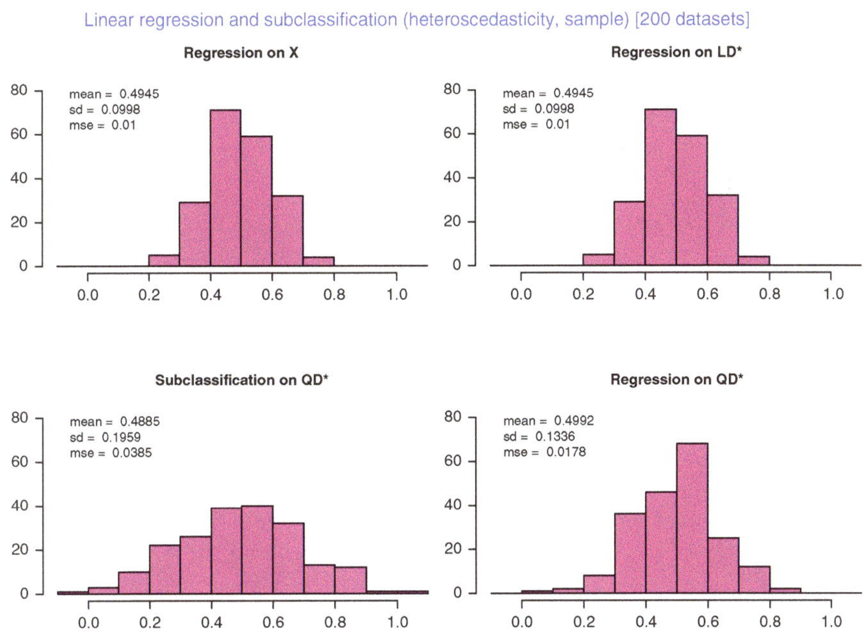

Fig. 3.7 Estimates of ACE by four different methods (clockwise): *1* Regression on sufficient covariate X. *2* Regression on sample linear discriminant LD*. *3* Regression on sample quadratic discriminant (propensity variable) QD*. *4* Subclassification on QD*

rather than on X, has absolutely no effect on the estimated ACE. LD* outperforms LD because the latter does not contain the response predictor. Regressions on LD, QD, and on QD* are roughly equal, because apart from X_1, the distributions of the remaining 19 variables are identical, with rather small multipliers. Thus, the two quadratic terms in QD are roughly the same, and QD $\approx \frac{1}{2}X_1$ works approximately as a function of a single variable X_1. Last comes subclassification on the quadratic PV, particularly when it is estimated.

4.4 *Propensity Analysis in Logistic Regression*

As already investigated, propensity analysis in linear regression is fairly straightforward. In many cases, however, response Y is not linear in X. We know that despite its name, generalised linear model (GLM) is not a linear model, because it is a non-linear function of the response that is linearly related to its predictors. Logistic regression is widely applied as a type of GLM if the response is binary. For example, doctors often record the outcome of a surgery on a patient as either 'cured' or 'not cured'. Next, a logistic model is used in our illustrative study.

4.4.1 Model Construction

For simplicity, suppose that $Y, T(1 \times 1)$ and $X(p \times 1)$ are all binary and components of X are mutually independent, The joint distribution of (F_T, X, T, Y) is constructed as follows:

$$X \mid F_T \sim Ber(\pi) \tag{3.31}$$

$$\text{logit}\{\text{P}_\emptyset(T \mid X)\} = c + a'X \tag{3.32}$$

$$\text{logit}\{\text{P}_f(Y \mid T, X)\} = d + \delta T + b'X, \tag{3.33}$$

for $f = 0, 1, \emptyset$; and π is $(p \times 1)$. Property 3 and $P_f(Y = 1 \mid T, X) \in (0, 1)$ are required such that (3.32) and (3.33) are well defined.

It is immediately seen that X is a strongly sufficient covariate and

$$\begin{aligned}
\text{ACE} &= \text{E}_\emptyset\{\text{E}_1(Y \mid X)\} - \text{E}_\emptyset\{\text{E}_0(Y \mid X)\} \\
&= \text{E}_\emptyset\{\text{P}_\emptyset(Y \mid T = 1, X)\} - \text{E}_\emptyset\{\text{P}_\emptyset(Y \mid T = 0, X)\} \\
&= \text{E}_\emptyset\left\{\frac{1}{1 + e^{-(d+\delta+b'X)}} - \frac{1}{1 + e^{-(d+b'X)}}\right\}.
\end{aligned}$$

If the parameters are set as follows:

$$p = 3, \quad \pi = (\pi_1, \pi_2, \pi_3)', \quad a = (a_1, a_2, 0)', \quad b = (0, b_2, b_3)', \tag{3.34}$$

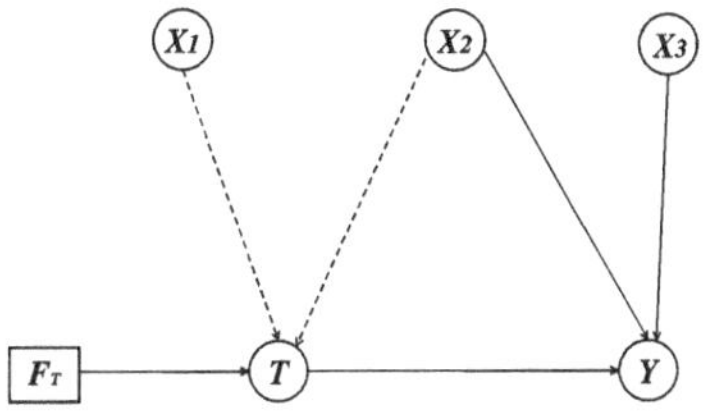

Fig. 3.8 DAG for the logistic model

then the response predictor is $b_2X_2 + b_3X_3$ and PV $= a_1X_1 + a_2X_2$. Figure 3.8 depicts the relationships of the variables in our model. Then we have that

$$\text{logit}\{P(Y \mid T, X_2, X_3)\} = \text{logit}\{P(Y \mid T, X)\} = d + \delta T + b_2X_2 + b_3X_3,$$

and

$$\begin{aligned} P(Y=1 \mid T, X_1, X_2) &= P(Y=1 \mid T, X_2) \\ &= \mathrm{E}\left\{\frac{1}{1+e^{-(d+\delta T+b_2X_2+b_3X_3)}} \mid T, X_2\right\} \\ &= \frac{\pi_3}{1+e^{-(d+\delta T+b_2X_2+b_3)}} + \frac{1-\pi_3}{1+e^{-(d+\delta T+b_2X_2)}} \end{aligned}$$

which does not depend on X_1. And we have that

$$\begin{aligned} \text{ACE} \quad = \quad & \pi_2\pi_3\left\{\frac{1}{1+e^{-(d+\delta+b_2+b_3)}} - \frac{1}{1+e^{-(d+b_2+b_3)}}\right\} \\ & + (1-\pi_2)\pi_3\left\{\frac{1}{1+e^{-(d+\delta+b_3)}} - \frac{1}{1+e^{-(d+b_3)}}\right\} \\ & + \pi_2(1-\pi_3)\left\{\frac{1}{1+e^{-(d+\delta+b_2)}} - \frac{1}{1+e^{-(d+b_2)}}\right\} \\ & + (1-\pi_2)(1-\pi_3)\left\{\frac{1}{1+e^{-(d+\delta)}} - \frac{1}{1+e^{-d}}\right\}, \end{aligned} \tag{3.35}$$

which is determined by $d, \delta, b_2, b_3, \pi_2$ and π_3. This extremely simple example, with only three components of X that are all binary, already results in a complicated form for ACE, which would be even worse for high dimensional X and various types of variables. Next, instead of simulation, we conduct propensity analysis on real data.

4.4.2 Propensity Analysis of Custodial Sanctions Study

We illustrate the method with the aid of a study involving 511 subjects sentenced to prison in 1980 by the California Superior Court, and 511 offenders sentenced to probation following conviction for certain felonies [2]. These probationers were matched to the prisoners on county of conviction, condition offence type and risk of imprisonment quantitative index, so as to bring into the final sample the most serious offenders on probation and the least serious offenders sentenced to prison. The structure of this study corresponds to the (partially matched) case–control design. In fact, this is analogous to the regression discontinuity designs where only observations near the cut-off of the risk score are included for causal effect analysis [13]. We were to compare the average causal effect of judicial sanction (probation or prison) on the probability of re-offence. We specify variables as follows.

- Treatment T: taking values 0 (probation) and 1 (prison);
- Response Y: occurrence of recidivism (re-offence);
- Pre-treatment variable X: including 17 carefully selected non-collinear variables that we can reasonably assume to make X a strongly sufficient covariate.

Simple random multiple imputation by bootstrapping (R package: mi) was applied to deal with missing data. We then considered two logistic regressions for the imputed data:

1. Y on (T, X), where X includes all the 17 variables.
2. Y on (T, EPS), where EPS is the propensity score estimated from the logistic regression of T on all the 17 variables. In selecting these variables, we took

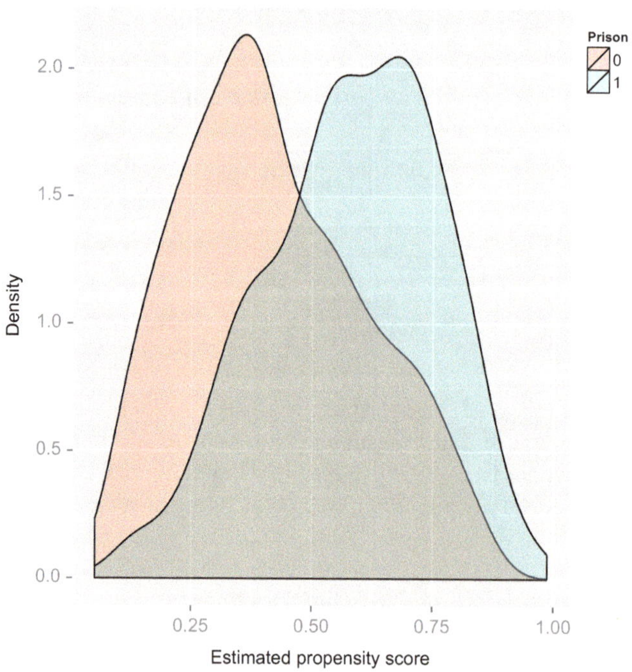

Fig. 3.9 Distribution density comparison of the estimated propensity score: prison vs. probation

Table 3.1 Coefficients of judicial sanction ('prison' with respect to 'probation') from logistic regressions: 1. Y on (T, X); 2. Y on (T, EPS)

Regression	Coefficient	Standard error	p-value
Y on (T, X)	−0.1631	0.1579	0.3014
Y on (T, EPS)	−0.1713	0.1503	0.2545

advantage of the possibility of trying various sets of covariates in the model, without inflating the type I error since these regressions do not involve the response information. The distribution densities of the two treatment groups are shown in Fig. 3.9, where we see a large overlapping area.

Shown in Table 3.1 are the results. In this case, regression on the full set of X and on the estimated PS makes little difference, since the summary statistics from the two approaches are quite similar. Although the negative values of both the coefficients imply reduced re-offence for the imprisonment, they are not statistically significant.

5 Double Robustness

Since the underlying response regression model (RRM): $Y \mid (X, T, F_T = \emptyset)$ and the propensity model (PM): $T \mid (X, F_T = \emptyset)$ are most likely unknown, one may specify parametric models based on previous experience. Moreover, as discussed in Sect. 3.3, a strongly sufficient covariate can be reduced by two alternative approaches from specified models, which enables estimating ACE by either method as follows:

1. Adjustment for response predictors from correctly specified RRM;
2. Adjustment for a PV (or PS) from correctly specified PM, either in response regression (if RRM is correctly specified), or otherwise, by non-parametric approaches, e.g., matching.

Due to lack of knowledge, it may well be that *at least one* model is misspecified. Little could be done if both models are wrong. Thus, our interest is to find a single estimator that produces a good estimate, given that at least one model is correct.

ACE is normally estimated from the observed data. Suppose there are n individuals in an observational study. Observations (x_i, t_i, y_i), where $i = 1, .., n$, are generated from the joint distribution of (X_i, T_i, Y_i) that are independent and identically distributed. The estimation of the ACE requires estimates of the expected response for both treatment groups assigned by intervention. We have already demonstrated that, within the decision-theoretic framework, ACE is identifiable from pure observational data if X is a strongly sufficient covariate. Here, X is again assumed to be strongly sufficient and thus satisfies Properties 1–3.

5.1 Augmented Inverse Probability Weighted Estimator

To construct the augmented inverse probability weighted (AIPW) estimator, we discuss two scenarios:

- **Correct RRM**: Suppose that we know the RRM. For convenience, we write $\mathrm{E}_t(Y)$ as μ_t, so
$$\mu_t = \mathrm{E}_\emptyset[\mathrm{E}_\emptyset(Y \mid X, T = t)] \tag{3.36}$$
since X is strongly sufficient. Hence, in observational studies, $\mathrm{E}_\emptyset(Y \mid X, T = t)$ is an unbiased estimator of μ_t, for $t = 0, 1$. Consequently, $\mathrm{E}_\emptyset(Y \mid X, T = 1) - \mathrm{E}_\emptyset(Y \mid X, T = 0)$ is an unbiased estimator of ACE.
- **Correct PM**: Consider that the PM is correct, i.e., $\pi(X) = \mathrm{P}_\emptyset(T = 1 \mid X)$.

Lemma 4. *Suppose that the propensity model is correct and that X is a strongly sufficient covariate. Then*

$$\mathrm{ACE} = \mathrm{E}_\emptyset\left\{\frac{T}{\pi(X)}Y\right\} - \mathrm{E}_\emptyset\left\{\frac{1-T}{1-\pi(X)}Y\right\}, \tag{3.37}$$

where $\mathrm{E}_\emptyset\left\{\frac{T}{\pi(X)}Y\right\} = \mu_1$ and $\mathrm{E}_\emptyset\left\{\frac{1-T}{1-\pi(X)}Y\right\} = \mu_0$.

Proof.

$$\begin{aligned}
\mathrm{E}_\emptyset\left\{\frac{T}{\pi(X)}Y\right\} &= \mathrm{E}_\emptyset\left\{\mathrm{E}_\emptyset(\frac{T}{\pi(X)}Y \mid X)\right\} = \mathrm{E}_\emptyset\left\{\frac{1}{\pi(X)}\mathrm{E}_\emptyset(TY \mid X)\right\} \\
&= \mathrm{E}_\emptyset\left\{\frac{1}{\pi(X)}\mathrm{E}_\emptyset(Y \mid X, T = 1)\mathrm{P}_\emptyset(T = 1 \mid X)\right\} \\
&= \mathrm{E}_\emptyset\left\{\mathrm{E}_\emptyset(Y \mid X, T = 1)\right\} = \mu_1 \qquad \text{by (3.36).}
\end{aligned}$$

It automatically follows that $\mathrm{E}_\emptyset\left\{\frac{1-T}{1-\pi(X)}Y\right\} = \mu_0$. By Lemma 4, we see that, under the observational regime, $\frac{T}{\pi(X)}Y$ and $\frac{1-T}{1-\pi(X)}Y$ are unbiased estimators of μ_1 and μ_0, respectively.

One may have noticed that the two terms for ACE in (3.37) are similar with the Horvitz–Thompson (HT) estimator for sample surveys [12]. They are, however, different in various aspects. The aim of HT estimator is to estimate the mean of a finite population $Y_1, \ldots, Y_N$, denoted by $\mu = N^{-1}\sum_{i=1}^{N} Y_i$, from a stratified sample of size n drawn without replacement. For $i = 1, \ldots, N$, let Δ_i be the binary sampling indicator ($\Delta_i = 1$: unit i is in sample; 0: unit i is not in sample), and π_i be the probability that unit i being drawn in the sample. Then HT estimator is given by

$$\hat{\mu}_{\mathrm{HT}} = N^{-1}\sum_{i=1}^{N}\frac{\Delta_i}{\pi_i}Y_i, \tag{3.38}$$

where π_i is pre-specified, and thus known in a sample survey design. But the propensity model $\pi(X)$ in (3.37) is normally unknown. Moreover, HT estimator is applied to estimate the mean of a finite population, while $\widehat{\text{ACE}}$ is used to estimate the mean of a superpopulation.[5] HT estimator depends on pre-specified sampling scheme, but observations involved in $\widehat{\text{ACE}}$ are generated from, and thus are dependent on, the joint distribution of (X, T, Y) in the observational regime. Nevertheless, both HT estimator and $\widehat{\text{ACE}}$ are formed by means of the inverse probability weights $1/\pi_i$ or $1/\pi(X)$. In fact, HT estimator is also termed the inverse probability weighted (IPW) estimator.

Sample surveys are closely related to missing data because the information is missing for those not sampled. So IPW estimator is frequently used in missing data models in the presence of partially observed response [1, 3, 14]. As counterfactuals are also regarded as missing data, IPW estimator can be used in the potential response framework with half observed information, to make causal inference of treatment effect under the assumptions of 'strongly ignorable treatment assignment': $(Y(0), Y(1)) \perp\!\!\!\perp T \mid X$ and 'no unobserved confounders' [1, 29].

5.1.1 Augmented Inverse Probability Weighted Estimator

From above discussion, there exists an unbiased estimator of ACE if either RRM or PM is correct. However, unknown RRM and PM makes it impossible to decide whether they are correct. Nevertheless, the augmented inverse probability weighted (AIPW) estimator can be constructed by combining the two models in the following alternative forms:

$$\begin{aligned}\hat{\mu}_{1,\text{AIPW}} &= m(X) + \frac{T}{\pi(X)}(Y - m(X)) \\ &= \frac{T}{\pi(X)}Y + \left[1 - \frac{T}{\pi(X)}\right] m(X),\end{aligned} \tag{3.39}$$

and similarly,

$$\begin{aligned}\hat{\mu}_{0,\text{AIPW}} &= m(X) + \frac{1-T}{1-\pi(X)}(Y - m(X)) \\ &= \frac{1-T}{1-\pi(X)}Y + \left[1 - \frac{1-T}{1-\pi(X)}\right] m(X),\end{aligned} \tag{3.40}$$

where $m(\cdot)$ and $\pi(\cdot)$ are arbitrary functions of X. As also indicated in its name, $\hat{\mu}_{t,\text{AIPW}}$ is the sum of the IPW estimator and an augmented term.

---

[5]In causal system, finite number of individuals in a study is called 'population', which can be regard as a sample from a larger 'superpopulation' of interest.

Lemma 5. *Suppose that X is a strongly sufficient covariate. The estimator $\hat{\mu}_{t,\text{AIPW}}$ has the property of* **double robustness**. *That is, $\hat{\mu}_{t,\text{AIPW}}$ is an unbiased estimator of the population mean given $T = t$ by intervention, if either $\pi(X) = p_{\emptyset}(T = 1 \mid X)$ or $m(X) = E_{\emptyset}(Y \mid X, T = t)$.*

Proof. By similarity, we only give proof of $\hat{\mu}_{1,\text{AIPW}}$. Consider the following two scenarios:

Scenario 1: $\pi(X) = p_{\emptyset}(T = 1 \mid X)$ and $m(X)$ is an arbitrary function of X.
It is easily seen that $\hat{\mu}_{1,\text{AIPW}}$ is unbiased, from the proof of Lemma 4. Since conditional on X, the last term in (3.39) vanishes when we take expectation of $\hat{\mu}_{1,\text{AIPW}}$ in the observational regime.
Scenario 2: $m(X) = \text{E}_{\emptyset}(Y \mid X, T = 1)$ and $\pi(X)$ is an arbitrary function of X.
By (3.39), we have that

$$\begin{aligned}
\text{E}(\hat{\mu}_{1,\text{AIPW}}) &= \text{E}\left[m(X) + \frac{T}{\pi(X)}(Y - m(X))\right] \\
&= \text{E}\left\{\text{E}\left[m(X) \mid X\right]\right\} + \text{E}\left\{\text{E}\left[\frac{T}{\pi(X)}(Y - m(X)) \mid X\right]\right\} \\
&= \text{E}[m(X)] + \text{E}\left\{\frac{\text{E}(TY \mid X) - m(X)\text{E}(T \mid X)}{\pi(X)}\right\} \\
&= \text{E}[m(X)] = \mu_1 \qquad\qquad \text{by (3.36).}
\end{aligned}$$

Indeed, if either $\pi(X) = p_{\emptyset}(T = 1 \mid X)$ or $m(X) = \text{E}_{\emptyset}(Y \mid X, T = 1)$, not necessarily both, $\hat{\mu}_{1,\text{AIPW}}$ is unbiased. Consequently,

$$\widehat{\text{ACE}}_{\text{AIPW}} = \hat{\mu}_{1,\text{AIPW}} - \hat{\mu}_{0,\text{AIPW}}.$$

Theorem 4. *Suppose that X is a strongly sufficient covariate. Then the AIPW estimator $\widehat{\text{ACE}}_{\text{AIPW}}$ is doubly robust.*

To prove Theorem 4, we simply apply the fact that both $\hat{\mu}_{1,\text{AIPW}}$ and $\hat{\mu}_{0,\text{AIPW}}$ are doubly robust, so is their difference.

5.2 *Parametric Models*

Suppose that we specify two parametric working models: the propensity working model $\pi(X; \alpha)$ and the response regression working model $m(T, X; \beta)$. Then by (3.39) and (3.40), we have, for the estimated $\text{E}_1(Y)$ and $\text{E}_0(Y)$, that

$$\hat{\mu}_{1,\text{AIPW}} = n^{-1}\left\{\sum_{i=1}^{n} \frac{T_i}{\pi(X_i; \hat{\alpha})} Y_i + \left[1 - \frac{T_i}{\pi(X_i; \hat{\alpha})}\right] m(1, X_i; \hat{\beta})\right\} \qquad (3.41)$$

and

$$\hat{\mu}_{0,\text{AIPW}} = n^{-1}\left\{\sum_{i=1}^{n} \frac{1-T_i}{1-\pi(X_i;\hat{\alpha})}Y_i + \left[1 - \frac{1-T_i}{1-\pi(X_i;\hat{\alpha})}\right] m(0, X_i; \hat{\beta})\right\} \quad (3.42)$$

respectively. Therefore, by (3.41) and (3.42), we have that

$$\begin{aligned}\widehat{\text{ACE}}_{\text{AIPW}} &= \hat{\mu}_{1,\text{AIPW}} - \hat{\mu}_{0,\text{AIPW}} \\ &= n^{-1}\left\{\sum_{i=1}^{n}\left[\frac{T_i}{\pi(X_i;\hat{\alpha})} - \frac{1-T_i}{1-\pi(X_i;\hat{\alpha})}\right](Y_i - m(T_i, X_i; \hat{\beta})\right\} ,(3.43)\end{aligned}$$

which is doubly robust, i.e., $\widehat{\text{ACE}}_{\text{AIPW}}$ is a consistent and asymptotically normal estimator of ACE if either of the working models is correctly specified.

5.2.1 Discussion

Kang and Schafer [14] state that there are various ways to construct an estimator which is doubly robust. In our view, they are essentially the same, i.e., it must be in the same (or similar) form of AIPW estimator which is constructed by combining RRM and PM. Other constructions proposed in [14] are just variations of AIPW estimator. For example, in (3.38), instead of using N as denominator for each unit, they use normalised weights $\sum_{i=1}^{N} \frac{\Delta_i}{\pi_i}$. Such normalised weights are especially useful for precision improvement in the case that subjects with very small probabilities of being sampled are actually drawn from the population. Because if N is used as the weight, these subjects will influence the estimated average response enormously, and consequently, result in poor precision.

Kang and Schafer [14] have also investigated the precision performance of an doubly robust estimator when both $\pi(X)$ and $m(X)$ are moderately misspecified. They state that ‘in at least some settings, two wrong models are not better than one’. This seems obvious because the performance of this estimator will depend on the degree of misspecification of both models. This can be easily analysed in theory but far more complicated in practice, as one cannot have a good control of specifying models $\pi(X)$ and $m(X)$ based on limited observed data and previous experience (if any). Therefore, it would be difficult to measure to what extent the specified models are different from the true ones.

5.3 *Precision of* $\widehat{\text{ACE}}_{\text{AIPW}}$

5.3.1 Known Propensity Score Model

We already see that $\widehat{\text{ACE}}_{\text{AIPW}}$ is an unbiased and doubly robust estimator of ACE. Then how can we choose an arbitrary function $m(X_i)$ to minimise the variance of

$\widehat{\text{ACE}}_{\text{AIPW}}$ given correct PM? Suppose that in an experiment, we know $\pi(X_i) = \text{P}(T_i = 1 \mid X_i)$. Then in terms of the variance, we have that

$$
\begin{aligned}
&\text{Var}(\widehat{\text{ACE}}_{\text{AIPW}}) \\
&= \text{Var}\left\{n^{-1}\left[\sum_{i=1}^{n}\left(\frac{T_i}{\pi(X_i)} - \frac{1-T_i}{1-\pi(X_i)}\right)(Y_i - m(X_i))\right]\right\} \\
&= n^{-2}\left\{\text{Var}\left[\sum_{i=1}^{n}\left(\frac{T_i}{\pi(X_i)} - \frac{1-T_i}{1-\pi(X_i)}\right)Y_i\right]\right. \\
&\quad +\text{Var}\left[\sum_{i=1}^{n}\left(\frac{T_i}{\pi(X_i)} - \frac{1-T_i}{1-\pi(X_i)}\right)m(X_i)\right] \\
&\quad \left. -2\text{Cov}\left[\sum_{i=1}^{n}\left(\frac{T_i}{\pi(X_i)} - \frac{1-T_i}{1-\pi(X_i)}\right)Y_i, \sum_{i=1}^{n}\left(\frac{T_i}{\pi(X_i)} - \frac{1-T_i}{1-\pi(X_i)}\right)m(X_i)\right]\right\} \\
&= n^{-2}\left\{\text{Var}(\widehat{\text{ACE}}_{\text{HT}}) + \text{E}\left[\sum_{i=1}^{n}\frac{m^2(X_i)}{\pi(X_i)(1-\pi(X_i))}\right]\right. \\
&\quad \left. -2\,\text{E}\left[\sum_{i=1}^{n}\frac{m(X_i)\mu_{1i}}{\pi(X_i)(1-\pi(X_i))} - \frac{m(X_i)(\mu_{1i}-\mu_i)}{(1-\pi(X_i))^2}\right]\right\} \\
&= n^{-2}\left\{\text{Var}(\widehat{\text{ACE}}_{\text{HT}}) + \text{E}\left[\sum_{i=1}^{n}\frac{m^2(X_i)}{\pi(X_i)(1-\pi(X_i))}\right.\right. \\
&\quad \left.\left. -2\sum_{i=1}^{n}\left\{\frac{\mu_{1i}}{\pi(X_i)(1-\pi(X_i))} - \frac{\mu_{1i}-\mu_i}{(1-\pi(X_i))^2}\right\}m(X_i)\right]\right\},
\end{aligned}
$$

where $\mu_{1i} = \text{E}_{\emptyset}(Y_i \mid X_i, T_i = 1)$ and $\mu_i = \text{E}_{\emptyset}(Y_i \mid X_i)$.

By minimising the quadratic function of $m(X_i)$ in the expectation, it follows that

$$
\begin{aligned}
m(X_i) &= [1-\pi(X_i)]\mu_{1i} + \pi(X_i)\mu_{0i} \\
&= [1-\pi(X_i)]\text{E}_{\emptyset}(Y_i \mid X_i, T_i = 1) + \pi(X_i)\text{E}_{\emptyset}(Y_i \mid X_i, T_i = 0), \quad (3.44)
\end{aligned}
$$

which minimises the variance of $\widehat{\text{ACE}}_{\text{AIPW}}$ among all functions of X_i. In fact, if either $\pi(X_i) = p_{\emptyset}(T_i = 1 \mid X_i)$ or (3.44) holds, $\widehat{\text{ACE}}_{\text{AIPW}}$ is unbiased, and thus is doubly robust.

Let $m_1(X_i)$ and $m_0(X_i)$ denote the regressions of Y on X_i for the two treatment groups in the observational regime. It is unnecessary to require that $m_1(X_i) = \text{E}_{\emptyset}(Y_i \mid X_i, T_i = 1)$ and that $m_0(X_i) = \text{E}_{\emptyset}(Y_i \mid X_i, T_i = 0)$. As long as $m(X_i)$ is specified as the sum of the weighted expectations as in the form of (3.44), $m(X_i)$ minimises the variance of the estimated ACE.

Same result is obtained in [26] as (3.44), by minimising a weighted mean squared error of $m(X_i)$. We now discuss an alternative approach provided in [26]. Let $\widetilde{Y}_i$ denote a weighted response in a form as follows:

$$\widetilde{Y}_i = \left[\left\{\frac{1}{\pi(X_i)} - 1\right\} T_i + \left\{\frac{1}{1-\pi(X_i)} - 1\right\} (1-T_i)\right] Y_i. \tag{3.45}$$

Then by (3.44), it follows that

$$\begin{aligned}
m(X_i) &= \frac{1-\pi(X_i)}{\pi(X_i)} \mathrm{E}_\emptyset(Y_i \mid X_i, T_i = 1)\mathrm{P}(T=1 \mid X) \\
&+ \frac{\pi(X_i)}{1-\pi(X_i)} \mathrm{E}_\emptyset(Y_i \mid X_i, T_i = 0)\mathrm{P}(T=0 \mid X) \\
&= \frac{1-\pi(X_i)}{\pi(X_i)} \mathrm{E}_\emptyset(T_i Y_i \mid X_i) + \frac{\pi(X_i)}{1-\pi(X_i)} \mathrm{E}_\emptyset[(1-T_i)Y_i \mid X_i] \\
&= \mathrm{E}_\emptyset\left\{\left[\frac{1-\pi(X_i)}{\pi(X_i)} T_i + \frac{\pi(X_i)}{1-\pi(X_i)} (1-T_i)\right] Y_i \mid X_i\right\} \\
&= \mathrm{E}_\emptyset\left\{\left[\left(\frac{1}{\pi(X_i)} - 1\right) T_i + \left(\frac{1}{1-\pi(X_i)} - 1\right) (1-T_i)\right] Y_i \mid X_i\right\} \\
&= \mathrm{E}_\emptyset(\widetilde{Y}_i \mid X_i),
\end{aligned}$$

where $m(X_i)$ is obtained by simply regressing $\widetilde{Y}_i$ on X_i, rather than regressing Y_i on both X_i and T_i. However, an obvious disadvantage of this approach is its low precision. When individuals with the PS close to 0 are actually in the treatment group and/or those with the PS close to 1 are actually assigned to the control group, the weights $1/\pi(X_i)$ or $1/(1-\pi(X_i))$ of these units will be very large, which leads to corresponding responses being highly influential, which is dangerous. In fact, it may be even worse than the HT estimator as we will see next.

To show the difference of these approaches, we have implemented Monte Carlo computations for four estimators of $\widehat{\mathrm{ACE}}_{\mathrm{AIPW}}$:

1. by (3.44) with $\mathrm{E}_\emptyset(Y_i \mid X_i, T_i = 1)$ and $\mathrm{E}_\emptyset(Y_i \mid X_i, T_i = 0)$ estimated by regressing Y_i on (X_i, T_i).
2. by (3.44) with $\mathrm{E}_\emptyset(Y_i \mid X_i, T_i = 1)$ and $\mathrm{E}_\emptyset(Y_i \mid X_i, T_i = 0)$ estimated by regressing Y_i on X_i for the treatment group and control group separately.
3. by Horvitz–Thompson approach, i.e. without covariate adjustment.
4. by regression of $\widetilde{Y}_i$ on X_i.

The results of simulated 100 datasets are shown in Fig. 3.10. The first two approaches give similar results. That is, we can estimate $\mathrm{E}_\emptyset(Y_i \mid X_i, T_i = 1)$ and $\mathrm{E}_\emptyset(Y_i \mid X_i, T_i = 0)$ either simultaneously from the response regression on the treatment and X, or separately from the response regression only on X for each of the two groups. As expected, the last approach generates several extreme estimates

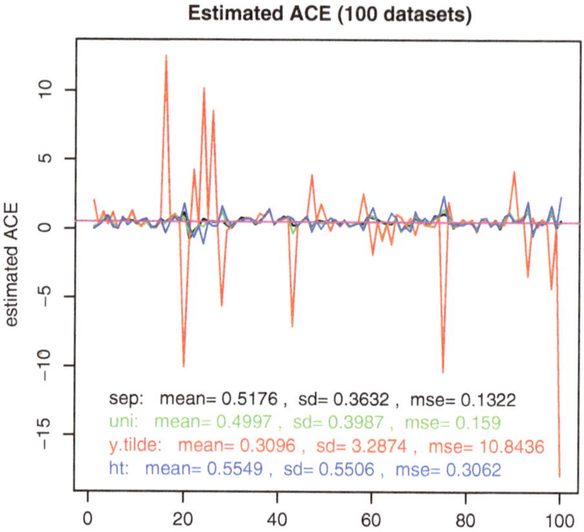

Fig. 3.10 Precision of the estimated ACE based on: (*1*) specified model for $E_{\emptyset}(Y_i \mid X_i, T_i)$; (*2*) specified models for $E_{\emptyset}(Y_i \mid X_i)$ separately for both groups; (*3*) Horvitz–Thompson estimator; (*4*) regression of $\widetilde{Y_i}$ on X_i

relative to others, which makes its variance even much larger than that of the HT estimator.

5.3.2 Known Response Regression Model

Suppose that $E_{\emptyset}(Y_i \mid X_i, T_i = 1)$ and $E_{\emptyset}(Y_i \mid X_i, T_i = 0)$ are both known but not the PM. Then the AIPW estimator can be constructed as:

$$\widehat{\text{ACE}}_{\text{AIPW}} = n^{-1}\left\{\sum_{i=1}^{n}\left[\frac{T_i}{g(X_i)} - \frac{1-T_i}{1-g(X_i)}\right](Y_i - m(X_i))\right\},$$

where

$$m(X_i) = (1 - g(X_i))\text{E}(Y_i \mid X_i, T_i = 1) + g(X_i)\text{E}(Y_i \mid X_i, T_i = 0),$$

and $g(X_i)$ is an arbitrary function of X_i.

So $\widehat{\text{ACE}}_{\text{AIPW}}$ is unbiased and its variance is computed as follows.

$$\begin{aligned}
&\text{Var}(\widehat{\text{ACE}}_{\text{AIPW}}) \\
&= \text{Var}\left\{n^{-1}\left[\sum_{i=1}^{n}\left(\frac{T_i}{g(X_i)} - \frac{1-T_i}{1-g(X_i)}\right)(Y_i - m(X_i))\right]\right\} \\
&= n^{-2}\text{Var}\left\{\sum_{i=1}^{n}\left(\frac{T_i}{g(X_i)} - \frac{1-T_i}{1-g(X_i)}\right)(Y_i - [(1-g(X_i))\mu_{1i} + g(X_i)\mu_{0i}])\right\}
\end{aligned}$$

$$= n^{-2}\text{Var}\left\{\sum_{i=1}^{n}(\mu_{1i} - \mu_{0i}) + \frac{T_i}{g(X_i)}(Y_i - \mu_{1i}) - \frac{1-T_i}{1-g(X_i)}(Y_i - \mu_{0i})\right\}$$

$$= n^{-2}\text{Var}\left\{\sum_{i=1}^{n}(\mu_{1i} - \mu_{0i})\right\}$$

$$+ n^{-2}\text{E}\left\{\text{Var}\left[\sum_{i=1}^{n}\frac{T_i}{g(X_i)}(Y_i - \mu_{1i}) - \frac{1-T_i}{1-g(X_i)}(Y_i - \mu_{0i}) \mid X_i\right]\right\}$$

$$> n^{-2}\text{Var}\left\{\sum_{i=1}^{n}(\mu_{1i} - \mu_{0i})\right\} = \text{Var}(\widehat{\text{ACE}}_{\text{RRM}}).$$

Hence, we conclude that, for each individual, if the conditional expectations of the response given X_i for both groups are known or correctly specified, then $\widehat{\text{ACE}}_{\text{AIPW}}$ will be less precise than the estimated ACE from the response regressions.

5.3.3 Discussion

If the PM is known, then the variance of $\widehat{\text{ACE}}_{\text{AIPW}}$ is minimised when $m(X_i)$ is specified as in (3.44)—where separate specification of $m_1(X_i)$ and $m_0(X_i)$ is not necessary. Rubin and van de Laan [26] have introduced a weighted response serving as an alternative, but we have shown, by simulations, that it could result in large variance of the estimated ACE and possibly larger than the HT estimator. In the case that the RRM is correctly specified, i.e., $m_1(X_i) = \text{E}_\emptyset(Y_i \mid X_i, T_i = 1)$ and $m_0(X_i) = \text{E}_\emptyset(Y_i \mid X_i, T_i = 0)$, then these two models rather than the AIPW estimator should be used to estimate ACE for higher precision of the estimator.

6 Summary

In this chapter, we have addressed statistical causal inference using Dawid's decision-theoretic framework within which assumptions are, in principle, testable. Throughout, the concept of sufficient covariate plays a crucial role. We have investigated propensity analysis in a simple normal linear model, as well as in logistic model, theoretically and by simulation. Adding weight to previous evidence [10, 11, 18, 28, 30], our results show that propensity analysis does little in improving estimation of the treatment causal effect, either unbiasedness or precision. However, as part of the augmented inverse probability weighted estimator that is doubly robust, correct propensity score model helps provide unbiased average causal effect.

Appendix: R Code of Simulations and Data Analysis

```
##################################################################
Figure 5: Linear regression (homoscedasticity)
------------------------------------------------------------------
1. Y on X;
2. Y on population linear discriminant / propensity variable LD;
3. Y on sample linear discriminant / propensity variable LD*;
4. Y on population linear predictor LP.
##################################################################

##  set parameters

p <- 2
delta <- 0.5
phi <- 1
n <- 20

alpha <- matrix(c(1,0), nrow=1)
sigma <- diag(1, nrow=p)
b <- matrix(c(0,1), nrow=p)


##  create a function to compute ACE from four linear regressions

ps <- function(r) {

  #  data for T, X and Y from the specified linear normal model

  set.seed(r)
  .Random.seed
  t <- rbinom(n, 1, 0.5)

  require(MASS)
  m <- rep(0, p)
  ex <- mvrnorm(n, mu=m, Sigma=sigma)
  x <- t%*%alpha + ex

  ey <- rnorm(n, mean=0, sd=sqrt(phi))
  y <- t*delta + x%*%b + ey

  #  calculate the true and sample linear discriminants

  ld.true <- x%*%solve(sigma)%*%t(alpha)
  pred <- x%*%b

  d1 <- data.frame(x, t)
  c <- coef(lda(t~.,d1))
  ld <- x%*%c

  #  extract estimated average causal effect (ACE)
  #  from the four linear regressions

  dhat.pred <- coef(summary(lm(y~1+t+pred)))[2]
  dhat.x <- coef(summary(lm(y~t+x)))[2]
  dhat.ld <- coef(summary(lm(y~t+ld)))[2]
  dhat.ld.true <- coef(summary(lm(y~t+ld.true)))[2]
```

```
  return(c(dhat.x, dhat.ld, dhat.ld.true, dhat.pred))
}


##  estimate ACE from 200 simulated datasets
##  compute mean, standard deviation and mean square error of ACE

g <- rep(0, 4)
for (r in 31:230) {
  g <- rbind(g, ps(r))
}
g <- g[-1,]

d.mean <- 0
d.sd <- 0
mse <- 0

for (i in 1:4) {
  d.mean[i] <- round(mean(g[,i]),4)
  d.sd[i] <- round(sd(g[,i]),4)
  mse[i] <- round((d.sd[i])^2+(d.mean[i]-delta)^2, 4)
}


##  generate Figure 5

par(mfcol=c(2,2), oma=c(1.5,0,1.5,0), las=1)
main=c("M0:  Y on (T, X=(X1, X2)')", "M3:  Y on (T, LD*)",
    "M1: Y on (T, LD=X1)", "M2:  Y on (T, LP=X2)")

for (i in 1:4){
  hist(g[,i], br=seq(-2.5, 2.5, 0.5), xlim=c(-2.5, 2.5), ylim=c(0,80),
       main=main[i], col.lab="blue", xlab="", ylab="",col="magenta")
  legend(-2.5,85, c(paste("mean = ",d.mean[i]), paste("sd = ",d.sd[i]),
       paste("mse = ",mse[i])), cex=0.85, bty="n")
}
mtext(side=3, cex=1.2, line=-1.1, outer=T, col="blue",
    text="Linear regression (homoscedasticity) [200 datasets]")

dev.copy(postscript,"lrpvpdecmbook.ps", horiz=TRUE, paper="a4")
dev.off()


############################################################################
Linear regression and subclassification (heteroscedasticity)
----------------------------------------------------------------------------

Figure 6:
1. Regression on population linear predictor LP;
2. Regression on population linear discriminant LD;
3. Regression on population quadratic discriminant / propensity variable QD;
4. Subclassification on QD.

Figure 7:
1. Regression on sample linear predictor LP*;
2. Regression on sample linear discriminant LD*;
3. Regression on sample quadratic discriminant / propensity variable QD*;
4. Subclassification on QD*.
############################################################################
```

```
## set parameters

p <- 20
d <- 0
delta <- 0.5
phi <- 1
n <- 500

a <- matrix(rep(0,p), nrow=1)
alpha <- matrix(c(0.5,rep(0,p-1)), nrow=1)
sigma1 <- diag(1, nrow=p)
sigma0 <- diag(c(rep(0.8, 10), rep(1.3, 10)), nrow=p)
b <- matrix(c(0, 1, rep(0,p-2)), nrow=p)


## create a function to compute ACE from eight approaches

ps <- function(r) {

  # data for T, X and Y from the specified linear normal model

  set.seed(r)
  .Random.seed
  pi <- 0.5
  t <- rbinom(n, 1, pi)
  n0 <- 0

  for (i in 1:n) {
    if (t[i]==0)
    n0 <- n0+1
  }

  t <- sort(t, decreasing=FALSE)
  mu1 <- a+alpha
  mu0 <- a

  require(MASS)
  m <- rep(0, p)
  ex0 <- mvrnorm(n0, mu=m, Sigma=sigma0)
  ex1 <- mvrnorm((n-n0), mu=m, Sigma=sigma1)

  a <- matrix(rep(a, n), nrow=n, byrow=TRUE)
  x0 <- a[(1:n0),] + t[1:n0]%*%alpha + ex0
  x1 <- a[(n0+1):n,] + t[(n0+1):n]%*%alpha + ex1
  x <- rbind(x0, x1)

  ey <- rnorm(n, mean=0, sd=sqrt(phi))
  d <- rep(d, n)
  y <- d + t*delta + x%*%b + ey

  # calculate linear discrimant, quadratic discrimant, for population
  # and for sample, extract estimated ACE from linear regressions

  ld <- x%*%solve(pi*sigma1+pi*sigma0)%*%t(alpha)
  d1 <- data.frame(x, t)
  c <- coef(lda(t~.,d1))
  ld.s <- x%*%c
```

```
  z1 <- x%*%(solve(sigma1)%*%t(mu1) - solve(sigma0)%*%t(mu0))
  z2 <- 0
  for (j in 1:n){
    z2[j] <-  - 1/2*matrix(x[j,], nrow=1)%*%(solve(sigma1)
      - solve(sigma0))%*%t(matrix(x[j,], nrow=1))
  }
  qd <- z1+z2

  dhat.x2 <- coef(summary(lm(y~1+t+x[,2])))[2]
  dhat.ld <- coef(summary(lm(y~1+t+ld)))[2]
  dhat.qd <- coef(summary(lm(y~1+t+qd)))[2]

  mn <- aggregate(d1, list(t=t), FUN=mean)
  m0 <- as.matrix(mn[1, 2:(p+1)])
  m1 <- as.matrix(mn[2, 2:(p+1)])
  v0 <- var(x0)
  v1 <- var(x1)

  c1 <- solve(v1)%*%t(m1)-solve(v0)%*%t(m0)
  z1.s <- x%*%c1
  c2 <- solve(v1)-solve(v0)
  z2.s <- 0
  for (i in 1:n){
    z2.s[i] <- -1/2*matrix(x[i,], nrow=1)%*%c2%*%t(matrix(x[i,], nrow=1))
  }
  qd.s <- z1.s+z2.s

  dhat.x <- coef(summary(lm(y~1+t+x)))[2]
  dhat.ld.s <- coef(summary(lm(y~1+t+ld.s)))[2]
  dhat.qd.s <- coef(summary(lm(y~1+t+qd.s)))[2]

  #  extract estimated ACE from subclassification

  d2 <- data.frame(cbind(qd, qd.s, y, t))

  tm1 <- vector("list", 2)
  tm0 <- vector("list", 2)
  te.qd <- 0

  for (k in 1:2) {
    d3 <- d2[, c(k,3,4)]
    d3 <- split(d3[order(d3[,1]), ], rep(1:5, each=100))

    tm <- vector("list", 5)
    for (j in 1:5) {
      tm[[j]] <-   aggregate(d3[[j]], list(Stratum=d3[[j]]$t), FUN=mean)
      tm1[[k]][j] <- tm[[j]][2,3]
      tm0[[k]][j] <- tm[[j]][1,3]
    }
    te.qd[k] <- sum(tm1[[k]] - tm0[[k]])/5
  }

#  return estimated ACE from the eight approaches

  return(c(dhat.x2, te.qd[1], dhat.ld, dhat.qd,
    dhat.x, te.qd[2], dhat.ld.s, dhat.qd.s))
}
```

```
##  estimate ACE from 200 simulated datasets
##  compute mean, standard deviation and mean square error of ACE

g <- rep(0, 8)
for (r in 31:230) {
   g <- rbind(g, ps(r))
}
g <- g[-1,]

d.mean <- 0
d.sd <- 0
d.mse <- 0

for (i in 1:8) {
  d.mean[i] <- round(mean(g[,i]),4)
  d.sd[i] <- round(sd(g[,i]),4)
  d.mse[i] <- round((d.sd[i])^2+(d.mean[i]-delta)^2, 4)
}


##  generate Figure 6

par(mfcol=c(2,2), oma=c(1.5,0,1.5,0), las=1)
main=c("Regression on LP=X2","Subclassification on QD",
    "Regression on  LD=5/9X1","Regression on QD")
for (i in 1:4){
  hist(g[,i], br=seq(-0.1, 1.1, 0.1), xlim=c(-0.1, 1.1), ylim=c(0,80),
      main=main[i], col.lab="blue", xlab="", , ylab="", col="magenta")
  legend(-0.2,85, c(paste("mean = ",d.mean[i]), paste("sd = ",d.sd[i]),
      paste("mse = ",d.mse[i])), cex=0.85, bty="n")
}
mtext(side=3, cex=1.2, line=-1.1, outer=T, col="blue",
    text="Linear regression and subclassification
    (heteroscedasticity) [200 datasets]")

dev.copy(postscript,"pslrsubtruebook.ps", horiz=TRUE, paper="a4")
dev.off()


##  generate Figure 7
main=c("Regression on X","Subclassification on QD*",
    "Regression on LD*",  "Regression on QD*")
for (i in 1:4){
  hist(g[,i+4], br=seq(-0.1, 1.1, 0.1), xlim=c(-0.1,1.1), ylim=c(0,80),
      main=main[i], col.lab="blue", xlab="", ylab="", col="magenta")
  legend(-0.2,85, c(paste("mean = ",d.mean[i+4]), paste("sd = ",d.sd[i+4]),
      paste("mse = ",d.mse[i+4])), cex=0.85, bty="n")
}
mtext(side=3, cex=1.2, line=-1.1, outer=T, col="blue",
    text="Linear regression and subclassification
    (heteroscedasticity, sample) [200 datasets]")

dev.copy(postscript,"pslrsubbook.ps", horiz=TRUE, paper="a4")
dev.off()
```

```
######################################################################
Figure 9 and Table 1: Propensity analysis of custodial sanctions study
----------------------------------------------------------------------
1. Y on all 17 variables X;
2. Y on estimated propensity score EPS.
######################################################################

##  read data, imputation by bootstrapping for missing data

dAll = read.csv(file="pre_impute_data.csv", as.is=T, sep=',', header=T)

set.seed(100)
.Random.seed
library(mi)
data.imp <- random.imp(dAll)


## estimate propensity score by logistic regression

glm.ps<-glm(Sentenced_to_prison~
                Age_at_1st_yuvenile_incarceration_y +
                N_prior_adult_convictions +
                Type_of_defense_counsel +
                Guilty_plea_with_negotiated_disposition +
                N_jail_sentences_gr_90days +
                N_juvenile_incarcerations +
                Monthly_income_level +
                Total_counts_convicted_for_current_sentence +
                Conviction_offense_type +
                Recent_release_from_incarceration_m +
                N_prior_adult_StateFederal_prison_terms +
                Offender_race +
                Offender_released_during_proceed +
                Separated_or_divorced_at_time_of_sentence +
                Living_situation_at_time_of_offence +
                Status_at_time_of_offense +
                Any_victims_female,
                data = data.imp, family=binomial)

summary(glm.ps)
eps <- predict(glm.ps, data = data.imp[, -1], type='response')
d.eps <- data.frame(data.imp, Est.ps = eps)


## Figure 9: densities of estimated propensity score (prison vs. probation)

library(ggplot2)

d.plot <- data.frame(Prison = as.factor(data.imp$Sentenced_to_prison),
    Est.ps = eps)
pdf("ps.dens.book.pdf")
ggplot(d.plot, aes(x=Est.ps, fill=Prison)) + geom_density(alpha=0.25) +
    scale_x_continuous(name="Estimated propensity score") +
    scale_y_continuous(name="Density")
dev.off()
```

```
## logistic regression of the outcome on all 17 variables

glm.y.allx<-glm(Recidivism~
                Sentenced_to_prison +
                Age_at_1st_yuvenile_incarceration_y +
                N_prior_adult_convictions +
                Type_of_defense_counsel +
                Guilty_plea_with_negotiated_disposition +
                N_jail_sentences_gr_90days +
                N_juvenile_incarcerations +
                Monthly_income_level +
                Total_counts_convicted_for_current_sentence +
                Conviction_offense_type +
                Recent_release_from_incarceration_m +
                N_prior_adult_StateFederal_prison_terms +
                Offender_race +
                Offender_released_during_proceed +
                Separated_or_divorced_at_time_of_sentence +
                Living_situation_at_time_of_offence +
                Status_at_time_of_offense +
                Any_victims_female,
                data = d.eps, family=binomial)

summary(glm.y.allx)


## logistic regression of the outcome on the estimated propensity score

glm.y.eps<-glm(Recidivism ~ Sentenced_to_prison + Est.ps,
  data = d.eps, family=binomial)
summary(glm.y.eps)
```

References

1. Bang, H., Robins, J.M.: Doubly robust estimation in missing data and causal inference models. Biometrics **61**, 962–972 (2005)
2. Berzuini, G.: Causal inference methods for criminal justice data, and an application to the study of the criminogenic effect of custodial sanctions. MSc Thesis in Applied Statistics, Birkbeck College, University of London (2013)
3. Carpenter, J.R., Kenward, M.G., Vansteelandt, S.: A comparison of multiple imputation and doubly robust estimation for analyses with missing data. J. R. Stat. Soc. Ser. A **169**, 571–584 (2006)
4. Dawid, A.P.: Conditional independence in statistical theory (with discussion). J. R. Stat. Soc. Ser. B **41**, 1–31 (1979)
5. Dawid, A.P.: Conditional independence for statistical operations. Ann. Stat. **8**, 598–617 (1980)
6. Dawid, A.P.: Causal inference without counterfactuals. J. Am. Stat. Assoc. **95**, 407–424 (2000)
7. Dawid, A.P.: Influence diagrams for causal modelling and inference. Int. Stat. Rev. **70**, 161–189 (2002)
8. Fisher, R.A.: Theory of statistical estimation. Proc. Camb. Philol. Soc. **22**, 700–725 (1925)
9. Guo, H., Dawid, A.P.: Sufficient covariates and linear propensity analysis. In: Teh, Y.W., Titterington, D.M. (eds.) Proceedings of the Thirteenth International Conference on Artificial Intelligence and Statistics (AISTATS), Chia Laguna, Sardinia, Italy, 13–15 May 2010. Journal of Machine Learning Research Workshop and Conference Proceedings, vol. 9, pp. 281–288 (2010)

10. Hahn, J.: On the role of the propensity score in efficient semiparametric estimation of average treatment effects. Econometrica **66**, 315–331 (1998)
11. Hirano, K., Imbens, G.W., Ridder, G.: Efficient estimation of average treatment effects using the estimated propensity score. Econometrica **71**, 1161–1189 (2003)
12. Horvitz, D.G., Thompson, D.J.: A generalization of sampling without replacement from a finite universe. J. Am. Stat. Assoc. **47**, 663–685 (1952)
13. Imbens, G.W., Lemieux, T.: Regression discontinuity designs: a guide to practice. J. Econ. **142**, 615–635 (2007)
14. Kang, J.D.Y., Schafer, J.L.: Demystifying double robustness: a comparison of alternative strategies for estimating a population mean from incomplete data. Stat. Sci. **22**, 523–539 (2007)
15. Mardia, K.V., Kent, J.T., Bibby, J.M.: Multivariate Analysis. Academic, New York (1979)
16. Pearl, J.: Causal diagrams for empirical research (with discussion). Biometrika **82**, 669–710 (1995)
17. Pearl, J.: Causality: Models, Reasoning, and Inference. Cambridge University Press, Cambridge (2000)
18. Robins, J.M., Mark, S.D., Newey, W.K.: Estimating exposure effects by modelling the expectation of exposure conditional on confounders. Biometrics **48**, 479–495 (1992)
19. Rosenbaum, P.R., Rubin, D.B.: The central role of the propensity score in observational studies for causal effects. Biometrika **70**, 44–55 (1983)
20. Rosenbaum, P.R., Rubin, D.B.: Reducing bias in observational studies using subclassification on the propensity score. J. Am. Stat. Assoc. **79**, 516–524 (1984)
21. Rubin, D.B.: Estimating causal effects of treatments in randomized and nonrandomized studies. J. Educ. Psychol. **66**, 688–701 (1974)
22. Rubin, D.B.: Assignment to treatment group on the basis of a covariate. J. Educ. Stat. **2**, 1–26 (1977)
23. Rubin, D.B.: Bayesian inference for causal effects: the role of randomization. Ann. Stat. **6**, 34–68 (1978)
24. Rubin, D.B.: Matched Sampling for Causal Effects. Cambridge University Press, Cambridge (2006)
25. Rubin, D.B., Thomas, N.: Characterizing the effect of matching using linear propensity score methods with normal distributions. Biometrika **79**, 797–809 (1992)
26. Rubin, D.B., van de Laan, M.J.: Covariate adjustment for the intention-to-treat parameter with empirical efficiency maximization. U.C.Berkeley Division of Biostatistics Working Paper 229 (2008)
27. Sekhon, J.: Multivariate and propensity score matching software with automated balance optimization: the matching package for R. J. Stat. Softw. **42**, 1–52 (2011)
28. Senn, S., Graf, E., Caputo, A.: Stratification for the propensity score compared with linear regression techniques to assess the effect of treatment or exposure. Stat. Med. **26**, 5529–5544 (2007)
29. Tang, Z.: Understanding OR, PS, and DR, Comment on "Demystifying double robustness: a comparison of alternative strategies for estimating a population mean from incomplete data" by Kang and Schafer. Stat. Sci. **22**, 560–568 (2007)
30. Winkelmayer, W.C., Kurth, T.: Propensity scores: help or hype? Nephrol. Dial. Transplant. **19**, 1671–1673 (2004)

Chapter 4
A Robustness Index of Propensity Score Estimation to Uncontrolled Confounders

Wei Pan and Haiyan Bai

Abstract As a crucial component of propensity score methods for reducing selection bias, propensity score estimation can only account for observed covariates. The behaviors of sensitivity and robustness of propensity score estimation to the impact of unobserved covariates or uncontrolled confounders have not been fully understood. This chapter introduces a new technique to assess the sensitivity and robustness of propensity score estimation to the impact of uncontrolled confounders. The sensitivity is defined as a change from a propensity score that is estimated from a propensity score model including all observed covariates to a potential propensity score that would be estimated from the propensity score model adding an unobserved covariate. The robustness is subsequently defined as the probability of the sensitivity would cross a pre-specified threshold. To assess the robustness, a reference distribution of the sensitivity is derived by borrowing information from observed covariates and further approximated to one of Pearson distributions. This procedure of assessment is illustrated with empirical data on substance abuse prevention for high-risk youth.

1 Introduction

Researchers often use non-randomized controlled trials (non-RCTs) or observational data to estimate treatment effects because RCTs are not always feasible [5, 35]. The use of non-RCTs or observational data poses a threat to the validity of causal inference due to selection bias in the treatment assignment. To tackle this problem, Rosenbaum and Rubin [30] theorized propensity score methods to mimic characteristics of RCTs by balancing the distributions of observed covariates between the treatment and comparison groups and, therefore, reduce selection bias.

W. Pan (✉)
Duke University School of Nursing, 307 Trent Dr., DUMC Box 3322, Durham, NC 27710, USA
e-mail: wei.pan@duke.edu

H. Bai
Department of Educational & Human Sciences, University of Central Florida, PO Box 161250, Orlando, FL 32816, USA
e-mail: haiyan.bai@ucf.edu

H. He et al. (eds.), *Statistical Causal Inferences and Their Applications in Public Health Research*, ICSA Book Series in Statistics, DOI 10.1007/978-3-319-41259-7_4

Over the past three decades, propensity score analysis has become increasingly popular in social, behavioral, and health research for making causal inferences based on non-RCTs or observational studies [2, 22].

Propensity score methods start with estimating propensity scores. Denote z as a treatment condition. Suppose one has N units (e.g., subjects). For each unit i ($i = 1, \ldots, N$), $z_i = 1$ indicates that the unit i is in the treatment group and $z_i = 0$ indicates that the unit i is in the comparison group. Suppose each unit i also has a covariate value vector $\mathbf{X}_i = (X_{i1}, \ldots, X_{iK})'$, where K is the number of covariates. Rosenbaum and Rubin [30] defined a propensity score for unit i as the probability of the unit being assigned to the treatment group, conditional on the covariate vector $\mathbf{X}_i$, that is, $e(\mathbf{X}_i) = \Pr\left(z_i = 1 \middle| \mathbf{X}_i\right)$. They also recommended using the logit of the propensity score, $\ln\left(\frac{e(\mathbf{X}_i)}{1-e(\mathbf{X}_i)}\right)$, to achieve normality.

2 Uncontrolled Confounders in Propensity Score Estimation

As a crucial component of propensity score methods for reducing selection bias, propensity score estimation can, however, only account for observed covariates in propensity score models. This is a major limitation of propensity score methods. Several techniques have been developed to address the potential impact of unobserved covariates or uncontrolled confounders on the estimation of propensity scores. For example, *sensitivity analysis* [6, 7, 31] has been adopted as the main technique for this purpose. Sensitivity analysis quantifies the impact of unobserved covariates under certain assumptions about unobserved covariates made by the researcher. This technique has been increasingly utilized [1, 9–11, 17, 28, 29, 33].

There are several variants of sensitivity analysis using specific techniques such as *marginal structural models* that examine the impact of residual confounding using the mean difference between treatment groups within each covariate stratum [3, 4, 14, 27]; *linear programming* that derives the upper and lower bounds for the causal effect on a binary outcome [15, 19]; *Bayesian sensitivity analysis* that corrects bias due to confounding, missing data, and classification error by including a bias model with relevant bias parameters [8, 20, 21]; *external adjustment* that simulates varying strengths of hypothetical associations between an unobserved covariate and observed covariates as well as between such a confounder and outcome [13]; and *propensity score-based approach* that uses inverse probability weighting [16, 36]. There are still a few other related strategies to directly or indirectly control for unobserved covariates. For example, *propensity score calibration* controls for one or more unobserved covariates in propensity score models by obtaining information on the unobserved covariates that were unobserved in the full data set but observed in a *subset* of the study population [12, 18, 37, 38]. *High-dimensional propensity score adjustment*, through including a substantial number (e.g., >20) of observed covariates in the propensity score model, indirectly controls for part of the confounding by unobserved covariates that are not included in the propensity score model and leaves little room for confounding by unobserved covariates that are correlated to the observed covariates [34, 39].

All the aforementioned approaches to sensitivity analysis of uncontrolled confounders, however, fell short of assessing the robustness of propensity score estimation to the impact of uncontrolled confounders. Robustness is about how sensitive is too sensitive. In other words, if adding an unobserved covariate in the propensity score model changes the propensity score estimates significantly, the propensity score model is too sensitive or not robust to the impact of the additional unobserved covariate. Following Pan and Frank [24] approach to assess robustness of confounders in linear models, the present study presents a new technique to assess not only sensitivity but also robustness of propensity score estimation to uncontrolled confounders by borrowing information from observed covariates.

3 Sensitivity and Robustness of Propensity Score Estimation

Denote $\mathbf{U} = (u_1, \ldots, u_N)'$ as an unobserved covariate, $\mathbf{X}_i = (X_{i1}, \ldots, X_{iK})'$ as the K observed covariates, and $\mathbf{X}_i^{(+u)} = (\mathbf{X}_i, \mathbf{U}) = (X_{i1}, \ldots, X_{iK}, u_i)'$, $i = 1, \ldots, N$. The sensitivity of propensity score estimation to the unobserved covariate is defined as a change from a propensity score $e(\mathbf{X}_i)$ that is estimated from a propensity score model including all the K observed covariates to a potential propensity score $e\left(\mathbf{X}_i^{(+u)}\right)$ that would be estimated from the propensity score model adding the unobserved covariate. That is, the theoretical definition of the sensitivity can be expressed as follows:

$$\Delta_i = e\left(\mathbf{X}_i^{(+u)}\right) - e\left(\mathbf{X}_i\right). \tag{4.1}$$

Here the problem is that the sensitivity index Δ_i can never be computed because $\mathbf{U}$ is unobserved and, thus, $e\left(\mathbf{X}_i^{(+u)}\right)$ is not calculable. By following McCandless et al. [21] and Pan and Frank [24] approaches, one could obtain the sensitivity index Δ_i by borrowing information from observed covariates, assuming the unobserved covariate has the same impact as that of the observed covariates on the propensity score estimation. In other words, the observed covariates are regarded as being representative of the "population" of all possible significant confounders or covariates, both observable and unobservable. Then, the sensitivity can be operationally estimated by treating each observed covariate as an unobserved covariate as follows:

$$\widehat{\Delta}_{ij} = e\left(\mathbf{X}_i\right) - e\left(\mathbf{X}_i^{(-j)}\right), \tag{4.2}$$

where $\mathbf{X}_i^{(-j)} = \left(X_{i1}, \ldots, X_{i(j-1)}, X_{i(j+1)}, \ldots, X_{iK}\right)', i = 1, \ldots, N$, for each covariate j $(j = 1, \ldots, K)$.

For each unit i ($i = 1, \ldots, N$), the estimated sensitivity indices $\widehat{\Delta}_{ij}$, $j = 1, \ldots, K$, constitute a sampling distribution or reference distribution of the sensitivity Δ_i. Then, the robustness of propensity score estimation to the unobserved covariate can be defined as the likelihood or probability that a sensitivity index Δ_i^* to a new unobserved covariate is not beyond the range of the estimated sensitivity indices $\widehat{\Delta}_{ij}$, $j = 1, \ldots, K$. That is, the theoretical definition of the robustness can be expressed as follows:

$$R_i = \Pr\left(\min_{1 \le j \le K} \widehat{\Delta}_{ij} \le \Delta_i^* \le \max_{1 \le j \le K} \widehat{\Delta}_{ij}\right). \tag{4.3}$$

For each unit i ($i = 1, \ldots, N$), if the robustness index R_i is larger than .95, which is analogous to a significance level of .05, one could claim that the propensity score estimation is robust to the impact of uncontrolled confounders for that unit. Then, for all the N units, if a majority of the R_i's (e.g., more than 80 %) are larger than .95, one could claim that the propensity score estimation is robust to uncontrolled confounders for the entire sample. If 50–80 % of the R_i's are larger than .95, one could caution that the propensity score estimation may be sensitive to uncontrolled confounders for the entire sample. If less than 50 % of the R_i's are larger than .95, one could conclude that the propensity score estimation is not robust to uncontrolled confounders for the entire sample. It is worth noting that the cut-off percentages (i.e., 50 % and 80 %) are arbitrary, and researchers can adopt more appreciate cut-off percentages for their own specific research areas.

Now the problem is that how to obtain the robustness index R_i without knowing the behavior of the sampling distribution or reference distribution of the sensitivity indices $\widehat{\Delta}_{ij}$. By following Pan and Frank [24, 25] approach, one could first approximate the shape of the distribution of the sensitivity indices $\widehat{\Delta}_{ij}$ to one of Pearson distributions [26] based on the first four moments of $\widehat{\Delta}_{ij}$. Once the following first four moments are calculated:

$$\widehat{\mu}'_{i1} = \frac{1}{K}\sum_{j=1}^{K}\widehat{\Delta}_{ij} \quad \text{and} \quad \widehat{\mu}_{im} = \frac{1}{K}\sum_{j=1}^{K}\left(\widehat{\Delta}_{ij} - \widehat{\mu}'_{i1}\right)^m, \; m = 2, 3, 4, \tag{4.4}$$

plug them into the Pearson distribution approximation to obtain an approximated distribution of $\widehat{\Delta}_{ij}$, denoted as $f_i\left(\widehat{\Delta}_{ij}\right)$. Then, the robustness index can be operationally estimated as follows:

$$\widehat{R}_i = \int_{\min_{1 \le j \le K} \widehat{\Delta}_{ij}}^{\max_{1 \le j \le K} \widehat{\Delta}_{ij}} f_i(t)dt. \tag{4.5}$$

The integration in Eq. 4.5 can be numerically evaluated using a SAS macro compiled by Pan and Boling [23].

4 An Empirical Example

The data for this empirical example were selected from a national database of 10,500 at-risk youth in a national evaluation of the High-Risk Youth Demonstration Grant Programs sponsored by the Substance Abuse and Mental Health Services Administration [32] focusing on prevention of substance use. The evaluation compared participants and non-participants of funded prevention programs over 18 months with respect to socio-demographic risk and protective factors. For demonstration purposes only, we sampled 547 youth who initiated substance use prior to entry to the national evaluation. Among the 547 youth in the sample, $n_T = 213$ were in the prevention (or treatment) group, and $n_C = 334$ were in the comparison group.

The outcome measure was a composite score of 30-Day substance use, including tobacco, alcohol, marijuana, and illicit drugs. There were 22 covariates in the study, including age, gender, race/ethnicity, family composition, family supervision, school prevention, community protection, neighborhood risk, family bonding, school bonding, self-efficacy, belief in self, self-control, social confidence, parental use attitudes, peer use attitudes, and peer use. Due to the focus on the methodological nature of this chapter, we would reference the detailed information about covariate selection to other resources such as SAMHSA [32].

To obtain the sensitivity indices $\widehat{\Delta}_{ij}$, each one of the 22 covariates was in turn treated as an unobserved covariate in a propensity score model (e.g., logistic regression model) to calculate $e(\mathbf{X}_i)$ and $e\left(\mathbf{X}_i^{(-j)}\right)$, $i = 1, 2, \ldots, 547; j = 1, 2, \ldots,$ 22. Figure 4.1 shows the empirical distribution of the sensitivity indices $\widehat{\Delta}_{311031,\,j}$ for Subject 311031, $j = 1, \ldots, 22$. Then, the smallest and largest sensitivity indices were identified for each unit. For instance, those two specific sensitivity indices for Subject 311031 were $\min_{1\le j\le 22} \widehat{\Delta}_{311031j} = -.58$ and $\max_{1\le j\le 22} \widehat{\Delta}_{311031j} = .30$.

The next step was to calculate the first four moments using Eq. 4.4 and, for example, the four moments for Subject 311031 are $\widehat{\mu}'_{311031,1} = -0.017$, $\widehat{\mu}_{311031,2} = 0.046$, $\widehat{\mu}_{311031,3} = -0.010$, and $\widehat{\mu}_{311031,4} = 3.003$, respectively. Then, for each subject i ($i = 1, \ldots, N$), using Pan and Boling [23] SAS macro, the reference distribution of the sensitivity indices was approximated to one of Pearson distributions using the first four moments. For Subject 311031, the reference distribution $f_{311031}(t)$ was approximated as a Type IV Pearson distribution (see Fig. 4.2). And, the same SAS macro also simultaneously computed the probability value for the robustness index $\widehat{R}_i$ (Eq. 4.5). For Subject 311031, the robustness index $\widehat{R}_{311031} = \int_{-.58}^{.30} f_{311031}(t)dt = .952$, suggesting that the propensity score estimation is robust to uncontrolled confounders for Subject 311031.

Figure 4.3 displays the empirical distribution of all the 547 robustness indices $\widehat{R}_i$'s, $i = 1, 2, \ldots, 547$. Among all the 547 robustness indices, 75 % of them are

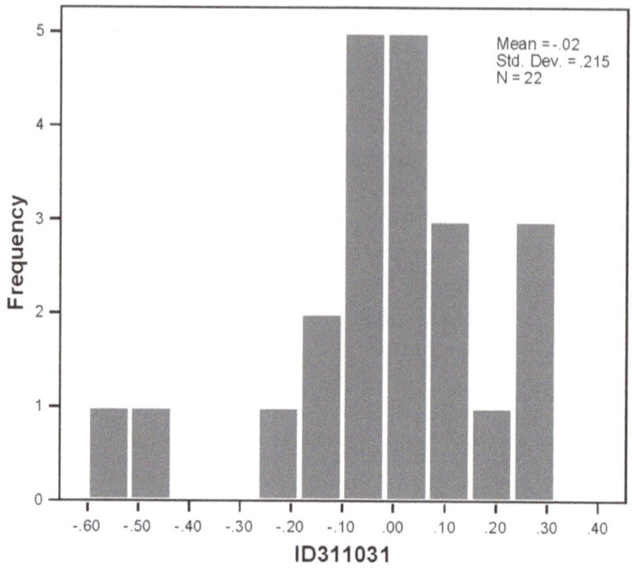

Fig. 4.1 The empirical distribution of the sensitivity indices $\widehat{\Delta}_{311031,\,j}$ for Subject 311031, $j = 1, \ldots, 22$

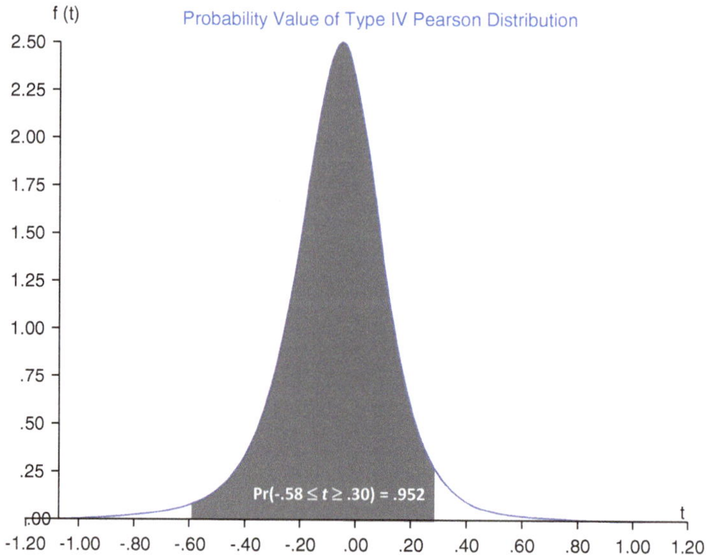

Fig. 4.2 The approximated Type IV Pearson distribution of the sensitivity for Subject 311031

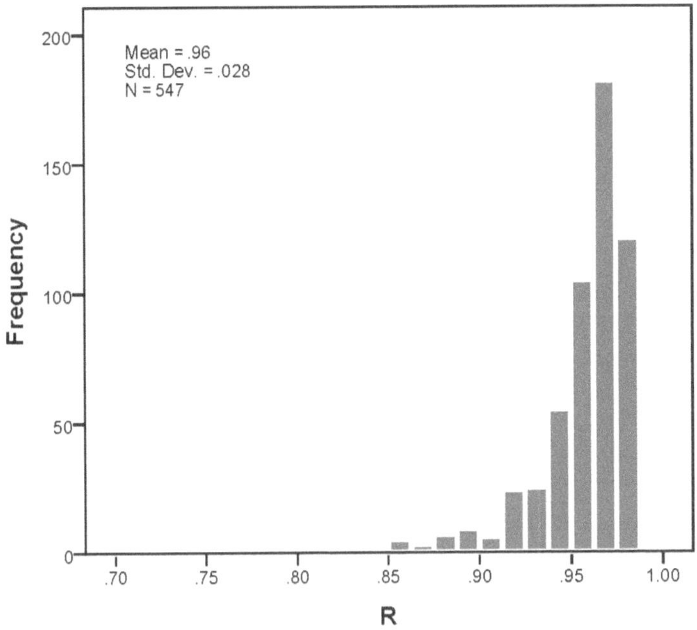

Fig. 4.3 The empirical distribution of robustness for the entire sample

larger than .95, indicating that the propensity score estimation may be sensitive to uncontrolled confounders for the entire sample. It suggests that more covariates should have been observed and controlled in the propensity score model.

5 Conclusion

The sensitivity and robustness of propensity score estimation to the impact of uncontrolled confounders are essential and a long-lasting issue in propensity score methods for making valid causal inference using non-RCTs or observational data. This chapter contributes to the literature on this topic through introducing a new technique to assess both the sensitivity and robustness by borrowing information from observed covariates, as did McCandless et al. [21] and Pan and Frank [24]. The following lists the steps for performing the assessment:

1. Run a logistic regression of the treatment condition z_i on all the K covariates $\mathbf{X}_i$ to obtain propensity score $e(\mathbf{X}_i)$, then run K logistic regressions of the treatment condition z_i, each on all but jth covariate, $\mathbf{X}_i^{(-j)}$, to obtain propensity scores $e\left(\mathbf{X}_i^{(-j)}\right), j = 1, \ldots, K; i = 1, \ldots, N.$

2. Calculate the sensitivity index $\widehat{\Delta}_{ij}$ for each subject i ($i = 1, \ldots, N$) to each observed covariate j ($j = 1, \ldots, K$) which is treated as an unobserved covariate, using Eq. 4.2.
3. Calculate the first four moments of $\widehat{\Delta}_{ij}$ using Eq. 4.4, $i = 1, \ldots, N; j = 1, \ldots, K$.
4. Find the smallest and largest sensitivity indices for each unit i, $\min_{1 \leq j \leq K} \widehat{\Delta}_{ij}$ and $\max_{1 \leq j \leq K} \widehat{\Delta}_{ij}$ ($i = 1, \ldots, N$).
5. Plug the first four moments, minimum, and maximum into Pan and Boling [23] SAS macro to obtain the reference distribution of the sensitivity indices as one of Pearson distributions and to obtain the robustness index $\widehat{R}_i$ for each unit i ($i = 1, \ldots, N$).
6. Calculate the proportion of the robustness indices $\widehat{R}_i$'s that are larger than .95 for all the N units.
7. Use the following suggested guideline to interpret the robustness of propensity score estimation to the impact of uncontrolled confounders:

   a. If more than 80 % of the robustness indices R_i's are larger than .95, one could claim that the propensity score estimation is robust to the impact of uncontrolled confounders for the entire sample.
   b. If 50–80 % of the robustness indices R_i's are larger than .95, one could caution that the propensity score estimation may be sensitive to the impact of uncontrolled confounders for the entire sample.
   c. If less than 50 % of the robustness indices R_i's are larger than .95, one could conclude that the propensity score estimation is not robust to the impact of uncontrolled confounders for the entire sample.

As mentioned earlier, the cut-off percentages (i.e., 50 % and 80 %) are arbitrary and researchers can adopt more appreciate cut-off percentages for their specific areas of research.

Some researchers may be uncomfortable with the use of observed covariates to generate a sampling distribution or reference distribution of sensitivity for uncontrolled confounders. We acknowledge that the reference distribution is only as valid as is the set of observed covariates representative of the "population" of all possible confounders or covariates, both observable and unobservable, which is, however, no different from any other statistical inference from a sample that must be representative of the population. In this light, the observed covariates represent important information by which the sensitivity and robustness are assessed. This strategy has been implemented in the literature (e.g., [21, 24]).

The number of observed covariates is another caveat to address. Of course, the more observed covariates, the more valid sensitivity and robustness indices we could obtain. In the situation where one can make a high-dimensional propensity score adjustment, one can also obtain good sensitivity and robustness indices. Nevertheless, how the observed covariates are representative of all possible covariates matters more than the number of the observed covariates.

References

1. Arah, O.A., Chiba, Y., Greenland, S.: Bias formulas for external adjustment and sensitivity analysis of unmeasured confounders. Ann. Epidemiol. **18**(8), 637–646 (2008). doi:10.1016/j.annepidem.2008.04.003
2. Bai, H.: A comparison of propensity score matching methods for reducing selection bias. Int. J. Res. Method Educ. **34**(1), 81–107 (2011). doi:10.1080/1743727X.2011.552338
3. Brumback, B.A., Hernán, M.A., Haneuse, S.J.P.A., Robins, J.M.: Sensitivity analyses for unmeasured confounding assuming a marginal structural model for repeated measures. Stat. Med. **23**(5), 749–767 (2004). doi:10.1002/sim.1657
4. Cole, S.R., Hernán, M.A., Margolick, J.B., Cohen, M.H., Robins, J.M.: Marginal structural models for estimating the effect of highly active antiretroviral therapy initiation on CD4 cell count. Am. J. Epidemiol. **162**(5), 471–478 (2005). doi:10.1093/aje/kwi216
5. Cook, T.D., Campbell, D.T.: Quasi-experimentation: Design & Analysis Issues for Field Settings. Rand McNally, Chicago (1979)
6. Cornfield, J., Haenszel, W., Hammond, E.C., Lilienfeld, A.M., Shimkin, M.B., Wynder, E.L.: Smoking and lung cancer: recent evidence and a discussion of some questions. J. Natl. Cancer Inst. **22**, 173–203 (1959)
7. Cornfield, J., Haenszel, W., Hammond, E.C., Lilienfeld, A.M., Shimkin, M.B., Wynder, E.L.: Smoking and lung cancer: recent evidence and a discussion of some questions. Int. J. Epidemiol. **38**(5), 1175–1191 (2009). doi:10.1093/ije/dyp289
8. Greenland, S.: Multiple-bias modelling for analysis of observational data. J. R. Stat. Soc. A. Stat. Soc. **168**(2), 267–306 (2005). doi:10.1111/j.1467-985X.2004.00349.x
9. Groenwold, R.H.H., Hak, E., Hoes, A.W.: Quantitative assessment of unobserved confounding is mandatory in nonrandomized intervention studies. J. Clin. Epidemiol. **62**(1), 22–28 (2009). doi:10.1016/j.jclinepi.2008.02.011
10. Groenwold, R.H.H., Hoes, A.W., Nichol, K.L., Hak, E.: Quantifying the potential role of unmeasured confounders: the example of influenza vaccination. Int. J. Epidemiol. **37**(6), 1422–1429 (2008). doi:10.1093/ije/dyn173
11. Groenwold, R.H.H., Nelson, D.B., Nichol, K.L., Hoes, A.W., Hak, E.: Sensitivity analyses to estimate the potential impact of unmeasured confounding in causal research. Int. J. Epidemiol. **39**(1), 107–117 (2010). doi:10.1093/ije/dyp332
12. Hsu, J.Y., Small, D.S.: Calibrating sensitivity analyses to observed covariates in observational studies. Biometrics **69**(4), 803–811 (2013). doi:10.1111/biom.12101
13. Huesch, M.D.: External adjustment sensitivity analysis for unmeasured confounding: an application to coronary stent outcomes, Pennsylvania 2004–2008. Health Serv. Res. **48**(3), 1191–1214 (2013). doi:10.1111/1475-6773.12013
14. Ko, H., Hogan, J.W., Mayer, K.H.: Estimating causal treatment effects from longitudinal HIV natural history studies using marginal structural models. Biometrics **59**(1), 152–162 (2003). doi:10.1111/1541-0420.00018
15. Kuroki, M., Cai, Z.: Formulating tightest bounds on causal effects in studies with unmeasured confounders. Stat. Med. **27**(30), 6597–6611 (2008). doi:10.1002/sim.3430
16. Li, L., Shen, C., Wu, A.C., Li, X.: Propensity score-based sensitivity analysis method for uncontrolled confounding. Am. J. Epidemiol. **174**(3), 345–353 (2011). doi:10.1093/aje/kwr096
17. Lin, D.Y., Psaty, B.M., Kronmal, R.A.: Assessing the sensitivity of regression results to unmeasured confounders in observational studies. Biometrics **54**(3), 948–963 (1998). doi:10.2307/2533848
18. Lunt, M., Glynn, R.J., Rothman, K.J., Avorn, J., Stürmer, T.: Propensity score calibration in the absence of surrogacy. Am. J. Epidemiol. **175**(12), 1294–1302 (2012). doi:10.1093/aje/kwr463
19. MacLehose, R.F., Kaufman, S., Kaufman, J.S., Poole, C.: Bounding causal effects under uncontrolled confounding using counterfactuals. Epidemiology **16**(4), 548–555 (2005). doi:10.2307/20486093

20. McCandless, L.C., Gustafson, P., Levy, A.: Bayesian sensitivity analysis for unmeasured confounding in observational studies. Stat. Med. **26**(11), 2331–2347 (2007). doi:10.1002/sim.2711
21. McCandless, L.C., Gustafson, P., Levy, A.: A sensitivity analysis using information about measured confounders yielded improved uncertainty assessments for unmeasured confounding. J. Clin. Epidemiol. **61**(3), 247–255 (2008). doi:10.1016/j.jclinepi.2007.05.006
22. Pan, W., Bai, H. (eds.): Propensity Score Analysis: Fundamentals and Developments. The Guilford Press, New York (2015)
23. Pan, W., Boling, J.: Computing and graphing probability Values of Pearson distributions: a SAS/IML macro. Paper presented at the 2013 Joint Statistical Meetings, Montreal, Canada, August 2013
24. Pan, W., Frank, K.A.: A probability index of the robustness of a causal inference. J. Educ. Behav. Stat. **28**(4), 315–337 (2003). doi:10.3102/10769986028004315
25. Pan, W., Frank, K.A.: An approximation to the distribution of the product of two dependent correlation coefficients. J. Stat. Comput. Sim. **74**(6), 419–443 (2004). doi:10.1080/00949650310001596822
26. Pearson, K.: Contributions to the mathematical theory of evolution. II. Skew variation in homogeneous material. Philos. Trans. R. Soc. Lond. A **186**, 343–414 (1895). doi:10.2307/90649
27. Robins, J.M.: Association, causation, and marginal structural models. Synthese **121**(1/2), 151–179 (1999). doi:10.2307/20118224
28. Robins, J.M., Rotnitzky, A., Scharfstein, D.O.: Sensitivity analysis for selection bias and unmeasured confounding in missing data and causal inference models. In: Halloran, M.E., Berry, D. (eds.) Statistical Models in Epidemiology, the Environment, and Clinical Trials, vol. 116. The IMA Volumes in Mathematics and its Applications, pp. 1–94. Springer, New York (2000). doi:10.1007/978-1-4612-1284-3_1
29. Rosenbaum, P.R., Rubin, D.B.: Assessing sensitivity to an unobserved binary covariate in an observational study with binary outcome. J. R. Stat. Soc. Ser. B (Methodol.) **45**(2), 212–218 (1983). doi:10.2307/2345524
30. Rosenbaum, P.R., Rubin, D.B.: The central role of the propensity score in observational studies for causal effects. Biometrika **70**(1), 41–55 (1983). doi:10.1093/biomet/70.1.41
31. Saltelli, A., Ratto, M., Andres, T., Campolongo, F., Cariboni, J., Gatelli, D., Saisana, M., Tarantola, S.: Global Sensitivity Analysis: The Primer. Wiley, West Sussex (2008)
32. SAMHSA: The National Cross-Site Evaluation of High-Risk Youth Programs. Substance Abuse and Mental Health Services Administration, U.S. Department of Health and Human Services, Rockville (2002)
33. Schneeweiss, S.: Sensitivity analysis and external adjustment for unmeasured confounders in epidemiologic database studies of therapeutics. Pharmacoepidemiol. Drug Saf. **15**(5), 291–303 (2006). doi:10.1002/pds.1200
34. Schneeweiss, S., Rassen, J.A., Glynn, R.J., Avorn, J., Mogun, H., Brookhart, M.A.: High-dimensional propensity score adjustment in studies of treatment effects using health care claims data. Epidemiology **20**(4), 512–522 (2009). doi:10.1097/EDE.0b013e3181a663cc
35. Shadish, W.R., Cook, T.D., Campbell, D.T.: Experimental and Quasi-experimental Designs for Generalized Causal Inference. Houghton Mifflin, Boston (2002)
36. Shen, C., Li, X., Li, L., Were, M.C.: Sensitivity analysis for causal inference using inverse probability weighting. Biom. J. **53**(5), 822–837 (2011). doi:10.1002/bimj.201100042
37. Stürmer, T., Schneeweiss, S., Avorn, J., Glynn, R.J.: Adjusting effect estimates for unmeasured confounding with validation data using propensity score calibration. Am. J. Epidemiol. **162**(3), 279–289 (2005). doi:10.1093/aje/kwi192
38. Stürmer, T., Schneeweiss, S., Rothman, K.J., Avorn, J., Glynn, R.J.: Performance of propensity score calibration—a simulation study. Am. J. Epidemiol. **165**(10), 1110–1118 (2007). doi:10.1093/aje/kwm074
39. Toh, S., García Rodríguez, L.A., Hernán, M.A.: Confounding adjustment via a semi-automated high-dimensional propensity score algorithm: an application to electronic medical records. Pharmacoepidemiol. Drug Saf. **20**(8), 849–857 (2011). doi:10.1002/pds.2152

Chapter 5
Missing Confounder Data in Propensity Score Methods for Causal Inference

Bo Fu and Li Su

Abstract Propensity score methods, including weighting, matching, or stratification, have been increasingly used to control potential confounding in observational studies and non-randomized trials to obtain causal effects of treatment or intervention. However, there are few studies to address the missing confounder data problem in propensity score estimation which is unique and different from most missing covariate data problems where the goal is parameter estimation. We will review existing methods for addressing missing confounder data in propensity score methods for causal inference and discuss the gap between current methodology developments in this area and the challenges in analyzing real observational data.

1 Introduction

In public health research, randomized clinical trials are often infeasible because of their size, time, budget, and ethical constraints and observational studies play an important role to evaluate treatment effects on long-term outcomes [9]. Because of the absence of randomization and the time-varying nature of medication initiation in such observational cohorts, it is crucial to adequately control potential confounding from various factors (both time-invariant and time-varying) in order to obtain causal effects of treatments and interventions. Overall, the ultimate goal in the design and analysis of observational studies is to mimic those of a randomized controlled trial. There has been rich literature on how to control potential confounding from baseline characteristics between treated and untreated patients, for example, using propensity

B. Fu (✉)
Administrative Data Research Centre for England & Institute of Child Health, University College London, London, UK
e-mail: b.fu@ucl.ac.uk

L. Su
MRC Biostatistics Unit, Cambridge Institute of Public Health, Cambridge, UK
e-mail: li.su@mrc-bsu.cam.ac.uk

H. He et al. (eds.), *Statistical Causal Inferences and Their Applications in Public Health Research*, ICSA Book Series in Statistics, DOI 10.1007/978-3-319-41259-7_5

score methods [25, 26], as well as on how to control time-varying confounders (possibly affected by prior treatment) using approaches such as marginal structural models [23].

Nevertheless, there are various important methodological issues that have not been adequately addressed by existing literature.

1. Since in practice a large number of measured confounders are commonly included in a propensity score model to minimize all possible confounding, missing data are almost unavoidable. These missing confounder data will create problems in propensity score estimation if the level of missingness is high [22].
2. Missing data also exist in important time-varying confounders, for example, due to selective measuring using invasive procedures. There is lack of work in the literature on how to apply the marginal structural model approach with such missing time-varying confounder data. Further uncertainties introduced by the missing data will also worsen the instability problem of estimating inverse probability of treatment weights (IPTW), which has led to criticisms of applying the marginal structural models in practice [14]. Estimation in marginal structural models can be unstable in certain situations, for example, when patients with unusual confounding histories go from being off treatment to on treatment and are assigned (unrealistically) with large IPTW. The problem is notable in event history data analysis, especially if the weights are assumed constant for the remaining event history [14].
3. Although propensity score methods are guaranteed to produce unbiased estimates on average across studies under correctly specified modeling assumptions, the bias in a particular study will depend on the balance achieved in that study [17] and different propensity score methods of balancing covariates may lead to very different treatment effect estimate [5]. In reports on studies using propensity score methods, there is lack of attention to assess the balance on measured confounders between treatment groups [34]. There is no consensus in statistical or medical literatures regarding choice of an appropriate balance measure, even in the absence of missing confounder data.
4. Most propensity score analyses assume that there is no unmeasured confounding. Sensitivity analysis of this untestable assumption is vital to assess uncertainty of observed findings due to unmeasured confounders. Similarly, sensitivity analysis is required when non-ignorable missing confounder data can possibly create imbalance between treated and untreated patients. Existing methods in this area are restricted to simple or very particular settings [32]. In particular, sensitivity analysis methods for unmeasured confounding are less developed for event history data.

In this chapter, we will first introduce two motivating examples from medical applications, and then review the current methodology developments for addressing the above methodological issues due to missing data and complex confounding when obtaining causal inference using observational data. The aim of this chapter is to discuss the existing gap between current methodological developments in this area and the challenges in analyzing real observational data, and to provide useful information and references for medical researchers and suggest important topics of future methodological research. We mainly focus on propensity score approaches as they are the commonly used causal inference method in the medical literature and are experiencing a tremendous increase of interest in medical research and many scientific areas (e.g., [1, 30, 34]).

2 Examples

2.1 Example 1: Long-Term Safety of Biologics Treatment in Rheumatoid Arthritis Patients

The introduction of biologics therapy in 2001 was a revolutionary step forward in the management of rheumatoid arthritis, offering patients a new and effective alternative to traditional drugs. However, randomized clinical trials were unable to address its long-term safety due to their relatively short follow-up and were under powered to detect differences in event rates of rarer outcomes. British Society for Rheumatology Biologics Register, the world's largest observational cohort of rheumatoid arthritis patients receiving biologics therapy, was thus established in 2001 to evaluate long-term adverse outcomes such as serious infection, malignancy, or death and was designed to follow up 12,000 rheumatoid arthritis patients treated with biologics and 4,000 rheumatoid arthritis patients with non-biologics [12, 13]. Many baseline covariates such as age, gender, disease duration, smoking status, comorbidities, disease activity score, health assessment questionnaire score have potential for confounding because of their significant differences between exposed and unexposed groups and their correlation with adverse event outcomes [12, 13]. However, there are substantially number of patients with missing values, e.g, 24 % of the unexposed patients did not have records for their smoking status.

Different propensity score methods (matching, weighting, stratification) were compared to control confounding and it was seen that they produced different estimates of relative risks [19]. This highlights the importance of developing confounder-balancing metrics to guide selection of the most appropriate propensity score method. In the presence of noticeable missing data in confounders, it would also be necessary and interesting to evaluate how different combinations of missing confounder data methods and propensity score methods affect estimation of causal effects of biologics on adverse outcomes.

Furthermore, most well-developed propensity score methods are based on assumptions of no unmeasured confounders and 'missing at random' in Rubin's taxonomy, which are often impractical in real situations. We need novel sensitivity analysis methods to check these assumptions and their inferential consequences in the context of missing confounder data in causal inference.

In addition, about 1/3 patients starting a biologics drug will stop it at a later observed time because of mild side effects, comorbidities, or inefficiency [13]. After changing from 'on exposure' to 'off exposure', the disease activity is less controlled and quite a number of patients will re-start the biologics treatment again in a future observed time or switch to a second-line biologics drug to control the disease progression. It is not clear how to handle this complex time-varying treatment process to adequately control for time-varying confounders (e.g., disease activity affected by prior exposure) in order to obtain causal effects of biologics on observational outcomes.

2.2 Example 2: Long-Term Survival After Pulmonary Endarterectomy (PEA) Surgery

Chronic thromboembolic pulmonary hypertension (CTEPH) is a life threatening condition that historically has a poor outcome with supportive medical treatment. Pulmonary endarterectomy (PEA) is the treatment of choice and offers the only chance of cure. Since there are limited data on the long-term survival after PEA, the UK national PEA cohort at Papworth Hospital provides a valuable opportunity to evaluate the causal effects of post-surgery drugs on the long-term survival after PEA. There were 880 CTEPH patients who had PEA at Papworth Hospital between 1997 and 2012. Following PEA, 45 patients remained on drug treatment, and 142 patients started post-surgery drugs between Month 1 and Month 126 after PEA. Preliminary analyses revealed that the initiation of post-surgery drugs was associated with previous right heart catheter measurements of pressures and hemodynamics in heart and lungs, which are key variables for monitoring CTEPH progression that are associated with patient survival. However, these right heart catheter variables are also affected by prior treatment. Thus a Cox regression by directly adjusting for confounding from the time-varying right heart catheter variables will reduce the causal effects of post-surgery drugs on long-term survival after PEA. Further, because right heart catheter is an invasive procedure, after the initial assessment after PEA and hospital discharge, patients were only going to have their right heart catheter measured if their mean pulmonary artery pressures were equal or greater than 30 mm/Hg in their immediate previous assessments, which led to missing data in the time-varying right heart catheter variables. This creates problems in dealing with time-varying confounding from these variables when estimating the probability of being treated over time in a weighted time-dependent Cox regression within the marginal structural model framework. There is very limited literature on missing confounder data in marginal structural models [21] and it remains unclear about the consequence of using complete records in marginal structural models under different missing data mechanisms.

3 Propensity Score Methods

Propensity score methods have become the standard techniques for the estimation of causal treatment effects from observational data. The propensity score is defined as the probability of receiving treatment conditional on measured confounders. Conditional on propensity score, treated and untreated patients have a similar distribution of measured confounders. Thus within similar levels of propensity score, a "virtual randomization" can be achieved to compare patients between treatment groups. Different methods of using estimated propensity score have been described in the literature, including stratification [26], matching [26], covariate adjustment [26], and weighting [25], and their performance has been compared by simulation studies in estimating odds ratio [7], risk difference [3], and hazard ratio for time-to-event outcomes [4], and by an empirical study in balancing confounders by checking residual confounding [19]. Marginal structural models have also been developed as an extension of the propensity score weighting method to tackle the time-varying confounding problem [23].

4 Missing Confounder Data in Propensity Score Estimation

Since a large number of measured confounders are commonly included in a propensity score model in practice, missing confounder data are almost unavoidable. Existing approaches to dealing with missing confounder data in propensity score estimation include:

1. **Using complete records only**. This common approach obviously will reduce the estimation efficiency when the missingness level is high as records with missing data in any single confounder are dropped. The generalizability of the estimated causal effects is also questionable [10].
2. **Pattern mixture models** [10, 28]. Observed data are split into groups defined by missing data patterns and propensity score estimation could be then done within patterns. This method ensures that the treated and untreated patients are balanced on the observed values of confounders and missing confounder patterns; but with many missing confounder patterns this approach may not be practical because the sample sizes within patterns can be very small. To alleviate this problem, ad-hoc algorithms to reduce the number of missing confounder patterns have been proposed [22].
3. **Use of missing value indicators** [10, 11, 29]. Missing indicators for partially observed confounders are created and the missing values are filled in by a chosen value [10] and [29]. Then both missing indicators and "filled-in" confounders are included in a propensity score model. If missing values are not filled in by a fixed value, then some restrictions are imposed in the propensity score model in order to obtain unique maximum likelihood estimates by Expectation Conditional Maximization algorithm [11]. This approach is

problematic in a general missing covariate data problem [15], but it might be reasonable in the propensity score estimation context, if it balances the observed values of confounders and missing confounder patterns.

4. **Multiple imputation**. Under various assumptions about the missing data mechanisms, multiple imputation methods have been applied to deal with missing confounder data in propensity score estimation. Essentially, the missing values are "filled in" several times before the actual propensity score estimation [20, 22]. Then the propensity score is estimated for each imputed dataset, and different propensity score methods can be used to obtain the final causal effects of treatments. It is not clear how the multiple imputation under the propensity score estimation scenario should differ from those developed for dealing with regular missing covariate data in the literature. For example, an unanswered interesting question concerns what should be combined across imputations—estimated treatment effects or estimated propensity score [20]. Nevertheless, any multiple imputation method will involve making unverifiable assumptions on the missing data mechanisms.
5. **Inverse probability weighting**. Inverse probability weighting (IPW) methods have also been proposed for tackling the missing confounder data problem in both the original propensity score estimation setting [35] and the marginal structural model setting [17], where the partially observed data are up-weighted to represent the full complete data. In particular, an improved IPW method through doubly robust estimation has been proposed [36]. These methods are currently restricted to the scenario of one single confounder with missing data. Again, the IPW approach relies on unverifiable assumptions on the missing data mechanism.

It is important to note that the missing confounder data problem in propensity score estimation is a unique missing data problem. D'Agostino and Rubin [11] emphasized: "It is important to note that our problem is different from most missing-data problems in which the goal is parameter estimation. We are not interested in obtaining one set of estimated parameters for a logistic regression.... Rather, parameters particular to each pattern of missing data serve only in intermediate calculations to obtain estimated propensity scores for each subject. Moreover, the propensity scores themselves serve only as devices to balance the observed distribution of covariates and patterns of missing covariates across the treated and control groups. Consequently, the success of the propensity score estimation is assessed by this resultant balance rather than by the fit of the models used to create the estimated propensity scores." Furthermore, in practice we are not able to assess the unverifiable assumption on the missing confounder data but we can assess the balance of observed values of confounders and missing confounder patterns between treated and untreated patients, after applying different missing data methods in propensity score estimation. In this sense, more sophisticated methods such as multiple imputation and IPW might not necessarily be superior to simple methods such as missing indicator methods in practice, as long as the same level of balance has been achieved. We aim to investigate the relative performance of these missing data methods as a topic of our future research.

Another interesting research problem is about the choice of propensity score methods with missing confounder data. In the absence of missing confounder data, it has been shown that both propensity score matching and IPTW using propensity score induce better balance on baseline confounders than stratification by propensity score and covariate adjustment using propensity score [5]. However, IPTW directly uses the estimated propensity score and thus is particularly sensitive to misspecification of the propensity score model or instability in the estimated propensity score [6]. This sensitivity is very likely when unverifiable assumptions are imposed on the missing confounder data, e.g., when applying multiple imputation and IPW methods. On the other hand, for propensity score matching methods the propensity score is not directly involved in estimating the treatment effects. As long as balance between treated and untreated patients is achieved in terms of observed values of confounders and missing confounder patterns, the unverifiable assumptions on the missing confounder data should have smaller impact on estimated treatment effects obtained through propensity score matching than through IPTW. Hence the question is *"Is propensity score matching more robust in this context than other propensity score methods such as IPTW using propensity score?"*.

The critical assumption in propensity score analyses is that of no unmeasured confounding. Specifically, in the missing confounder data scenario, we assume that no other variables influencing treatment assignment given the observed values of confounders and missing confounder patterns [11]. In other words, we allow the missingness itself to be predictive about which treatment is received; but given the missing confounder patterns, the actual missing values of the confounders do not impact the treatment assignment. This, of course, is an assumption we cannot verify using the observed data. Therefore, analyses are required to check the sensitivity of the observed findings to the missing values of the confounders. These sensitivity analyses for missing confounder data essentially should be similar to those sensitivity analyses developed to examine an unmeasured confounder, therefore similar strategies can be applied. However, since we ensure that the observed values of confounders and missing confounder patterns are balanced between treated and untreated patients, in order to alter inferences about the treatment effects, the hidden bias due to actual missing values of the confounders will probably need to be larger in magnitude than the hidden bias due to unmeasured confounders for which we have absolutely no control [24].

5 Assessing Balance for Confounders with Missing Data

There is no consensus in the statistical or medical literatures regarding choice of an appropriate balance measure for propensity score methods and a variety of balance measures are available including mean differences, Kolmogorov-Smirnov distance [8], Levy distance [8], overlapping coefficient [8], Mahalanobis distance [16], C-statistics [5, 30], L1 metric [18]. Particularly in the presence of missing confounder data, assessing balance will not be straightforward [29], either in terms of requiring a measure balancing both observed distribution and missing data pattern for each confounder or in how to summarize across confounders. Thus methodological development in this area is needed.

6 Sensitivity Analysis for Unmeasured Confounders and Missing Confounder Data

To address unmeasured confounding, propensity score calibration can be carried out using external validation data if available [31] or one may use instrumental variable analysis [2]. The latter has limited feasibility if it is not possible to identify instruments. An alternative approach is to formulate a specific model for the bias and consider the sensitivity analysis of estimated treatment effects to plausible assumptions about unknown bias parameters [27]. Existing sensitivity analysis techniques are restricted to simple or very particular settings [32]. There are limited sensitivity analysis methods devoted to event history data as well [33]. Recently, a general framework for sensitivity analysis that is applicable to event history data was developed but requires specification of a large number of bias parameters [32, 33]. Methodological developments in sensitivity analysis for unmeasured confounders would be also useful for the case of missing confounder data, which also requires sensitivity analysis on unverifiable assumptions.

References

1. Ali, M., Groenwold, R., Klungel, O.: Covariate selection and assessment of balance in propensity score analysis in the medical literature: a systematic review. J. Clin. Epidemiol. **68**(2), 112–121 (2015)
2. Angrist, J. D., Imbens, G.W., Rubin, D. B.: Identification of causal effects using instrumental variables (with discussion). J. Am. Stat. Assoc. **91**, 444–472 (1996)
3. Austin, P.C.: The performance of different propensity score methods for estimating difference in proportions (risk differences or absolute risk reductions) in observational studies. Stat. Med. **29**, 2137–2148 (2010)
4. Austin, P.C.: The performance of different propensity score methods for estimating marginal hazard ratios. Stat. in Med. **32**(16), 2837–2849 (2013)
5. Austin, P.C.: The relative ability of different propensity score methods to balance measured covariates between treated and untreated subjects in observational studies. Med. Decis. Mak. **29**, 661–677 (2009)
6. Austin, P.C.: Balance diagnostics for comparing the distribution of baseline covariates between treatment groups in propensity-score matched samples. Stat. Med. **28**, 3083–3107 (2009)
7. Austin, P.C., Grootendorst, P., Anderson, G.M.: A comparison of the ability of different propensity score models to balance measured variables between treated and untreated subjects: a Monte Carlo study. Stat. Med. **26**(4), 734–753 (2007)
8. Belitser, S.V., Martens, E.P., Pestman, W.R., Groenwold, R.H.H., Boer, A., Klungel, O.H.: Measuring balance and model selection in propensity score methods. Pharmacoepidemiol. Drug Saf. **20**, 1115–1129 (2011)

9. Concato, J., et al.: Randomized, controlled trials, observational studies, and the hierarchy of research designs. N. Engl. J. Med. **342**(25), 1887–1892 (2000)
10. D'Agostino, R., et al.: Examining the impact of missing data on propensity score estimation in determining the effectiveness of SMBG. Health Serv. Outcome Res. Methodol. **2**, 291–315 (2011)
11. D'Agostino, R.B., Rubin, D.B.: Estimating and using propensity scores with partially missing data. J. Am. Stat. Assoc. **95**(451), 749–59 (2000)
12. Dixon, W., Watson, K.D., Lunt, M., Hyrich, K.L., British Society for Rheumatology Biologics Register Control Centre Consortium, Silman, A.J., Symmons, D.P., on behalf of the British Society for Rheumatology Biologics Register: Serious infection following anti-tumor necrosis factor alpha therapy in patients with rheumatoid arthritis: lessons from interpreting data from observational studies. Arthritis Rheum. **56**, 2896–2904 (2007)
13. Fu, B., Lunt, M., et al.: A threshold hazard model for estimating serious infection risk following anti-tumor necrosis factor therapy in rheumatoid arthritis patients. J. Biopharm. Stat. **23**(2), 461–476 (2013)
14. Gran, J.M., Roysland, K., Wolbers, M., Didelez, V., Sterne, J., Ledergerber, B., Furrer, H., von Wyl, V., Aalen, O.: A sequential Cox approach for estimating the causal effect of treatment in the presence of time-dependent confounding applied to data from the Swiss HIV cohort study. Stat. Med. **29**, 2757–68 (2010)
15. Groenwold, R.H., White, I.R., Donders, A.R.T., Carpenter, J.R., Altman, D.G., Moons, K.G.: Missing covariate data in clinical research: when and when not to use the missing-indicator method for analysis. Can. Med. Assoc. J. **184**(11), 1265–1269 (2012)
16. Gu, X.S., Rosenbaum, P.R.: Comparison of multivariate matching methods: structures, distances, and algorithms. J. Comput. Graph. Stat. **2**, 405–420 (1993)
17. Hirano, K., Imbens, G.W., Ridder, G.: Efficient estimation of average treatment effects using the estimated propensity score. Econometrica. **71**, 1161–1189 (2003)
18. Iacus, S.M., King, G., Porro, G.: Multivariate matching methods that are monotonic imbalance bounding. J. Am. Stat. Assoc. **106**, 345–361 (2011)
19. Lunt, M., et al.: Different methods of balancing covariates leading to different effect estimates in the presence of effect modification. Am. J. Epidemiol. **169**(7), 909–917 (2009)
20. Mitra, R., Reiter, J.P.: A comparison of two methods of estimating propensity scores after multiple imputation. Stat. Methods Med. Res. **25**(1), 188–204 (2016)
21. Moodie, E., Delaney, J., Lefebvre, G., Platt, R.: Missing confounding data in marginal structure models: a comparison of inverse probability weighting and multiple imputation. Int. J. Biostat. **4**, 1557–4679 (2008)
22. Qu, Y., Lipkovich, I.: Propensity score estimation with missing values using a multiple imputation missingness pattern (MIMP) approach. Stat. Med. **28**, 1402–414 (2009)
23. Robins, J.M., Hernán, M.A., Brumback, B.: Marginal structural models and causal inference in epidemiology. Epidemiology. **11**, 550–60 (2000)
24. Rosenbaum, P.R.: Observational Studies. Springer, New York (2002)
25. Rosenbaum, P.R.: Model-based direct adjustment. J. Am. Stat. Assoc. **82**, 387–94 (1987)
26. Rosenbaum, P.R., Rubin, D.B.: Assessing sensitivity to an unobserved binary covariate in an observational study with binary outcome. J. R. Stat. Soc. Ser. B **45**, 212–218 (1983)
27. Rosenbaum, P., Rubin, D.: The central role of the propensity score in observational studies for causal effect. Biometrika **70**, 41–55 (1983)
28. Rosenbaum, P.R., Rubin, D.B.: Reducing bias in observational studies using subclassification on the propensity score. J. Am. Stat. Assoc. **79**, 516–524 (1984)
29. Stuart, E.A.: Matching methods for causal inference. Stat. Sci. **25**(1), 1–21 (2010)
30. Stürmer, T., Joshi, M., Glynn, R.J., Avorn, J., Rothman, K.J., Schneeweiss, S.: A review of the application of propensity score methods yielded increasing use, advantages in specific settings, but not substantially different estimates compared with conventional multivariable methods. J. Clin. Epidemiol. **59**, 431–437 (2006)

31. Stürmer, T., Schneeweiss, S., Avorn, J., et al.: Adjusting effect estimates for unmeasured confounding with validation data using propensity score calibration. Am. J. Epidemiol. **162**(3), 279–289 (2005)
32. VanderWeele, T.J., Arah, O.A.: Bias formulas for sensitivity analysis of unmeasured confounding for general outcomes, treatments, and confounders. Epidemiology. **22**(1), 42–52 (2011)
33. VanderWeele, T.J.: Unmeasured confounding and hazard scales: sensitivity analysis for total, direct, and indirect effects. Eur. J. Epidemiol. **28**(2), 113–117 (2013)
34. Weitzen, S., et al.: Principles for modelling propensity scores in medical research: a systematic literature review. Pharmacoepidemiol. Drug Saf. **13**(12), 841–853 (2004)
35. Williamson, E., Morley, R., Lucas, A., Carpenter, J.: Propensity scores: from naive enthusiasm to intuitive understanding. Stat. Methods Med. Res. **21**(3), 273–93 (2012)
36. Williamson, E.J., Forbes, A., Wolfe, R.: Doubly robust estimators of causal exposure effects with missing data in the outcome, exposure or a confounder. Stat. Med. **31**(30), 4382–400 (2012)

Chapter 6
Propensity Score Modeling and Evaluation

Yeying Zhu and Lin (Laura) Lin

Abstract In causal inference for binary treatments, the propensity score is defined as the probability of receiving the treatment given covariates. Under the ignorability assumption, causal treatment effects can be estimated by conditioning on/adjusting for the propensity scores. However, in observational studies, propensity scores are unknown and need to be estimated from the observed data. Estimation of propensity scores is essential in making reliable causal inference. In this chapter, we first briefly discuss the modeling of propensity scores for a binary treatment; then we will focus on the estimation of the generalized propensity scores for categorical treatment variables with more than two levels and continuous treatment variables. We will review both parametric and nonparametric approaches for estimating the generalized propensity scores. In the end, we discuss how to evaluate the performance of different propensity score models and how to choose an optimal one among several candidate models.

1 Propensity Score Modeling for a Binary Treatment

The potential outcomes framework [23] has been a popular framework for estimating causal treatment effects. An important quantity to facilitate causal inference has been the propensity score [22], defined as the probability of receiving the treatment given a set of measured covariates. In observational studies, propensity scores are unknown and need to be estimated from the observed data. Consistent estimation of propensity scores is essential in making reliable causal inference. In this section, we briefly review the modeling of propensity scores for a binary treatment variable.

We first define some notations. Let Y denote the response of interest, T be the treatment variable, and X be a p-dimensional vector of baseline covariates. The data can be represented as (Y_i, T_i, X_i), $i = 1, \ldots, n$, a random sample from (Y, T, X). In addition to the observed quantities, we further define $Y_i(t)$ as the potential outcome

Y. Zhu (✉) • L. (Laura) Lin
Department of Statistics & Actuarial Science, University of Waterloo, Waterloo, ON, Canada
e-mail: yeying.zhu@uwaterloo.ca; linlin.laura@gmail.com

H. He et al. (eds.), *Statistical Causal Inferences and Their Applications in Public Health Research*, ICSA Book Series in Statistics, DOI 10.1007/978-3-319-41259-7_6

if subject i were assigned to treatment level t. Here, T is a random variable and t is a specific level of T. In the case of a binary treatment, let $T = 1$ if treated and $T = 0$ if untreated. The propensity score is then defined as $r(X) \equiv P(T = 1|X)$. The quantities we are interested in estimating are usually the average treatment effect (ATE):

$$\text{ATE} = E[Y(1) - Y(0)],$$

and the average treatment effect among the treated (ATT):

$$\text{ATT} = E[Y(1) - Y(0)|T = 1].$$

1.1 Parametric Approaches

In the causal inference literature, propensity score for a binary treatment variable is usually estimated by logistic regression. Using logistic regression to estimate propensity scores can be easily implemented in R. However, logistic regression is not without drawbacks. First of all, a parametric form of $r(X)$ needs to be specified. Consistent estimation of ATE and ATT relies on the correct logistic regression model. In most cases, only including main effects into the model is not adequate, but it is also hard to determine which interaction terms should be included, especially when the vector of covariates is high-dimensional. In addition, logistic regression is not resistant to outliers [11, 18]. In particular, Kang and Schafer [11] show when the logistic regression model is mildly misspecified, propensity score-based approaches lead to large bias and variance of the estimated treatment effects.

Other parametric approaches for estimating propensity scores include Probit regression modeling and linear discriminant analysis, both of which assume normality. However, through a simulation study, Zhu et al. [31] found that these parametric models give very similar treatment effect estimates.

1.2 Machine Learning Techniques

Due to the above-mentioned drawbacks of parametric approaches for modeling propensity scores, more recent literature advocates using machine learning algorithms to model propensity scores [13, 24]. Since in causal inference, propensity scores are auxiliary in the sense that one usually is not interested in interpreting or making inference for the propensity score model, the nonparametric black-box algorithms can be directly used to estimate the propensity scores. Examples are classification and regression trees (CART, [2]) and its various extensions, such as pruned CART, bagged CART, random forests (RF [1]), and boosting [16]. Other classification methods that can indirectly yield class probability estimates include support vector machines (SVM) and K-nearest neighbors (KNN), etc. R packages are readily available, such as *rpart* for CART; *randomForest* for RF, *twang* or *gbm* package for boosting models, and *e1071* for SVM. A detailed review of each approach for estimating propensity scores can be found in [31]. In a simulation study, Zhu et al. found there is a trade-off between bias and variance among parametric and nonparametric approaches. More specifically, parametric methods tend to yield lower bias but higher variance than nonparametric methods for estimating ATE and ATT.

1.3 Propensity Score Modeling via Balancing Covariates

Recently, a new propensity score modeling approach termed covariate balance propensity scores is proposed by Imai and Ratkovic [8], which also assumes a logistic regression model, i.e.,

$$r(X) \equiv r_\beta(X) = \frac{1}{1+\exp\{-\beta' X\}}. \tag{6.1}$$

Then, β is solved by satisfying the following condition:

$$\mathrm{E}\left\{\frac{T\widetilde{X}}{r_\beta(X)} - \frac{(1-T)\widetilde{X}}{1-r_\beta(X)}\right\} = 0, \tag{6.2}$$

where $\widetilde{X}$ is a function of X specified by the researcher. If setting $\widetilde{X} = \frac{dr_\beta(X)}{d\beta}$, one solves the maximum likelihood estimator (MLE) of β because Eq. (6.2) is the score function for MLE. However, if setting $\widetilde{X} = X$, one aims to achieve optimal balance in the first order of the covariates, because this balancing condition implies the weighted mean value of each covariate is the same between the treatment and the control group. If letting $\widetilde{X} = \frac{dr_\beta(X)}{d\beta}$ and $\widetilde{X} = X$ at the same time, there will be more equations than unknown parameters to solve and a generalized method of moments [5] is employed for estimation. The above balancing condition is for the estimation of ATE. For estimating ATT, the balancing condition becomes

$$\mathrm{E}\left\{T\widetilde{X} - \frac{r_\beta(X)(1-T)\widetilde{X}}{1-r_\beta(X)}\right\} = 0. \tag{6.3}$$

The advantage of this approach is that, by achieving better balance in the covariates, it is less susceptible to model misspecification of the propensity scores, compared to logistic regression.

A related issue is whether we should achieve balance in all the measured covariates in a study or a subset of the available covariates. This is a variable selection issue. Zhu et al. [32] have shown through a simulation study that one should aim to achieve balance in the real confounders, i.e. covariates related to both the treatment variable and the outcome variable, as well as the covariates related only to the outcome variable. Adding additional balancing condition on covariates that are only related to the treatment variable may increase the bias and variance of the estimated treatment effects.

2 Propensity Score Modeling for a Multi-level Treatment

In most of the causal inference literature based on potential outcomes framework, researchers have focused on binary treatments. Imbens [10] extended this framework to more general case by defining the generalized propensity score, which is the conditional probability of being assigned to a particular treatment group given the observed covariates. In the past decade, a few studies (e.g., [9, 12, 28]) have extended the propensity score-based approaches to multi-level treatments. Compared with binary treatments, there are two important issues specific to the causal inference with multi-level treatments. The first issue is to define the parameters of interest and to determine whether the parameters are identifiable. As discussed by Imbens [10] and Tchernis et al. [28], for a multi-level treatment, the following parameters may be of interest: (1) the average causal effect of treatment t relative to k, i.e., $E[Y(t) - Y(k)]$; (2) the average causal effect of treatment t relative to k among those who receive treatment t, i.e., $E[Y(t) - Y(k)|T = t]$ or (3) the average causal effect of treatment t relative to all other treatments among those who receive treatment t, i.e., $E[Y(t) - Y(\bar{t})|T = t]$, where $\bar{t}$ refers to other treatment groups except group t. In any of the three definitions, the multi-level treatment variable is dichotomized; in this sense, causal inference with multiple treatments is essentially an extension of the binary case. Therefore, matching, stratification, or inverse probability weighting methods can be employed to estimate the targeted causal effects in a similar way as in binary treatments. The second issue is that in many studies, the treatments are correlated: the odds ratio of receiving one treatment against the other is affected by whether a third treatment is taken into consideration or not. Tchernis et al. [28] pointed out in a simulation study that if the treatments are correlated, ignoring correlations while estimating propensity scores will lead to biased estimation of the causal effect. The commonly used multinomial logistic regression model does not account for correlation. Therefore, the nested logit model or multinomial probit model has been suggested for modeling propensity scores to allow specification of a correlation matrix among treatments. Due to developments in machine learning methods, nonparametric algorithms such as random forests or boosting algorithms can be easily implemented to estimate propensity scores for multiple treatments.

We define some additional notations here. Let T_i be the treatment status for the ith subject, so $T_i = t$ if subject i was observed under treatment $t \in \{1, \ldots, M\}$, where there are M total treatment groups. We further define an indicator variable, indicating membership of a particular treatment group t, as $A_i(t) = I(T_i = t)$, $t \in \{1, \ldots, M\}$. According to Imai and Van Dyk [9], the generalized propensity score is defined as $r(t|X) \equiv Pr(T = t|X)$, for $t = 1, \ldots, M$.

2.1 Parametric Approaches

In this section, we describe multinomial logistic regression (MLR), which is an extension of logistic regression to cases where the treatment variable has more than two levels. We now assume an underlying multinomial distribution with a probability of inclusion into each treatment group and use maximum likelihood to find the estimates of the regression parameters. The exact steps are as follows:

1. We assume the following model for the generalized propensity scores:

$$r(t|X)_{\mathrm{MLR}} = \frac{1}{1+\sum_{s=2}^{M} e^{\beta_s' X}} \quad \text{for} \quad t=1$$

and

$$r(t|X)_{\mathrm{MLR}} = \frac{e^{\beta_t' x}}{1+\sum_{s=2}^{M} e^{\beta_s' X}} \quad \text{for} \quad t=2,\ldots,M$$

2. We maximize the multinomial likelihood function with respect to all the β's:

$$L(\beta) = \prod_{i=1}^{n}\prod_{t=1}^{M} r_i(t|X)^{A_i(t)}$$

where $r_i(t|X)$ follows the model as defined in Step 1. Equivalently, we maximize the log likelihood function:

$$l(\beta) = \sum_{i=1}^{n}\sum_{t=1}^{M} A_i(t)\log(r_i(t|X)).$$

3. The solution $\hat{\beta}_s$ for $s = 2,\ldots,M$ is substituted into the model to obtain the estimates for the generalized propensity score.

While MLR is a seemingly simple way to estimate the generalized propensity score, there is the question of variable selection and which interactions to be included. In addition, Tchernis et al. [28] pointed out that MLR does not take into account the correlation among treatments in the sense that for two treatment levels $t \neq s$, we have

$$\frac{r(t|X)_{\mathrm{MLR}}}{r(s|X)_{\mathrm{MLR}}} = e^{(\beta_t-\beta_s)'X},$$

which does not depend on the information of other treatment levels. This assumption could be violated in real applications, which makes an MLR model not suitable for estimating the generalized propensity scores.

In R, to fit an MLR model, we can use the package *nnet* [29].

2.2 Machine Learning Techniques

In this section, we are going to introduce two machine learning approaches for the modeling of generalized propensity scores: generalized boosted model (GBM) and random forests (RF).

GBM uses an iterative procedure that adds together many simple regression trees to approximate the propensity score function. A regression tree algorithm divides the dataset into two non-overlapping regions based on one of the covariates. Then, it recursively divides each of those regions into two smaller regions, where each split is based on one of the covariates [2]. Note that the splits may occur on a different covariate or the same covariate each time. The splits are chosen so that the prediction error is minimized. After the allowed number of splits have occurred, for each region of the dataset, the estimated response value equals the average response values of the data points within the region.

Now we describe the GBM method for binary treatments, then we extend the procedure to multi-level treatments. McCaffrey et al. [16] provides a detailed algorithm for estimating propensity scores using GBM. In the binary case, let $g(X) = \log[r(X)/(1-r(X))]$ and the maximum likelihood function can be rewritten as

$$l(g) = \sum_{i=1}^{n} T_i g(X_i) - \log\{1 + \exp[g(X_i)]\}. \quad (6.4)$$

To maximize $l(g)$ in (6.4), $g(X)$ is updated at each iteration with $g(X) + h(X)$ where $h(X)$ is the fitted value from a regression tree which models $\gamma_i = T_i - 1/\{1 + \exp[-g(X_i)]\}$, the largest increase in (6.4). To avoid overfitting, a shrinkage parameter α is introduced so the update is $g(X) + \alpha h(X)$, where α is usually a small value, such as 0.0001. This iterative estimation procedure can be tuned to yield propensity scores that achieve optimal balance in covariate distribution between the treatment and control groups. The key is to stop the algorithm at the optimal number of trees when a certain balance statistic (e.g., average standardized absolute mean difference in the covariates) is minimized. Interactions are automatically included when multi-level splits are allowed in regression trees and since splits are automatically determined by the algorithm based on a criterion, variable selection is automatically done [16].

McCaffrey et al. [17] extended this algorithm to the multi-level treatment case. We first note that while estimating the generalized propensity score for a particular treatment level t, we are interested in the probability that each subject is assigned to a particular treatment t as opposed to any other treatment. So essentially we have two groups: those assigned to treatment t (equivalent to the treatment group in the binary case), and those that were not assigned to treatment t (equivalent to the control group in the binary case). Then we can fit a GBM that balances the covariates between the treatment t group and the entire sample [17]. We do this for each of the M treatments to obtain the generalized propensity scores $\hat{r}(t|X)$. The estimation of the generalized propensity scores for multi-level treatment can be realized in the R package *twang* [19].

The downside to this method is that by fitting separate GBMs for all M treatment groups, it is not guaranteed that the generalized propensity scores for each treatment group will add up to 1. McCaffrey et al. [17] justified that estimating the ATE only requires the propensity scores for the particular treatment groups involved, so as long as the estimated generalized propensity scores are not biased, they do not need to add up to 1.

Next, we are going to introduce RF model for estimating the generalized propensity scores. An RF model [1] is built on a collection of classification trees, fitted on bootstrap samples of the original dataset. Classification trees are different from regression trees in that classification trees predict the class label for each input vector of covariates and use nonparametric information criteria, such as Entropy, misclassification rate, or Gini Index, for splitting at each node. The random forest classification tree finds the best split from only a random subsample of the covariates at each node. Then the estimated generalized propensity score for treatment t is the fraction of votes for t from the collection of the random forest classification trees. The specific random forest algorithm for estimating the generalized propensity score is

1. Draw a random sample with replacement of size n (size of dataset), called a bootstrap sample, from the dataset.
2. Fit a random forest classification tree to the bootstrap sample.
3. Repeat steps 1 and 2 a large number, B, times and obtain a collection of B classification trees (usually, $B = 500$).
4. For a given vector of covariates X, predict the class label from each fitted tree. The estimated generalized propensity score is then

$$\hat{r}(t|X)_{\text{RF}} = \frac{\text{number of trees that voted for class } t}{B}$$

An issue with this method is that it is possible for none of the trees to vote for a particular treatment, resulting with an estimated generalized propensity score of 0 for that treatment. Another possibility is that all the trees vote for one treatment, resulting with an estimated generalized propensity score of 1 for that treatment. In both cases, the positivity assumption, i.e., $0 < r(t|X) < 1$ for all X and t, is violated. In addition, since inverse probability weighting and double-robust estimation involve the reciprocal of the estimated propensity score or one minus the estimated propensity score, an estimated score close to 1 or 0 may result in extreme weights. This issue has been frequently discussed in the literature (e.g., [11, 14, 31]). One way to deal with this issue is to trim extreme weights to a percentile. For example, the inverse probability weights higher than the 95th percentile are set to the 95th percentile. Lee et al. [14] showed that trimming extreme weights gain little benefit in terms of bias, standard error, and 95 % confidence interval coverage, and trimming beyond the optimal level increases bias. Another way to deal with extreme weights is to use a weighted average between a parametric model (such as an MLR model) and RFs as the generalized propensity score estimator [31]. This so-called data-adaptive matching score is

$$\hat{r}(t|X)_{\text{DAMS}} = \lambda\hat{r}(t|X)_{\text{MLR}} + (1-\lambda)\hat{r}(t|X)_{\text{RF}} \quad (6.5)$$

where

$$\lambda = \frac{\hat{r}(t|X)_{\mathrm{MLR}}{}^{A(t)}[1-\hat{r}(t|X)_{\mathrm{MLR}}]^{1-A(t)}}{\hat{r}(t|X)_{\mathrm{MLR}}{}^{A(t)}[1-\hat{r}(t|X)_{\mathrm{MLR}}]^{1-A(t)}+\hat{r}(t|X)_{\mathrm{RF}}{}^{A(t)}[1-\hat{r}(t|X)_{\mathrm{RF}}]^{1-A(t)}} \tag{6.6}$$

As explained by Zhu et al. [31], the intuition of this approach comes from the fact there is a trade-off in bias and variance between parametric and nonparametric approaches. By combining, both bias and variance of the estimated causal effects will be reduced. The choice of λ in (6.6) gives more weight to the estimate that is closer to the observed value of $A(t)$, so it trims extreme weights to more reasonable values without ad hoc adjustment. In addition, it would not attain 0 or 1 as a possible value due to the MLR component.

3 Propensity Score Estimation for a Continuous Treatment

Finally, we are going to focus on the case when the treatment variable is continuous. In this case, we are interested in estimating the so-called dose–response function: $\mu(t) = E[Y_i(t)]$. We assume $Y_i(t)$ is well defined for $t \in \tau$, where $\tau = [t_0, t_1]$.

To draw causal inference, we assume the ignorability assumption:

$$f(t|Y(t), X) = f(t|X), \quad \text{for} \quad t \in \tau,$$

where $f(t|\cdot)$ refers to the conditional density. In other words, we assume the vector of covariates X include all the real confounders that may jointly affect the treatment and the potential outcomes.

In the continuous treatment case, the generalized propensity score is defined as $r(t|X) \equiv f_{t|X}(t|X)$, which is the conditional density of the treatment level t conditioning on the covariates [10]. The ignorability assumption also implies

$$f(t|Y(t), r(t|X)) = f(t|r(t|X)), \quad \text{for} \quad t \in \tau.$$

That is, to adjust for confounding, it is sufficient to condition on the generalized propensity scores instead of conditioning on the vector of covariates. In the literature, Robins et al. [20, 21] propose inverse probability weighting based on the marginal structural model to estimate the dose–response function. To obtain consistent estimation, the inverse probability weight for subject i is

$$w_i = \frac{r(T_i)}{r(T_i|X_i)} \quad \text{for} \quad i = 1, \ldots, n. \tag{6.7}$$

However, the estimation of the conditional probability function (generalized propensity score) in the denominator is a non-trivial problem because when X is high-dimensional, the traditional nonparametric approach for estimating conditional density (e.g., [4]) suffers from curse of dimensionality.

3.1 Parametric Approaches

Robins et al. [21] proposed a two-step approach to estimate $r(T_i|X_i)$. The treatment variable T is assumed to follow a parametric model:

$$T = X^{'}\beta + \epsilon, \quad \epsilon \sim N(0, \sigma^2). \tag{6.8}$$

The generalized propensity score can be estimated by first regressing T_i on X_i, $i = 1, \ldots, n$, and get $\hat{T}_i$ and $\hat{\sigma}$; Then, the residuals $\hat{\epsilon}_i = T_i - \hat{T}_i$, $i = 1, \ldots, n$, are calculated and $r(T_i|X_i)$ can be approximated by

$$\hat{r}(T_i|X_i) \approx f(\hat{\epsilon}_i) \approx \frac{1}{\sqrt{2\pi}\hat{\sigma}} \exp\left\{-\frac{\hat{\epsilon}_i^{\,2}}{2\hat{\sigma}^2}\right\}, \quad i = 1, \ldots, n,. \tag{6.9}$$

To be noticed, if T does not follow a normal distribution (which can be checked based on data), we can always employ nonparametric density estimation approaches, such as Kernel density estimation to estimate $r(T_i|X_i)$ using residuals $\hat{\epsilon}_i, i = 1, \ldots, n$.

3.2 Machine Learning Techniques

In practice, to ensure there is no unmeasured confounders, researchers usually collect a large number of covariates. In the case when X is high-dimensional, the parametric model (6.8) may not be true. A more general approach is to assume

$$T = m(X) + \epsilon, \quad \epsilon \sim N(0, \sigma^2). \tag{6.10}$$

where $m(X) = E(T|X)$ and we employ a nonparametric approach to estimate the mean function.

In [30], we advocate a machine learning algorithm, boosting, to estimate $m(X)$. The boosting model for a continuous response variable can be represented as

$$m(X) = \sum_{m=1}^{M}\sum_{j=1}^{K_m} c_{mj} I\{X \in R_{mj}\}, \tag{6.11}$$

where M is the total number of trees, K_m is the number of terminal nodes for the mth tree, R_{mj} is the indicator of rectangular region in the feature space spanned by X, and c_{mj} is the predicted constant in region R_{mj}. K_m and R_{mj} are determined by optimizing some nonparametric information criterion, such as Entropy, misclassification rate, or Gini Index. c_{mj} is simply the average value of T_i in the training data that falls in the region R_{mj}. Details about how to construct a classification/regression tree can be found in [2].

In boosting, M is a tuning parameter. If M is too large, the model tends to overfit and results in a large variance and if M is too small, bias will occur. In [30], we propose an innovative criterion to determine the value of M. Notice in the inverse probability weighting approach, if subject i receives a weight w_i as in (6.7), it means the subject will be replicated $w_i - 1$ times in the weighted pseudo sample. In the weighted sample, if the propensity scores are correctly estimated, the treatment assignment and the covariates are supposed to be unconfounded under the ignorability assumption [21]. Therefore, a reasonable criterion is to stop the algorithm at the number of trees such that the treatment assignment and the covariates are independent (unconfounded) in the weighted sample. Based on this idea, we propose the following procedure to determine the optimal number of trees in [30]:

1. Calculate $\hat{r}(T_i|X_i)$ using boosting with M trees. Then, calculate

$$w_i = \frac{\hat{r}(T_i)}{\hat{r}(T_i|X_i)} \quad \text{for} \quad i = 1, \ldots, n.$$

   where $\hat{r}(T_i)$ is estimated by normal density.
2. For the jth covariate, denoted as X^j, calculate the weighted correlation coefficient between T and X^j using weights $w_i, i = 1, \ldots, n$ obtained in the first step and denote it as $\bar{d}_j$;
3. Average the absolute value of $\bar{d}_j$ over all the covariates and get the average absolute correlation coefficient ($AACC$).

For each value of $M = 1, 2, \ldots, 20{,}000$, calculate $AACC$ and find the optimal number of trees that lead to the smallest $AACC$ value. In step 2, we employ a bootstrapping approach to obtain the weighted correlation coefficient. Also, we advocate distance correlation coefficient [26, 27] over other correlation metrics. The reason is that the distance correlation takes values between zero and one and it equals zero if and only if T and X^j are independent, regardless of the type of X^j. The R code for calculating $AACC$ is displayed in the Appendix of [30]. After the value of M is determined, the generalized propensity score is estimated by (6.11). More details of this approach can be found in [30].

4 Propensity Score Evaluation

Given the buffet of methods available to researchers, it is important to select the best one among all the candidate propensity score models. On the other hand, it is commonly accepted that there is no uniformly best procedure for all the datasets. In this section, we briefly talk about how to evaluate a propensity score model and how to choose an optimal one among several candidate models. We are going to focus on the binary treatment case. One way to evaluate the performance of different propensity score models is to see how close the estimates are to the true propensity scores using simulations. However, Hirano et al. [7] and Lunceford and Davidian [15] showed that conditioning on the estimated propensity score rather than the true propensity score can yield smaller variance of the estimated causal effects. That is, even when the propensity score is estimated more accurately, it does not necessarily yield better causal inference estimates.

4.1 Evaluation by Checking Balance

One commonly accepted practice is to check balance after the propensity scores are estimated. The underlying idea is that if the propensity score is correctly estimated, the covariates should be distributed almost the same among different treatment groups. There are many ways to evaluate balance in the covariates and it also depends on the particular approach employed to estimate the causal treatment effect. For example, in inverse probability weighting, we may look at the absolute standardized mean difference (ASMD) in the covariates. For a single covariate X, the standardized mean difference is defined as

$$d = \frac{\bar{X}^w_{\text{treated}} - \bar{X}^w_{\text{control}}}{\sqrt{(s^2_{\text{treated}} + s^2_{\text{control}})/2}}, \tag{6.12}$$

where s_{treated} is the standard deviation of X in the treatment group and s_{control} is the standard deviation of X in the control (untreated) group; $\bar{X}^w_{\text{treated}}$ is the weighted average of X in the treatment group and $\bar{X}^w_{\text{control}}$ is the weighted average of X in the control group. When estimating ATE,

$$\bar{X}^w_{\text{treated}} = \frac{\sum_{i=1}^n X_i T_i / \hat{r}_i}{\sum_{i=1}^n T_i / \hat{r}_i},$$

where $\hat{r}_i = \hat{r}(T_i|X_i), i = 1, \ldots, n$ and

$$\bar{X}^w_{\text{control}} = \frac{\sum_{i=1}^n X_i(1 - T_i)/(1 - \hat{r}_i)}{\sum_{i=1}^n (1 - T_i)/(1 - \hat{r}_i)}.$$

When estimating ATT,

$$\bar{X}^w_{\text{treated}} = \frac{\sum_{i=1}^n X_i T_i}{\sum_{i=1}^n T_i},$$

and

$$\bar{X}^w_{\text{control}} = \frac{\sum_{i=1}^n X_i(1 - T_i)\hat{r}_i/(1 - \hat{r}_i)}{\sum_{i=1}^n (1 - T_i)\hat{r}_i/(1 - \hat{r}_i)}.$$

In some literature, the denominator in (6.12) is replaced by s_{treated}. We then look at the mean/mediation/maximum value of the ASMD among the covariates and the propensity score model that leads to the smallest value is usually claimed as the best model.

Other criteria to evaluate the balance in the covariates include Kolmogorov–Smirnov statistic [17], t-test statistic [6], and c statistic. Recently, an innovative prognostic score-based balance measurement has been proposed by Stuart et al. [25], which accounts for the information in the outcome variable while checking balance. The approach works as follows: first, a model of the outcome on the covariates is fitted and the predicted outcome if untreated is calculated for each subject in the study, which is termed the prognostic score. Then, the weighted ASMD in the prognostic score is calculated as a measure of balance. The authors show in a comprehensive simulation study that this measurement outperforms the other balance measurements, such as mean/median/maximum ASMD and KS statistic, in the sense that it is highly correlated with the bias in the estimated causal treatment effect.

4.2 Evaluation Based on a Two-Stage Procedure

In the propensity score-based approaches, we may treat the estimation of propensity scores as the first stage and the estimation of causal treatment effect using matching, stratification or inverse probability weighting as the second stage. The estimated propensity score can be treated as the input into the second stage. While evaluating a propensity score model, we should focus on the quality of the estimates in the second stage rather than the first stage. The two-stage causal inference procedure also fits the model structure discussed by Brookhart and van der Laan [3]. We denote the causal effect as ψ, which is the parameter of interest, and the propensity score as η, which is the nuisance parameter. Assuming we have K different candidate models for estimating η, we aim to choose the optimal one in terms of estimating ψ. Denote the resulting estimates of ψ from the K candidate models as $\hat{\psi}_1(X),\ldots,\hat{\psi}_K(X)$, and assume there exists an approximately unbiased but highly variable estimate of ψ, denoted as $\hat{\psi}_0(X)$. The model used to estimate η in $\hat{\psi}_0(X)$ is regarded as the reference model. To account for the fact that there is a trade-off between bias and variance while estimating ψ, the authors proposed a cross-validation criterion for selecting the optimal estimator of the nuisance parameter among the K candidate models. Let X_v^0 be the training sample and X_v^1 be the testing sample in the vth iteration of the Monte-Carlo cross-validation, the criterion function is defined as follows:

$$C_v(k) = \frac{1}{V}\sum_{v=1}^{V}(\hat{\psi}_k(X_v^0) - \hat{\psi}_0(X_v^1))^2.$$

The optimal model for estimating propensity scores is then chosen to be the one which leads to the smallest C_v among the K models. Brookhart and van der Laan [3] proved that the optimal model selected by the Monte Carlo cross-validation criteria leads to the smallest mean square error of the parameter of interest. This approach has been adopted to compare different propensity score models in [33], in which an over-fitted logistic regression model using all the available covariates is treated as the reference propensity score model to obtain $\hat{\psi}_0(X)$.

References

1. Breiman, L.: Random forests. Mach. Learn. **45**(1), 5–32 (2001)
2. Breiman, L., Friedman, J.H., Olshen, R.A., Stone C.J.: Classification and Regression Trees Chapman & Hall/CRC, Boca Raton, FL (1984)
3. Brookhart, M.A., van der Laan, M.J.: A semiparametric model selection criterion with applications to the marginal structural model. Comput. Stat. Data Anal. **50**(2), 475–498 (2006)
4. Hall, P., Wolff, R.C.L., Yao, Q.: Methods for estimating a conditional distribution function. J. Am. Stat. Assoc. **94**(445), 154–163 (1999)
5. Hansen, L.P.: Large sample properties of generalized method of moments estimators. Econometrica **50**(4), 1029–1054 (1982)
6. Hirano, K., Imbens, G.W.: Estimation of causal effects using propensity score weighting: an application to data on right heart catheterization. Health Serv. Outcome Res. Methodol. **2**(3), 259–278 (2001)
7. Hirano, K., Imbens, G.W., Ridder, G.: Efficient estimation of average treatment effects using the estimated propensity score. Econometrica **71**(4), 1161–1189 (2003)
8. Imai, K., Ratkovic, M.: Covariate balancing propensity score. J. R. Stat. Soc. Ser. B (Stat Methodol.) **76**(1), 243–263 (2014)
9. Imai, K., Van Dyk, D.A.: Causal inference with general treatment regimes. J. Am. Stat. Assoc. **99**(467), 854–866 (2004)
10. Imbens, G.W.: The role of the propensity score in estimating dose-response functions. Biometrika **87**(3), 706–710 (2000)
11. Kang, J.D.Y., Schafer, J.L.: Demystifying double robustness: a comparison of alternative strategies for estimating a population mean from incomplete data. Stat. Sci. **22**(4), 523–539 (2007)
12. Lechner, M.: Program heterogeneity and propensity score matching: an application to the evaluation of active labor market policies. Rev. Econ. Stat. **84**(2), 205–220 (2002)
13. Lee, B.K., Lessler, J., Stuart, E.A.: Improving propensity score weighting using machine learning. Stat. Med. **29**(3), 337–346 (2010)
14. Lee, B.K., Lessler, J., Stuart, E.A.: Weight trimming and propensity score weighting. PLoS ONE **6**(3), e18174 (2011)
15. Lunceford, J.K., Davidian, M.: Stratification and weighting via the propensity score in estimation of causal treatment effects: a comparative study. Stat. Med. **23**(19), 2937–2960 (2004)
16. McCaffrey, D.F., Ridgeway, G., Morral, A.R.: Propensity score estimation with boosted regression for evaluating causal effects in observational studies. Psychol. Methods **9**(4), 403–425 (2004)
17. McCaffrey, D.F., Griffin, B.A., Almirall, D., Slaughter, M.E., Ramchand, R., Burgette, L.F.: A tutorial on propensity score estimation for multiple treatments using generalized boosted models. Stat. Med. **32**(19), 3388–3414 (2013)
18. Pregibon, D.: Resistant fits for some commonly used logistic models with medical applications. Biometrics **38**(2), 485–498 (1982)
19. Ridgeway, G., McCaffrey, D., Morral, A., Burgette, L., Griffin, B.A.: Toolkit for weighting and analysis of nonequivalent groups: a tutorial for the twang package. R vignette. RAND, 2015.

20. Robins, J.M.: Association, causation, and marginal structural models. Synthese **121**(1), 151–179 (1999)
21. Robins, J.M., Hernán, M.Á., Brumback, B.: Marginal structural models and causal inference in epidemiology. Epidemiology. **11**(5), 550–560 (2000)
22. Rosenbaum, P.R., Rubin, D.B.: The central role of the propensity score in observational studies for causal effects. Biometrika **70**(1), 41–55 (1983)
23. Rubin, D.B.: Estimating causal effects of treatments in randomized and nonrandomized studies. J. Educ. Psychol. **66**(5), 688–701 (1974)
24. Setoguchi, S., Schneeweiss, S., Brookhart, M.A., Glynn, R.J., Cook, E.F.: Evaluating uses of data mining techniques in propensity score estimation: a simulation study. Pharmacoepidemiol. Drug Saf. **17**(6), 546–555 (2008)
25. Stuart, E.A., Lee, B.K., Leacy, F.P.: Prognostic score–based balance measures can be a useful diagnostic for propensity score methods in comparative effectiveness research. J. Clin. Epidemiol. **66**(8), S84–S90 (2013)
26. Székely, G.J., Rizzo, M.L.: Brownian distance covariance. Ann. Appl. Stat. **32**(8), 1236–1265 (2009)
27. Székely, G.J., Rizzo, M.L., Bakirov, N.K.: Measuring and testing dependence by correlation of distances. Ann. Stat. **35**(6), 2769–2794 (2007)
28. Tchernis, R., Horvitz-Lennon, M., Normand, S.L.T.: On the use of discrete choice models for causal inference. Stat. Med. **24**(14), 2197–2212 (2005)
29. Venables, W.N., Ripley, B.D.: Modern Applied Statistics with S, 4th edn. Springer, New York (2002). ISBN 0-387-95457-0
30. Zhu, Y., Coffman, D.L., Ghosh, D.: A boosting algorithm for estimating generalized propensity scores with continuous treatments. J. Causal Inference **3**(1), 25–40 (2015)
31. Zhu, Y., Ghosh, D., Mitra, N., Mukherjee, B.: A data-adaptive strategy for inverse weighted estimation of causal effect. Health Serv. Outcome Res. Methodol. **14**(3), 69–91 (2014)
32. Zhu, Y., Schonbach, M., Coffman, D.L., Williams, J.S.: Variable selection for propensity score estimation via balancing covariates. Epidemiology **26**(2), e14–e15 (2015)
33. Zhu, Y., Ghosh, D., Coffman, D.L., Savage, J.S.: Estimating controlled direct effects of restrictive feeding practices in the 'early dieting in girls' study. J. R. Stat. Soc.: Ser. C: Appl. Stat. **65**(1), 115–130 (2016)

Chapter 7
Overcoming the Computing Barriers in Statistical Causal Inference

Kai Zhang and Ding-Geng Chen

Abstract The massive development in statistical causal inference to the era of big data commonly seen in public health applications can be always hindered due to the computational barriers. In this chapter we discuss a practical concern on computing barriers in statistical causal inference with example in optimal pair matching and consequently offer a novel solution by constructing a stratification tree based on exact matching and propensity scores. We demonstrate the implementation of this novel method with a large observational study from Philadelphia obstetric unit closure from 1995 to 2003 with 59 observed covariates in each of the 132,786 birth deliveries and 5,998,111 potential controls. Algorithms and R program code are also provided for interested readers.

1 Statistical Causal Inference and Optimal Pair Matching

In standard statistical modelling, such as the typical regression, estimation, and hypothesis testing techniques, we estimate parameters of a statistical distribution from samples drawn of that distribution. With the estimated parameters for this distribution, we can then make statistical inferences for the associations among variables as well as estimate the probabilities of past and future events with new evidence or new measurements. The processes in statistical modelling can be legitimized and substantiated at the same experimental conditions which are static in the process of statistical design and data collection. This static experimental condition is always a debating topic in the standard statistical modelling.

K. Zhang (✉)
Department of Statistics and Operations Research, University of North Carolina, Chapel Hill, NC, USA
e-mail: zhangk@email.unc.edu

D.-G. Chen
School of Social Work & Department of Biostatistics, Gilling School of Global Public Health, University of North Carolina, Chapel Hill, NC, USA
e-mail: dinchen@email.unc.edu

H. He et al. (eds.), *Statistical Causal Inferences and Their Applications in Public Health Research*, ICSA Book Series in Statistics, DOI 10.1007/978-3-319-41259-7_7

Relaxing the static assumptions in statistical modeling, causal inference goes one step further which can infer not only the probabilities under static conditions, but also the dynamics under the changing conditions by treatments or external interventions. This distinction implies that there is a fundamental difference between causal and associational concepts. In standard statistical modelling, the estimated distribution function cannot tell us how that distribution would differ if external conditions were to change, such as from observational to experimental setup. This information change must be provided by causal assumptions which identify relationships that remain invariant when external conditions change.

The fundamental problem in causal inference is often defined by the counterfactual. This counterfactual can be illustrated in a simple example: "I have diabetes", then "I take Metformin", and " My diabetes got cured". The question now is whether we can infer the cure of diabetes from taking metformin. Is it true that I am cured because I took Metformin? It is impossible to know for sure. This could be certain only if I could have also observed what happened to me if I had not taken the Metformin. But this control condition is impossible to observe for one single individual. This is the so-called counterfactual.

To reach causal statements in causal inference, the ideal situation is to have identical twins randomly assigned to different treatments. The comparison of the effects from the treatments in these cases establishes causal conclusions because all covariates here, measured or unmeasured, are identical. However, in practice, such situation is very rare, particularly in observational studies. Instead, in observational studies the distributions of covariates in the treatment group are often quite different from those in the control group, making it difficult to reach causal conclusions.

Pair matching is one of the most important methods in observational studies to overcome the above difficulty. Intuitively, by finding one subject from the control group for each individual in the treatment group based on similarity of covariates, the pair matching approach is like creating "artificial twins" from the treatment group and the control group. For an excellent reference on pair matching methods, see [3]. Among the basic pair matching methods, the optimal pair matching method (OPM) in [1] aims at minimizing the total rank-based Mahalanobis distance between the treatment group and the control group. The optimal pair matching method has many advantages. For example, compared to the propensity score matching, which only balances the overall distribution of the covariates in two groups, the OPM also balances the covariates within each matched pair. Therefore, the OPM often forms closer pairs in terms of covariate balance and is a popular choice for matching. For algorithms and examples of the usage of OPM, see Sect. 8.5 in [3].

The better balance obtained by the OPM comes, however, at a cost of high computing expenses. For instance, the optimal pairs are formed based on a distance matrix of rank-based Mahalanobis distances from every individual in the treatment group to every individual in the control group. Calculating and storing these distances can be time and space consuming, let alone the optimization algorithm in creating the matched pairs. Although the computing expense in OPM is manageable for small studies, in large observational studies, such expense can be prohibitive.

As an example, in a large observational study, Zhang et al. [5] investigated the effect of massive obstetric unit closures in Philadelphia to the health care quality of the mothers and babies. In this study, the treatment group are the birth delivery records in Philadelphia from 1995 to 2003, and the control group are those in California, Missouri, and the rest of Pennsylvania. The data consist of 59 observed covariates in each of the 132,786 birth delivery records in Philadelphia and 5,998,111 potential control ones. For this massive dataset, if we use OPM directly, we would have to create a $132,786 \times 5,998,111$ distance matrix by calculating 8×10^{11} rank-based Mahalanobis distances [2] based on 59 covariates. The cost from this direct calculation and the storage much exceeds the capacity of standard software, such as the 2.10.0 version of R, which was used in this study in 2009. The data size also exceeds the algorithm for optimal pair matching: The R package `optmatch` stops to work if the size of the distance matrix exceeds about 9×10^6 (the limit for more current R version 3.1.2 is about 10^7).

2 Constructing a Stratification Tree Based on Exact Matching and Propensity Scores

One way to overcome the computing barrier for optimal matching problems is to take advantage of the structure of the data. In particular, there are two general observations:

1. Individuals with similar propensity scores are more likely to have close covariates, and in general the individuals in the treatment group have higher propensity scores than the ones in the control group.
2. Some covariates are of more importance than others for the field of research.

Guided by these considerations, one can stratify the data into small subclasses with a tree structure and match within each subclass.

In what follows, we describe the construction of the stratification tree. In general, decision on whether or not the stratification is needed at each node is based on several scientific, statistical, and computational criteria, while the stratification process can be done by estimated propensity scores and by exact matching of important variables.

At the root of the stratification tree, the entire data is regarded as a stratum. The algorithm then runs through the following steps.

1. **Checking statistical criteria**. We first check if the stratum makes statistical sense. For example, we ask if there is any treated observation in the stratum. If yes, we shall proceed with the stratification and matching steps. Otherwise, we shall ignore the stratum.
2. **Checking matching feasibility**. In this step we check whether the stratum is feasible for the OPM algorithm to get matched pairs. For example, we check whether the size of the distance matrix is below a preset tolerance, for example

9×10^6. If yes, then we can proceed matching within the stratum. Otherwise, we shall further split the stratum. There are two methods of stratification: propensity score stratification (PSS) and important variables stratification (IVS).

3. **Propensity scores stratification (PSS)**. In this stratification process, we first fit a logistic regression to get propensity scores for each individual. We then rank the propensity scores from high to low and start stratification from the top. A subclass keeps recruiting people until both (1) there are more control units than the treated ones and (2) the size of distance matrix reaches the preset tolerance in Step 2. The key idea behind PSS is based on [4] that stratification on estimated propensity scores can effectively reduce bias and unbalance. If the logistic regression encounters difficulty for some reason, such as when the stratum is too large for the logistic regression, the stratum will be split by the important variable stratification (IVS) described below.
4. **Checking the number of strata after propensity**. In [4], the authors recommend five subclasses from PSS. Indeed, since individuals with similar propensity scores may have very different covariates, and since a large number of strata may lead to difficult interpretations, a discretion on the number of strata from PSS is needed. In the algorithm, we check if the number of strata from PSS is below a preset tolerance bound, which should be a compromise between exact matching and propensity score matching and should be advised by field experts. If the number of strata is small, we proceed with matching within each subclass. Otherwise, we disregard the PSS and consider the following important variable stratification (IVS).
5. **Important Variables Stratification (IVS)**. To stratify the data by important variables, we first set a list of priority and a set of ranges for interval splitting. These order of importance and intervals should be advised by field experts before the study. For example, in [5], the variable "Mom's age" has three stratification intervals $(0, 18], (18, 34]$, and $(34, \infty)$ based on medical considerations. Thus, a stratum reaching the IVS step based on "Mom's age" will be split into the three intervals above so that treated and control units are exactly matched in each of the three strata from IVS. The algorithm then repeats from Step 1 for each of the three strata. The key idea behind this process is the exact matching idea described in Chap. 9 in [3]. If there is no more variables for IVS but matching is not feasible in the current data, the algorithm stops and reports an error.

When all strata are of a size that is feasible for matching, the stratification process is complete, and matched pairs are formed within each subclass. The aggregated pairs from all subclasses then form the matched pairs of treated and control individuals for the entire study. The flowchart in Fig. 7.1 describes the complete algorithm.

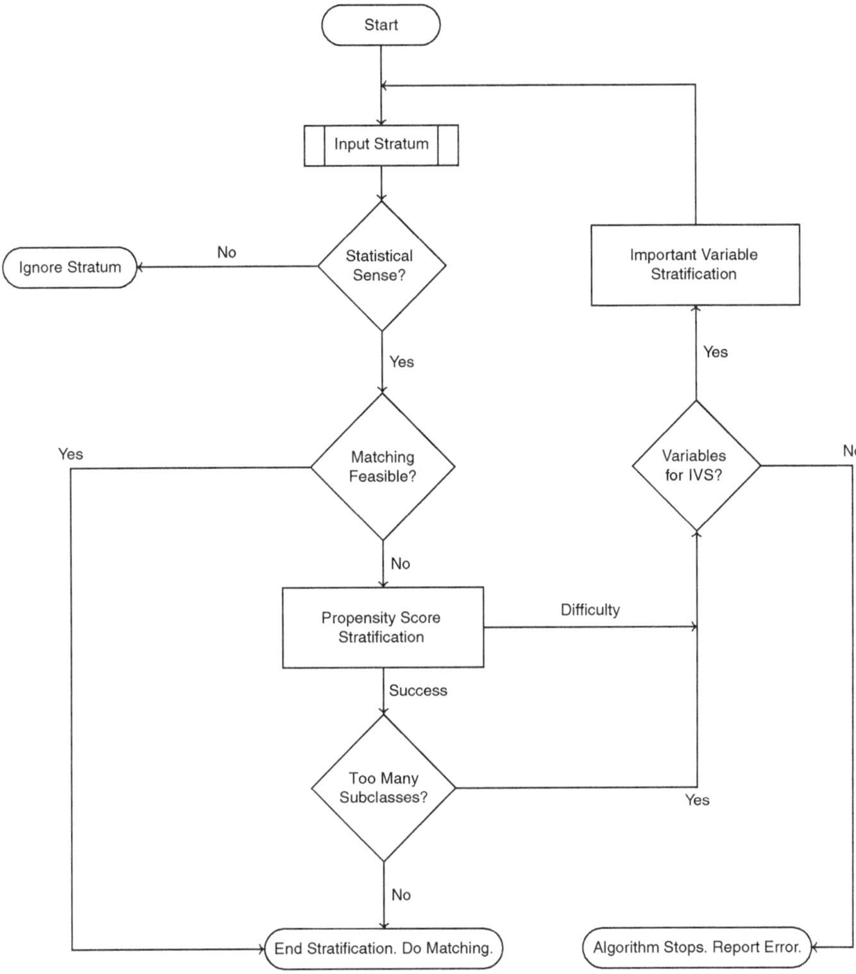

Fig. 7.1 Flowchart of the complete stratification process

3 Example: Creating a "Control-Philadelphia"

We take the 1995 data in [5] for example in illustrating the algorithm. There are 14,768 treated units and 681,743 control ones in this dataset, and the data size is beyond the 2.10.0. version of software R in 2009 for logistic regression and OPM. To apply the algorithm described in Sect. 2, we set the following argument as input: The tolerance size of the distance matrix within each subclass is set to be 9×10^6. The tolerance number of the subclasses is set to be 5. The stratification variables and stratification intervals suggested by the doctors, listed by the priority from high to low, are

1. "Gestation Age" with intervals $(0,33]$, $(33,36]$, $(36,38]$, $(38,40]$, and $(40,\infty)$.
2. "Mom's Age" with intervals $(0,18]$, $(18,34]$, and $(34,\infty)$.
3. "Mom's Education" with categories "Less than High School", "High School Degree", "College Degree", and "More than College".

We shall only explain Step 3 of PSS here since other steps are straightforward. For this step, we use the delivery records with gestational ages more than 40 weeks as an example. There are 1791 treated units and 137,463 control ones. To subclassify this stratum, we fit the logistic regression with all covariates but "Gestation Age" to get estimated propensity scores for each individual. We then sort the estimated propensity scores from high to low and present a few of them in Table 7.1. We form the first subclass by searching in the treated group for the lowest estimated propensity score p_1, above which (1) the number of treated propensity scores is less than the number of control propensity scores—so there are enough control units to pair with treated ones, and (2) the product of the number of treated and control units—which is going to be the size of the distance matrix in OPM—is less than a threshold required by certain software. The observations with a estimated propensity score higher than p_1 are collected to be the first subclass. The detailed R function of this PSS step is provided in Appendix.

In this stratum, for the first 312 treated units there are not enough control units above their propensity scores so pair matching cannot be done. From the 313-th treated unit on the pool of control units is large enough to match for each treated unit. However, from the 1045-th treated units on, the product of the sizes of treated and control pools exceeds the 9×10^6 tolerance bound. Therefore, $p_1 = 0.0388$, and the first subclass consists of 1044 treated units and 8614 control units.

In summary, by going through the process described in Sect. 2, the resulting tree of strata for the 1995 data is shown in Fig. 7.2. At the beginning step, the stratum is so large that even logistic regression cannot be fit. After the split based on "Gestation Age", four of the five strata can be divided into a few subclasses for which matching is feasible. However, the $(38,40]$ stratum is still too large. Its PSS will result in 12 subclasses, which is above the limit of 5. Thus, this stratum is further split by "Mom's Age". We further stratify the "Mom's age" group of $(18,34]$ by "Mom's Education" since it is too large too. The resulting tree has 10 ending nodes of strata with the numbers of subclasses $2,2,5,2,2,5,4,1,2,4$.

After the stratification process is complete, OPM is performed within each of the 29 subclasses. Table 7.2 from [5] shows that the covariate balance in terms of standardized differences before and after matching. It can be seen that before matching, the distributions of covariates are quite different between the Philadelphia group and the control group. Many covariates have a standardized difference greater than 0.2. For example, on average Philadelphia mothers were younger and their prenatal care started later, Philadelphia babies were lighter in weight, and

Table 7.1 Illustration of PSS in [5]

Rank of treated propensity score	Estimated treated propensity score	Number of control units above	Enough control?	Size of dist. matrix	Exceeding OPM limit?	Matching feasible?
1	0.973	0	No	0	No	No
⋮	⋮	⋮	⋮	⋮	⋮	⋮
312	0.314	308	No	96,096	No	No
313	0.312	315	Yes	98,595	No	Yes
⋮	⋮	⋮	⋮	⋮	⋮	⋮
1043	0.0389	8594	Yes	8,963,542	No	Yes
1044	0.0388	8614	Yes	8,993,016	No	Yes
1045	0.0387	8629	Yes	9,017,305	Yes	No
⋮	⋮	⋮	⋮	⋮	⋮	⋮

The first subclass is formed by the top 1044 treated units and 8614 control units

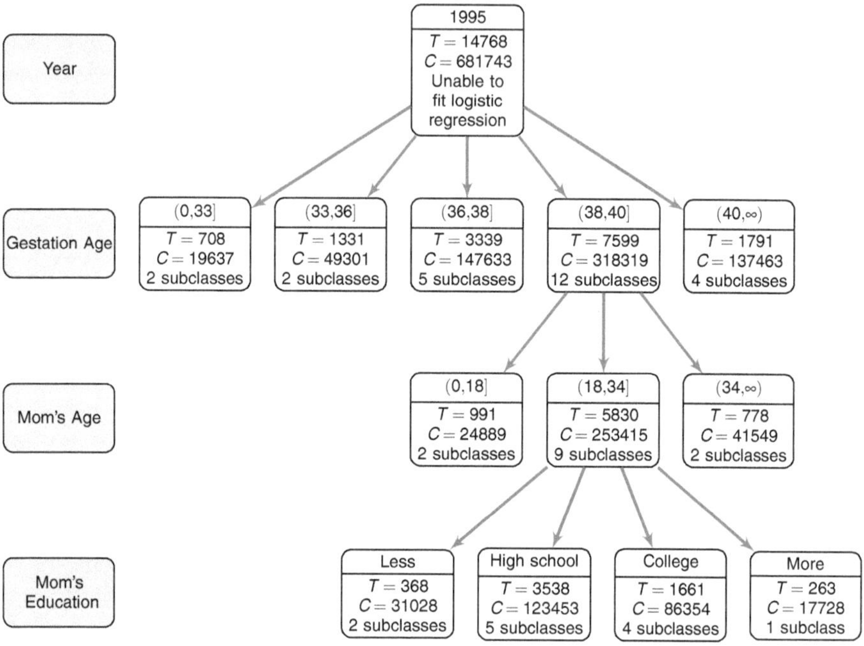

Fig. 7.2 The stratification tree from the algorithm. Each node represents a resulting stratum from IVS. Within each node, we list the number of treated units, the number of control units, and the number of subclasses from PSS if feasible

Table 7.2 Covariate balance before and after matching

Sample size	5,998,111 potential controls	132,786 Philadelphia births	132,786 matched controls	Absolute standardized difference	
Covariate	Covariate mean or proportion			Before	After
	Mom's neighborhood (Zip code)				
Income (K$)	46	30	30	**1.16**	0.04
Income missing	0.00	0.00	0.00	0.06	0.00
Poverty (zip-fr)	0.15	0.25	0.23	**0.91**	0.13
Poverty missing	0.00	0.00	0.00	0.06	0.00
High school (zip-fr)	0.74	0.68	0.69	**0.37**	0.07
HS missing	0.00	0.00	0.00	0.06	0.00
College (zip-fr)	0.22	0.15	0.15	**0.51**	0.01
College missing	0.00	0.00	0.00	0.06	0.00
	Mom				
Mom's age	28	26	26	**0.21**	0.01
Parity	2.10	2.20	2.20	0.07	0.03
Parity missing	0.00	0.01	0.01	0.09	0.04

(continued)

Table 7.2 (continued)

Sample size	5,998,111 potential controls	132,786 Philadelphia births	132,786 matched controls	Absolute standardized difference	
Covariate	Covariate mean or proportion			Before	After
Prenatal care (month started)	2.40	2.70	2.60	**0.22**	0.04
PC missing	0.02	0.11	0.08	**0.37**	0.11
	Mom's education				
Below 8th grade	0.10	0.02	0.02	**0.32**	0.02
Some high school	0.17	0.21	0.20	0.11	0.04
HS graduate	0.30	0.38	0.40	0.17	0.05
Some college	0.20	0.19	0.19	0.02	0.01
College graduate	0.13	0.09	0.10	0.11	0.01
More than college	0.09	0.06	0.06	0.11	0.00
Missing	0.01	0.04	0.04	0.17	0.04
	Mom's race				
White	0.71	0.31	0.32	**0.87**	0.03
Black	0.07	0.42	0.46	**0.88**	0.11
Asian	0.07	0.03	0.03	0.18	0.03
Other	0.12	0.06	0.05	**0.20**	0.05
Missing	0.02	0.17	0.14	**0.52**	0.13
	Mom's health insurance				
Government	0.40	0.40	0.39	0.01	0.02
Other insurance	0.57	0.58	0.60	0.02	0.04
Uninsured	0.03	0.01	0.01	0.11	0.04
Missing	0.00	0.01	0.00	0.11	0.06
	Baby				
Birth weight (g)	3345	3179	3189	**0.26**	0.02
Birth weight missing	0.00	0.00	0.00	0.04	0.03
Gestational age (weeks)	39	38	38	0.14	0.01
Gestational age missing	0.05	0.01	0.01	**0.22**	0.02
Small at gestational age	0.09	0.14	0.12	0.16	0.05

For Zip Code data, zip-fr means the fraction of the Zip Code with this attribute. An absolute standardized difference in mean of 0.2 or greater is in **bold**

Philadelphia families had less income. Such discrepancy made causal conclusions difficult from direct comparisons. However, after matching, the covariates were well balanced: The standardized differences were all below 0.2, and the average of each covariate was close.

4 Summary

As shown in this chapter, the optimal pair matching method can achieve good balance among covariates for causal conclusions in large observational studies. However, it has to be used with care to avoid the high computation cost it can incur. In the era of Big Data, such large studies can be more and more often. Thus, it is important to develop and consider efficient ways in OPM to facilitate statistical analysis in observational studies.

In the Philadelphia obstetric unit closure study [5], a stratification tree method was applied to split the data into small subclasses where matching is computationally feasible. The construction process of the tree structure was based on an integration of propensity score stratification and important variable stratification. The underlying ideas are that propensity scores [4] and exact matching [3] are important ways to balance covariates. In the Philadelphia obstetric unit closure study, the difference between covariates is substantially reduced after the matching based on the stratification tree, which in turn facilitates the establishment of causal statements.

In general, stratification provides efficient ways to perform OPM for large datasets. Since the overall goal is to achieve the balance in covariates between matched pairs for causal conclusion, the stratification process, especially important quantities such as the points of split, should be carefully designed with the advice from field experts. The resulting stratification strategy should be a good compromise between covariate balance and computation costs.

The R code for the tree stratification algorithm described in this chapter is available upon request for interested readers. An R package of this algorithm is also under development.

Acknowledgements Zhang's research is partially supported by NSF DMS-1309619, DMS-1613112, and IIS-1633212. Chen's research was supported in part by NIH grants from the Eunice Kennedy Shriver National Institute of Child Health and Human Development (NICHD, R01HD075635, PIs: Xinguang. Chen and Ding-Geng Chen). This material was also partially based upon work supported by the NSF under Grant DMS-1127914 to the Statistical and Applied Mathematical Sciences Institute. Any opinions, findings, and conclusions or recommendations expressed in this material are those of the author(s) and do not necessarily reflect the views of the National Science Foundation.

Zhang thanks Dylan S. Small for very helpful suggestions.

Appendix: R Code for Propensity Score Stratification

The following R function `opt_pstrat` implements the PSS algorithm in Step 3 described above and forms the subclasses. The function takes three arguments as inputs:

1. `indicator`: This argument takes a binary vector which takes value 1 for treated units and 0 for control ones.
2. `pscore`: This argument takes a vector of propensity scores of each unit.
3. `sizemax`: This argument takes a preset tolerance level on the size of the distance matrix. The default value is 9,000,000.

The function `opt_pstrat` returns with the following values:

1. `flag`: This output returns 1 for successful stratification and 2 otherwise.
2. `cutoffs`: This output returns the cutoff points where the subclasses are split.
3. `t.pstrata`: This output returns a vector listing the number of treated units in each subclass formed.
4. `c.pstrata`: This output returns a vector listing the number of control units in each subclass formed.
5. `prodsize.pstrata`: This output returns a vector listing the size of distance matrix in each subclass formed.

```
opt_pstrat <- function(indicator,pscore,sizemax=9000000){

    cutoffs <- max(pscore)

    t.strata <- NULL

    c.strata <- NULL

    size.strata <- NULL

    indicator_iter <- indicator

    pscore_iter <- pscore

    num_strata_formed <- 0

    while(sum(indicator_iter)>0 & sum(1-indicator_iter)>0){

        n <- length(indicator_iter)

        t_ind <- which(indicator_iter==1)

        n_treated <- sum(indicator_iter)

        treated_pscore_iter <- pscore_iter[t_ind]

        t_geq_t <- n_treated+1-rank(treated_pscore_iter,
        ties.method="min")

        c_geq_t <- n+1-rank(pscore_iter,ties.method="min"
        )[t_ind]-t_geq_t

        matchable <- c_geq_t >= t_geq_t

        if (sum(matchable)>0){
```

```
        matchable_set <- t_ind[c_geq_t >= t_geq_t]
    }else{
        print("No way to stratify: c_geq_t < t_geq_t."
    ); stop}
    size.dist <- t_geq_t*c_geq_t
    if (min(size.dist[matchable])>sizemax){
        print("No way to stratify: min(size.dist)>sizemax.");
        return(list(flag=2))
    }
    cutoff.size.ind <- which(size.dist==max(size.dist[
    matchable][size.dist[matchable]<sizemax]))[1]
    cutoff <- pscore_iter[t_ind[cutoff.size.ind]]
    cutoffs <- c(cutoffs,cutoff)
    t.strata <- c(t.strata,sum(treated_pscore_iter>=cutoff))
    c.strata <- c(c.strata,sum(pscore_iter>=cutoff)-sum(
    treated_pscore_iter >= cutoff))
    size.strata <- c(size.strata,sum(treated_pscore_iter>=
    cutoff)*(sum(pscore_iter>=cutoff)-sum(
    treated_pscore_iter>=cutoff)))
    num_strata_formed <- num_strata_formed+1
    print(num_strata_formed)
    indicator_iter <- indicator_iter[pscore_iter<cutoff]
    pscore_iter <- pscore_iter[pscore_iter<cutoff]
}
if (sum(indicator_iter)==0){
    print("Stratification Finished: Treated Units Used Up.")
    return(list(flag=1,num_pstrata=num_strata_formed,
    cutoffs=rev(cutoffs),t.pstrata=rev(t.strata),c.pstrata=
    rev(c.strata),prodsize.pstrata=rev(size.strata)))
}
if (sum(!indicator_iter)==0){
    print("Stratification Finished: Control Units Used Up.
```

```
            Cannot Form New Strata.")
            return(list(flag=2,num_pstrata=num_strata_formed,
            cutoffs=rev(cutoffs),t.pstrata=rev(t.strata),c.pstrata=
            rev(c.strata),prodsize.pstrata=rev(size.strata)))
        }
}
```

As described in the main text, the function `opt_pstrat` is applied when each stratum goes through Step 3. The outputs of this function provide useful information on whether to further split or match within the subclasses.

References

1. Hansen, B.B., Klopfer, S.O.: Optimal full matching and related designs via network flows. J. Comput. Graph. Stat. **15**, 609–627 (2006)
2. Rosenbaum, P.R.: Observational Studies. Springer Series in Statistics. Springer, New York (2002)
3. Rosenbaum, P.R.: Design of Observational Studies. Springer, New York (2010)
4. Rosenbaum, P.R., Rubin, D.B.: Reducing bias in observational studies using subclassification on the propensity score. J. Am. Stat. Assoc. **79**, 516–524 (1984)
5. Zhang, K., Small, D.S., Lorch, S., Srinivas, S., Rosenbaum, P.R.: Using split samples and evidence factors in an observational study of neonatal outcomes. J. Am. Stat. Assoc. **106**, 511–524 (2011)

Part III
Causal Inference in Randomized Clinical Studies

Chapter 8
Semiparametric Theory and Empirical Processes in Causal Inference

Edward H. Kennedy

Abstract In this paper we review important aspects of semiparametric theory and empirical processes that arise in causal inference problems. We begin with a brief introduction to the general problem of causal inference, and go on to discuss estimation and inference for causal effects under semiparametric models, which allow parts of the data-generating process to be unrestricted if they are not of particular interest (i.e., nuisance functions). These models are very useful in causal problems because the outcome process is often complex and difficult to model, and there may only be information available about the treatment process (at best). Semiparametric theory gives a framework for benchmarking efficiency and constructing estimators in such settings. In the second part of the paper we discuss empirical process theory, which provides powerful tools for understanding the asymptotic behavior of semiparametric estimators that depend on flexible nonparametric estimators of nuisance functions. These tools are crucial for incorporating machine learning and other modern methods into causal inference analyses. We conclude by examining related extensions and future directions for work in semiparametric causal inference.

1 Introduction

Causality and counterfactual questions lie at the heart of many if not most scientific endeavors. Counterfactual questions are about what *would have happened* in some system had it undergone a particular change. For example: How would the distribution of patient outcomes differ had everyone versus no one received some medical treatment? Which rule for treatment assignment would maximize outcomes if it were implemented in the population?

In fact many scientific questions are causal even if they are not framed using explicitly causal language and notation. For example, standard regression analyses

E.H. Kennedy (✉)
University of Pennsylvania, Philadelphia, PA, USA
e-mail: kennedye@mail.med.upenn.edu

H. He et al. (eds.), *Statistical Causal Inferences and Their Applications in Public Health Research*, ICSA Book Series in Statistics, DOI 10.1007/978-3-319-41259-7_8

are often explained in implicitly causal terms, e.g., when regression coefficients are portrayed as representing the expected difference in outcome if all covariates were held constant, except for one covariate whose value was increased by one. In contrast, without causal assumptions, these coefficients can only represent the expected difference in outcome for two units who happen to have the same covariate values, except for one covariate whose values happen to differ by one; manipulation of the covariate cannot be allowed without invoking causal assumptions.

In this chapter we give a review of semiparametric theory and empirical processes as they arise in causal inference problems. These include very powerful methodological tools that can be especially useful in causal settings.

In Sect. 2 we give an introduction to causal inference, following Robins [37, 41, 59], van der Laan [57, 59, 60], and others. In order to answer causal questions with observed data, we need causal assumptions. Sometimes these causal assumptions can hold by virtue of the study design (e.g., in randomized trials), while at other times the assumptions we need are untestable and need to be justified based on subject matter expertise (e.g., in standard observational studies). In either case, as we discuss in detail in Sect. 2.1, it is important to have a clearly defined study question (with a corresponding causal parameter of interest). It is similarly important to be precise about the assumptions that are required to estimate the causal parameter of interest with observed data. This is the enterprise of identification, which we discuss briefly in Sect. 2.2.

After a causal parameter of interest has been precisely defined and identified (i.e., expressed in terms of observed data), then estimation and inference for that parameter is essentially a purely statistical problem. Classical maximum likelihood approaches can in theory be used to estimate such identified causal parameters, but typically require unrealistic parametric assumptions about the entire data-generating process. In contrast, semiparametric methods allow parts of the data-generating process to be completely unrestricted, e.g., if they are unknown or involve nuisance functions that are not of particular interest to the study question. Thus, if investigators have a good understanding of the treatment assignment process, for example, this information can be incorporated into a semiparametric analysis, and no assumptions might be needed about the outcome process. This is particularly useful in causal inference settings since the outcome process is often complex and difficult to model, while investigators may have some information about the treatment mechanism (e.g., by surveying doctors about how they prescribe some treatment).

Alternatively, in many cases investigators may not have much information available about any part of the data-generating process. Then it will often be most reasonable to use a nonparametric model, which does not make any parametric assumptions at all about the data-generating process. A nonparametric model can be viewed as a special case of a semiparametric model, so the theory reviewed in this chapter covers these settings as well as those where treatment is assigned according to some known process.

In Sect. 3 we review semiparametric theory, following foundational work by numerous authors, including Begun et al. [4], Bickel et al. [7], Pfanzagl [33], van der Vaart [64, 65], Robins [37, 41, 59], van der Laan [57, 59, 60], and many others [21, 53]. We start in Sect. 3.1 with a general introduction to semiparametric models, and discuss influence functions as representations of estimators in such models in Sect. 3.2. Then in Sect. 3.3 we introduce the notion of tangent spaces and a related space where influence functions reside, give an example illustrating basic semiparametric theory for estimation of the average treatment effect in Sect. 3.4, and wrap up by discussing links to general missing data problems in Sect. 3.5.

Semiparametric theory gives us efficiency benchmarks in models where parts of the data-generating process are unrestricted, and tells us how to construct potentially efficient estimators. However, in order to understand the asymptotic behavior of such semiparametric estimators, particularly when flexible nonparametric methods are used to estimate nuisance functions, we need empirical process theory. This is the topic of Sect. 4. The field of empirical processes is vast, so we only discuss parts that especially relate to estimation of nuisance functions. Our review follows important work by Andrews [1, 2], Pollard [35, 36], van der Vaart [64, 65, 67], Wellner [48, 67], and others [21, 60]. We start by giving the motivation for empirical process theory in semiparametric problems in Sect. 4.1, discuss Donsker classes and examples in Sects. 4.2 and 4.3, and illustrate with an analysis of the doubly robust estimator of the average treatment effect in Sect. 4.4.

We close the chapter in Sect. 5 by considering extensions and future directions for work in semiparametric causal inference.

2 Setup

In this section we briefly introduce the basic setup of a typical causal inference problem. We focus on two essential components of causal inference: first, formulating a clearly defined parameter of interest, and second, exploring how and whether this target parameter is identified with observed data. These issues are very important and provide a crucial foundation for semiparametric causal inference; however, we give only a brief treatment since the main goal of this chapter is to discuss semiparametric theory and empirical processes. Much of the discussion here is inspired by pioneering work by Robins [37, 41, 59], van der Laan [57, 59, 60], and colleagues.

2.1 The Target Parameter

An important first step in any scientific pursuit is to have a clearly defined goal. In a statistical analysis, this includes giving a precise expression for a parameter of interest, which we will refer to as *the target parameter*.

The target parameter is the main feature of interest in the analysis, and ideally is decided upon based on collaborative discussion between scientific investigators and the statistician or analyst. In practice, however, the target parameter is sometimes defined only in vague terms, or is chosen based on convenience rather than scientific interest. In causal inference problems, the target parameter is typically formulated in terms of hypothetical interventions and corresponding counterfactual data, which represent the data that would have been observed under some intervention. In this chapter we mostly rely on the potential outcome framework, due to Neyman [28] and Rubin [46, 47], but note that alternative frameworks based on structural equation models and graphs [30, 31], or decision theory [10] can also be useful.

For example, in some population of units (e.g., patients), let $Y \in \mathbb{R}$ denote a random variable representing an outcome of interest (e.g., blood pressure, or an indicator for whether a heart attack occurred), and let $A \in \{0, 1\}$ denote a binary treatment (e.g., receipt of a statin), whose effect is in question. Then it may be of interest to estimate the average causal effect, i.e., how the expected outcome would have differed had everyone in the population taken treatment versus if no one in the population had taken treatment. This quantity can be represented notationally as follows. Let Y^a denote the potential outcome that would have been observed (for a particular unit in the population) had that unit taken treatment level $A = a$. For a binary treatment, for example, this notation gives rise to two potential outcomes, Y^1 and Y^0, which are the outcomes that would have been observed for a particular unit under treatment ($A = 1$) and control ($A = 0$), respectively. Then the *average causal effect* in the population can be defined as

$$\psi = \mathbb{E}(Y^1 - Y^0). \tag{8.1}$$

Of course, different contrasts may instead be of interest under this hypothetical intervention; for example, if the outcome is binary, then one may be more concerned with the risk ratio $\mathbb{E}(Y^1)/\mathbb{E}(Y^0) = \mathbb{P}(Y^1 = 1)/\mathbb{P}(Y^0 = 1)$, or with the odds ratio $\{\mathbb{P}(Y^1 = 1)/\mathbb{P}(Y^1 = 0)\}/\{\mathbb{P}(Y^0 = 1)/\mathbb{P}(Y^0 = 0)\}$. Alternatively, one may care more about how the effect of treatment changes with some other variable. Or some other entirely different intervention may be of interest; for example, one may want to learn what the mean outcome would have been if treatment had been assigned via some rule based on other variables [9, 24], or how outcomes would have changed under treatment versus control if a mediating variable (a variable occurring subsequent to treatment, but prior to outcome) was fixed at some value [51, 69].

We will consider a number of different types of causal parameters and hypothetical interventions in subsequent sections, but a full taxonomy is beyond the scope of this chapter. The main point is that it is necessary to have a clear definition of the target parameter (i.e., the object one wants to learn about using data) when working in the semiparametric framework. In fact, regardless of framework or philosophical perspective, a clearly defined target parameter is necessary in order to meaningfully address estimation bias or variance relative to any meaningful standard.

2.2 Identification

Once a target parameter is clearly defined based on some hypothetical intervention, the next step is to explore how and whether it can be *identified* (i.e., expressed uniquely in terms of a distribution for observed data). This step translates the causal question of interest into a statistical problem defined in terms of observed data.

For example, suppose that in a population of interest we actually get to observe potential outcomes under the received treatment for each unit, i.e.,

$$A = a \implies Y = Y^a. \tag{C1}$$

Condition (C1) is called "consistency" [68] and holds if potential outcomes are defined uniquely by a unit's own treatment and not others' (i.e., no interference), and also not by the way treatment is administered (i.e., no different versions of treatment). Also suppose that there exists some set of observed covariates L that render treatment independent of potential outcomes when conditioned upon, i.e.,

$$A \perp\!\!\!\perp Y^a \mid L, \tag{C2}$$

where $\perp\!\!\!\perp$ denotes statistical independence. Condition (C2) is often called "no unmeasured confounding," "exchangeability," or "ignorability," and holds if treatment is externally randomized, or if treatment decisions are made based only on covariates L. Finally suppose that, regardless of covariate value, each unit has a non-zero chance to receive treatment level $A = a$, i.e.,

$$p(A = a \mid L = l) \geq \delta > 0 \text{ whenever } p(L = l) > 0, \tag{C3}$$

where $p(\cdot)$ denotes densities with respect to an appropriate dominating measure. Condition (C3) is called "positivity" and means treatment is not assigned deterministically [32]. Then, if Conditions (C1)–(C3) hold for treatment value a, it follows that

$$p(Y^a = y \mid L = l) = p(Y = y \mid L = l, A = a). \tag{8.2}$$

Therefore we can express the conditional distribution of the potential outcome Y^a given L in terms of observed data; thus, we can also identify the conditional distribution given any subset of L, including the null set, by simply marginalizing. In particular if Conditions (C1)–(C3) hold for $a = 0, 1$, then the average causal effect ψ from (8.1) can be written as

$$\psi = \int_{\mathscr{L}} \Big\{\mathbb{E}(Y \mid L = l, A = 1) - \mathbb{E}(Y \mid L = l, A = 0)\Big\}\, dP(L = l). \tag{8.3}$$

The above identification result is an example of the g-computation formula, which was first proposed for general time-varying treatments by Robins [37, 43]. Numerous alternative identification schemes are also available, for example based on instrumental variables [3, 16]. The literature on causal identification is extensive, and includes graphical criteria [30, 31], bounds [23], and many other topics.

In this chapter we focus on settings where the target causal parameter (call it ψ) is identified, and thus can be written in terms of the distribution P of the observed data. In the next section we illustrate ideas with the average causal effect ψ defined in Eq. (8.1), and defined by Eq. (8.3) under Conditions (C1)–(C3); although we focus on simple average effects, the general logic is similar for other parameters.

3 Semiparametric Theory

In this section we give a general review of semiparametric theory, using as a running example the common problem of estimating an average causal effect. Our review draws on foundational work in general semiparametric theory by Begun et al. [4], Bickel et al. [7], Pfanzagl [33, 34], and van der Vaart [64, 65], among others [21, 26], as well as further developments for missing data and causal inference problems by Robins [37–39, 41, 59], van der Laan [57, 59, 60], and colleagues [15, 53].

3.1 Semiparametric Models

Standard semiparametric theory generally considers the following setting. We observe an independent and identically distributed sample $(Z_1, \ldots, Z_n)$ distributed according to some unknown probability distribution P_0 on the Borel σ-field $\mathscr{B}$ for some sample space $\mathscr{Z}$. The general goal is estimation and inference for some target parameter $\psi_0 = \psi(P_0) \in \mathbb{R}^p$, where $\psi = \psi(P)$ can be viewed as a map from a probability distribution to the parameter space (assumed to be Euclidean here). In our running example where ψ is the average causal effect defined in (8.3) (after imposing identifying assumptions), the observed data consist of an independent and identically distributed sample of $Z = (L, A, Y)$ where L denotes covariates, A is a binary treatment, and Y is the outcome of interest. Here we suppose the distribution P_0 has density given by

$$p(z) = p(y \mid l, a)p(a \mid l)p(l) \tag{8.4}$$

with respect to some dominating measure. In general we write $p(X = t)$ for the density of X at t, but when there is no ambiguity we let $p(x) = p(X = x)$.

A *statistical model* $\mathscr{P}$ is a set of possible probability distributions, which is assumed to contain the observed data distribution P_0. In a parametric model, $\mathscr{P}$ is assumed to be indexed by a finite-dimensional real-valued parameter $\theta \in \mathbb{R}^q$, e.g., we may have $\mathscr{P} = \{P_\theta : \theta \in \mathbb{R}^q\}$ with $\psi \subseteq \theta$. For example, if Z is a scalar random variable, one might assume it is normally distributed with unknown mean and variance, $Z \sim N(\mu, \sigma^2)$, in which case the model is indexed by $\theta = (\mu, \sigma^2) \in \mathbb{R} \times \mathbb{R}^+$. *Semiparametric models* are simply sets of probability distributions that cannot be indexed by only a Euclidean parameter, i.e., models that are indexed by an infinite-dimensional parameter. Semiparametric models can vary widely in the amount of structure they impose; for example, they can range from *nonparametric models* for which $\mathscr{P}$ consists of all possible probability distributions, to simple regression models that characterize the regression function parametrically but leave the residual error distribution unspecified.

In semiparametric causal inference settings it is common to impose some structure on the treatment mechanism (e.g., with a parametric model) leaving the outcome mechanism unspecified. This is because the outcome mechanism is often a complex natural process outside of human control, whereas the treatment mechanism is known in randomized trials, and can be well understood in some observational settings (for example, when a medical treatment is assigned in a standardized way, which is communicated by physicians to researchers). In our running example, one may wish to do inference for the average causal effect ψ under a parametric model for the treatment mechanism, leaving everything else unspecified, so that

$$p(z; \eta, \alpha) = p(y \mid l, a; \eta_y) p(a \mid l; \alpha) p(l; \eta_l), \tag{8.5}$$

where $\alpha \in \mathbb{R}^q$ but $\eta = (\eta_y, \eta_l)$ represents an infinite-dimensional parameter that does not restrict the distribution of the outcome given covariates and treatment $p(y \mid l, a)$ or the marginal covariate distribution $p(l)$.

Of course it is not always the case that there is substantive information available about the treatment mechanism; in many observational studies, neither the exposure nor the outcome process is under human control, and both processes may be equally complex (e.g., in studies where the treatment or exposure is itself a disease or other medical condition). In such cases it is often more appropriate to consider inference for ψ under a nonparametric model that makes no parametric assumptions about the distribution P. As we will see in Sect. 4.4, in order to obtain usual root-n rates of convergence in nonparametric models, we will still require some conditions on how well we can estimate the nuisance functions.

Another way semiparametric models arise in causal settings is through parametric assumptions about high-level treatment effects. For example, suppose we were not interested in the average causal effect $\mathbb{E}(Y^1 - Y^0)$ but in how this effect varied with a subset of covariates $V \subset L$, i.e., the goal was to estimate $\gamma(v) = \mathbb{E}(Y^1 - Y^0 \mid V = v)$. Letting $W = L \setminus V$ so that $L = (V, W)$, it is straightforward to show that this conditional effect is also identified under Conditions (C1)–(C3) as in (8.3), except replacing $dP(l)$ with $dP(w \mid v)$. If V includes a continuous variable or has many strata, it may be desirable to make parametric assumptions to reduce the dimension of $\gamma(v)$ (or in rare cases, there may be substantive knowledge about the parametric form of the effect modification), and thus one may want to assume $\gamma(v) = \gamma(v;\psi)$ for $\psi \in \mathbb{R}^p$. Such assumptions are not always easily encoded directly in the distribution $p(z)$, but can still be employed in conjunction with parametric assumptions about the treatment mechanism, for example, or in otherwise nonparametric models. An alternative approach is to use nonparametric *working models* [25], where instead of assuming $\gamma(v) = \gamma(v;\psi)$ we define our target parameter as a projection of $\gamma(v)$ onto the model $\gamma(v;\psi)$ (using, for example, a weighted least squares projection).

3.2 Influence Functions

In the previous subsection we discussed the concept of a semiparametric model (in which part of the distribution P is allowed to have unrestricted or infinite-dimensional components) and gave some examples. Now we begin to discuss estimation and inference in such models. This requires the concept of the *influence function*, which is a foundational object of statistical theory that allows us to characterize a wide range of estimators and their efficiency.

Let $\mathbb{P}_n = n^{-1}\sum_i \delta_{Z_i}$ denote the empirical distribution of the data, where δ_z is the Dirac measure that simply indicates whether $Z = z$. This means, for example, that empirical averages can be written as $n^{-1}\sum_i f(Z_i) = \int f(z)\, d\mathbb{P}_n = \mathbb{P}_n\{f(Z)\}$. An estimator $\hat{\psi} = \hat{\psi}(\mathbb{P}_n)$ is asymptotically linear with influence function φ if the estimator can be approximated by an empirical average in the sense that

$$\hat{\psi} - \psi_0 = \mathbb{P}_n\{\varphi(Z)\} + o_p(1/\sqrt{n}), \tag{8.6}$$

where φ has mean zero and finite variance (i.e., $\mathbb{E}\{\varphi(Z)\} = 0$ and $\mathbb{E}\{\varphi(Z)^{\otimes 2}\} < \infty$). Here $o_p(1/\sqrt{n})$ employs the usual stochastic order notation so that $X_n = o_p(1/r_n)$ means $r_n X_n \xrightarrow{p} 0$ where $\xrightarrow{p}$ denotes convergence in probability.

Importantly, by the classical central limit theorem, an estimator $\hat{\psi}$ with influence function φ is asymptotically normal with

$$\sqrt{n}(\hat{\psi} - \psi_0) \rightsquigarrow N\Big(0,\ \mathbb{E}\{\varphi(Z)^{\otimes 2}\}\Big), \tag{8.7}$$

where $\rightsquigarrow$ denotes convergence in distribution. Thus if we know the influence function for an estimator, we know its asymptotic distribution, and we can easily construct confidence intervals and hypothesis tests, for example. Also, the efficient influence function for an asymptotically linear estimator is almost surely unique (i.e., unique up to measure zero sets) [53], so in a sense the influence function contains all information about an estimator's asymptotic behavior (up to $o_p(1/\sqrt{n})$ error).

Consider our running example where ψ is the average causal effect defined in Eqs. (8.1) and (8.3). Suppose we are in a randomized trial setting where the propensity score $\pi(l) = p(A = 1 \mid L = l)$ is known. A simple inverse-probability weighted estimator is given by

$$\hat{\psi}_{ipw} = \mathbb{P}_n \left\{ \frac{AY}{\pi(L)} - \frac{(1-A)Y}{1-\pi(L)} \right\} . \tag{8.8}$$

(Note that $\mathbb{E}(\hat{\psi}_{ipw}) = \psi_0$ by iterated expectation.) The influence function for the estimator $\hat{\psi}_{ipw}$ is clearly given by

$$\varphi_{ipw}(Z) = \frac{AY}{\pi(L)} - \frac{(1-A)Y}{1-\pi(L)} - \psi_0 \tag{8.9}$$

since $\hat{\psi}_{ipw} - \psi_0 = \mathbb{P}_n\{\varphi_{ipw}(Z)\}$ exactly, without any $o_p(1/\sqrt{n})$ approximation error.

Now suppose we are in an observational study setting where the propensity score $\pi(l)$ needs to be estimated, and suppose we do so with a correctly specified parametric model $\pi(l;\alpha)$, with $\alpha \in \mathbb{R}^q$, so that the estimator $\hat{\alpha}$ solves some estimating equation $\mathbb{P}_n\{S(Z;\hat{\alpha})\} = 0$. Then the inverse-probability-weighted estimator $\hat{\psi}^*_{ipw}$ is given by (8.8) above, except with the estimated propensity score $\pi(L;\hat{\alpha})$ replacing the true propensity score $\pi(L)$. We can find the corresponding influence function by standard estimating equation techniques [49]. Specifically, we have that $\hat{\theta} = (\hat{\psi}^*_{ipw}, \hat{\alpha}^{\mathrm{T}})^{\mathrm{T}}$ solves $\mathbb{P}_n\{m(Z;\hat{\theta})\} = 0$ where $m(z;\theta) = \{\varphi_{ipw}(Z;\psi,\alpha), S(Z;\alpha)^{\mathrm{T}}\}^{\mathrm{T}}$ are the stacked estimating equations for ψ and α, with the influence function for known propensity score given by $\varphi_{ipw}(Z;\psi,\alpha) = AY/\pi(L;\alpha) - (1-A)Y/\{1 - \pi(L;\alpha)\} - \psi$. Then under standard regularity conditions [27, 53, 64] we have

$$\hat{\theta} - \theta_0 = \mathbb{P}_n \left[\mathbb{E}\left\{ \frac{\partial m(Z;\theta_0)}{\partial \theta} \right\}^{-1} m(Z;\theta_0) \right] + o_p(1/\sqrt{n}), \tag{8.10}$$

which after evaluating and rearranging implies that the influence function for $\hat{\psi}^*_{ipw}$ when the propensity score $\pi(l;\alpha)$ is estimated is

$$\varphi^*_{ipw}(Z) = \varphi_{ipw}(Z;\psi_0,\alpha_0) - \mathbb{E}\left\{ \frac{\partial \varphi_{ipw}(Z;\psi_0,\alpha_0)}{\partial \alpha^{\mathrm{T}}} \right\} \mathbb{E}\left\{ \frac{\partial S(Z;\alpha_0)}{\partial \alpha} \right\}^{-1} S(Z;\alpha_0).$$

Surprisingly, even if the propensity score is known, it can be shown [53] that the inverse-probability-weighted estimator $\hat{\psi}^*_{ipw}$ based on an estimated propensity score is at least as efficient as the inverse-probability-weighted estimator $\hat{\psi}_{ipw}$ that uses the known propensity score. In other words, the variance of the influence function $\varphi^*_{ipw}(Z)$ is less than or equal to the variance of the influence function $\varphi_{ipw}(Z)$ for known propensity score. Thus the propensity score should be estimated from the data (according to a correct model, of course) even when it is known; discarding information can actually yield better efficiency.

So far we have seen that, given an estimator $\hat{\psi}$, we can learn about its asymptotic behavior by considering its influence function $\varphi(Z)$. But we can also use influence functions to find or construct estimators. Suppose we are given a candidate influence function $\varphi(Z; \psi, \eta)$ that depends on the target parameter ψ as well as a nuisance parameter η as in the previous examples. Then we can construct an estimator by solving the estimating equation $\mathbb{P}_n\{\varphi(Z; \psi, \hat{\eta})\} = 0$ in ψ, where $\hat{\eta}$ is some estimate of the nuisance parameter. Under standard regularity conditions, along with some additional conditions on the nuisance estimation, the corresponding estimator will itself be asymptotically linear with an influence function related to $\varphi(Z; \psi_0, \eta_0)$ depending on the form of the function φ and how the nuisance parameter η is estimated (as in the previous example). Other approaches for constructing estimators based on a particular influence function are also possible [60, 61].

There is a deep connection between (asymptotically linear) estimators for a given model and the influence functions under that model. In some sense, if we know one then we know the other. Thus if we can find all the influence functions for a given model, we can characterize all asymptotically linear estimators for that model.

3.3 *Tangent Spaces*

In this subsection we discuss the fundamental problem of how to find influence functions for a given semiparametric model, by characterizing the space in which influence functions reside. As noted previously, once we have solved this problem we can characterize valid estimators under our model. In particular, we can use influence functions to construct estimators and explore their efficiency.

To ease notation, consider the case where the target parameter is a scalar, so that $\psi \in \mathbb{R}$. As discussed in the previous subsection, influence functions φ are functions of the observed data Z with mean zero and finite variance. These influence functions reside in the Hilbert space $L_2(P)$ of measurable functions $g : \mathscr{Z} \to \mathbb{R}$ with $Pg^2 = \int g^2 \, dP = \mathbb{E}\{g(Z)^2\} < \infty$, equipped with covariance inner product $\langle g_1, g_2 \rangle = P(g_1 g_2)$. The space of influence functions will be a subspace of this Hilbert space. A Hilbert space is a complete inner product space, and can be viewed as a generalization of usual Euclidean space; it provides a notion of distance and direction for spaces whose elements are potentially infinite-dimensional functions.

A fundamentally important subspace of $L_2(P)$ in semiparametric problems is the *tangent space*. First we will discuss the tangent spaces for parametric models. For parametric models indexed by real-valued parameter $\theta \in \mathbb{R}^{q+1}$, the tangent space $\mathscr{T}$ is defined as the linear subspace of $L_2(P)$ spanned by the score vector, i.e.,

$$\mathscr{T} = \{b^{\mathrm{T}} S_\theta(Z; \theta_0) : b \in \mathbb{R}^{q+1}\}, \tag{8.11}$$

where $S_\theta(Z; \theta_0) = \partial \log p(z; \theta)/\partial\theta|_{\theta=\theta_0}$. If we can decompose $\theta = (\psi, \eta)$, then we can equivalently write $\mathscr{T} = \mathscr{T}_\psi \oplus \mathscr{T}_\eta$ for

$$\mathscr{T}_\psi = \{b_1 S_\psi(Z; \theta_0) : b_1 \in \mathbb{R}\}\,,\ \mathscr{T}_\eta = \{b_2^{\mathrm{T}} S_\eta(Z; \theta_0) : b_2 \in \mathbb{R}^q\}, \tag{8.12}$$

where $S_\psi(Z;\theta_0) = \partial \log p(z;\theta)/\partial\psi|_{\theta=\theta_0}$ is the score function for the target parameter, and similarly $S_\eta(Z;\theta_0) = \partial \log p(z;\theta)/\partial\eta|_{\theta=\theta_0}$ is the score for the nuisance parameter ($A \oplus B$ denotes the direct sum $A \oplus B = \{a+b : a \in A, b \in B\}$). In the above formulation, the space $\mathscr{T}_\eta$ is called the *nuisance tangent space*. Influence functions for ψ reside in the *orthogonal complement of the nuisance tangent space*, denoted by $\mathscr{T}_\eta^\perp = \{g \in L_2(P) : P(gh) = 0 \text{ forany } h \in \mathscr{T}_\eta\}$. In such parametric settings, this orthogonal space $\mathscr{T}_\eta^\perp$ can be written as

$$\begin{aligned}\mathscr{T}_\eta^\perp &= \{g \in L_2(P) : g = h - \Pi(h \mid \mathscr{T}_\eta),\ h \in L_2(P)\}\\ &= \{g \in L_2(P) : g = h - P(hS_\eta^{\mathrm{T}})P(S_\eta S_\eta^{\mathrm{T}})^{-1}S_\eta,\ h \in L_2(P)\},\end{aligned} \tag{8.13}$$

where $\Pi(g \mid S)$ denotes projections of g on the space S, i.e., $P[h\{g - \Pi(g \mid S)\}] = 0$ for all $h \in S$. The subspace of influence functions is the set of elements $\varphi \in \mathscr{T}_\eta^\perp$ that satisfy $P(\varphi S_\psi) = 1$. The *efficient influence function* is the influence function with the smallest covariance $P(\varphi^2)$, and is given by $\varphi_{\text{eff}} = P(S_{\text{eff}}^2)^{-1}S_{\text{eff}}$, where S_{eff} is the *efficient score*, given by $S_{\text{eff}} = S_\psi - \Pi(S_\psi \mid \mathscr{T}_\eta)$.

Thus if we can characterize the nuisance tangent space and its orthogonal complement, then we can characterize influence functions. In fact, one can show that all regular asymptotically linear estimators have influence functions φ that reside in $\mathscr{T}_\eta^\perp$ with $P(\varphi S_\psi) = 1$, and conversely any element in this space corresponds to the influence function for some regular asymptotically linear estimator [53]. Thus characterizing the nuisance tangent space allows us to also characterize all regular asymptotically linear estimators. (Recall that a regular estimator is one whose limiting distribution is insensitive to local changes to the data generating process, as defined, for example, in [53, 64] and elsewhere.)

We have seen that in parametric models the tangent space is defined as the span of the score vector S_θ. However, in semiparametric models, the nuisance parameter is infinite-dimensional and cannot be indexed by a real-valued parameter, so we cannot define scores in the usual way, since this requires differentiation with respect to the nuisance parameter. How can we extend the concept of the tangent space to semiparametric settings?

Constructing tangent spaces in semiparametric models requires a technical device called a *parametric submodel*. A parametric submodel $\mathscr{P}_\epsilon$ indexed by real-valued parameter ϵ is a set of distributions contained in the larger model $\mathscr{P}$, which also contains the truth (i.e., $P_0 \in \mathscr{P}_\epsilon$); typically, we have $\mathscr{P}_\epsilon = \{P_\epsilon : \epsilon \in \mathbb{R}\}$ with $P_\epsilon|_{\epsilon=0} = P_0$. Thus a parametric submodel needs to respect the semiparametric model $\mathscr{P}$ and also needs to equal the true distribution at $\epsilon = 0$. A typical example of a parametric submodel is given by

$$p_\epsilon(z) = p_0(z)\{1 + \epsilon g(z)\}, \tag{8.14}$$

where $\mathbb{E}\{g(Z)\} = 0$ and we have $\sup_z |g(z)| < M$ and $|\epsilon| < 1/M$ so that $p_\epsilon(z) \geq 0$. We will often index the parametric submodel by the function g, and so let $P_\epsilon = P_{\epsilon,g}$. Note again that parametric submodels like the one above are a technical device for constructing tangent spaces and analyzing semiparametric models, rather than a usual model whose parameters we want to estimate from data (since P_ϵ depends on the true distribution P_0, it cannot be used as a model in the usual sense) [53].

One intuition behind parametric submodels can be expressed in terms of efficiency bounds as follows [64]. First note that it is an easier problem to estimate ψ under the parametric submodel $\mathscr{P}_\epsilon \in \mathscr{P}$ than it is to estimate ψ under the entire (larger) semiparametric model $\mathscr{P}$. Therefore the efficiency bound under the larger model $\mathscr{P}$ must be larger than the efficiency bound under any parametric submodel. In fact we can define the efficiency bound for semiparametric models as the supremum of all such parametric submodel efficiency bounds.

Now that we have defined parametric submodels, how can they be used to construct tangent spaces? Just as the tangent space is defined as the linear span of the score vector in parametric models, in semiparametric models the tangent space $\mathscr{T}$ is defined as the (closure of the) linear span of scores of the parametric submodels. In other words, we first define scores on the parametric submodels P_ϵ with $S_\epsilon(z) = \partial \log p_\epsilon(z)/\partial\epsilon|_{\epsilon=0}$, and then construct parametric submodel tangent spaces as described earlier for standard parametric models, i.e., $\mathscr{T}_\epsilon = \{b^{\mathrm{T}} S_\epsilon(Z) : b \in \mathbb{R}\}$. Note that for parametric submodels like the one defined in (8.14) we have

$$S_\epsilon(z) = g(z)/\{1 + \epsilon g(z)\}|_{\epsilon=0} = g(z), \tag{8.15}$$

so that the functions g indexing the parametric submodels are set up to equal the parametric submodel scores. The closure $\mathscr{T}$ of the parametric submodel tangent spaces $\mathscr{T}_\epsilon$ is the minimal closed set that contains them; roughly speaking, $\mathscr{T}$ is the union of all the spaces $\mathscr{T}_\epsilon$ along with their limit points. Similarly, the nuisance tangent space $\mathscr{T}_\eta$ for a semiparametric model is the set of scores in $\mathscr{T}$ that do not vary the target parameter ψ, i.e.,

$$\mathscr{T}_\eta = \{g \in \mathscr{T} : \partial\psi(P_{\epsilon,g})/\partial\epsilon|_{\epsilon=0} = 0\}. \tag{8.16}$$

Importantly, in nonparametric models the tangent space is the whole Hilbert space of mean zero functions. For more restrictive semiparametric models the tangent space will be a proper subspace.

Now that we are equipped with definitions of tangent spaces and nuisance tangent spaces in semiparametric models, we can define influence functions, efficient influence functions, and efficient scores in much the same way we did before with parametric models.

Specifically, the subspace of influence functions is the set of elements $\varphi \in \mathscr{T}_\eta^\perp$ that satisfy $P(\varphi S_\psi) = 1$. The efficient influence function is the influence function with the smallest covariance $P(\varphi_{\mathrm{eff}}^2) \leq P(\varphi^2)$ for all φ; it is given by $\varphi_{\mathrm{eff}} = P(S_{\mathrm{eff}}^2)^{-1} S_{\mathrm{eff}}$, where S_{eff} is the efficient score defined as the projection of the score onto the tangent space, i.e., $S_{\mathrm{eff}} = \Pi(S_\psi \mid \mathscr{T}_\eta^\perp) = S_\psi - \Pi(S_\psi \mid \mathscr{T}_\eta)$ as before. The efficient influence function can also be defined as the projection of any influence function φ onto the tangent space, $\varphi_{\mathrm{eff}} = \Pi(\varphi \mid \mathscr{T})$ for any influence function φ, which is also a pathwise derivative of the target parameter in the sense that $P(\varphi S_\epsilon) = \partial\psi(P_\epsilon)/\partial\epsilon|_{\epsilon=0}$.

3.4 Efficient Influence Function for Average Treatment Effect

As an illustration, return to our example involving the average treatment effect $\psi = \mathbb{E}(Y^1 - Y^0) = \mathbb{E}\{\mu(L,1) - \mu(L,0)\}$, where we let $\mu(l,a) = \mathbb{E}(Y \mid L = l, A = a)$ denote the outcome regression function. Also let $\pi(l) = P(A = 1 \mid L = l)$ denote the propensity score as before. In this subsection, we will show using the results from previous subsections that, under a nonparametric model where the distribution P is unrestricted, the efficient influence function for ψ is given by $\varphi(Z;\psi,\eta) = m_1(Z;\eta) - m_0(Z;\eta) - \psi$, where

$$m_a(Z;\eta) = m_a(Z;\pi,\mu) = \frac{I(A=a)\{Y - \mu(L,a)\}}{a\pi(L) + (1-a)\{1-\pi(L)\}} + \mu(L,a) \tag{8.17}$$

with $\eta = (\pi, \mu)$ the nuisance function for this problem.

We will show this result by checking that the proposed efficient influence function φ is a pathwise derivative in the sense that $\partial\psi(P_\epsilon)/\partial\epsilon|_{\epsilon=0} = P(\varphi S_\epsilon)$.

Here we let $p_\epsilon(z) = p(z;\epsilon)$ denote a parametric submodel with parameter $\epsilon \in \mathbb{R}$. For notational simplicity let $f'_\epsilon(t;0) = \{\partial f(z;\epsilon)/\partial\epsilon\}|_{\epsilon=0}$ for any function f of ϵ and z, and also let $\ell(v \mid w;\epsilon) = \log p(v \mid w;\epsilon)$ for any partition $(V,W) \subseteq Z$, so that, for example, scores on the parametric submodels are denoted by $S_\epsilon(z) = \ell'_\epsilon(z;0)$. Then by definition from (8.3) we have

$$\ell'_\epsilon(z;\epsilon) = \ell'_\epsilon(y \mid l,a;\epsilon) + \ell'_\epsilon(a \mid l;\epsilon) + \ell'_\epsilon(l;\epsilon). \tag{8.18}$$

First consider the term $\partial\psi(P_\epsilon)/\partial\epsilon|_{\epsilon=0} = \psi'_\epsilon(0)$. By definition we have $\psi = \int\int\{y\,dP(y \mid l,a=1) - y\,dP(y \mid l,a=0)\}\,dP(l)$, so that

$$\begin{aligned}\psi'_\epsilon(\epsilon) = &\int\int\{y\ell'_\epsilon(y \mid l,a=1;\epsilon)\,dP(y \mid l,a=1;\epsilon)\\ &-y\ell'_\epsilon(y \mid l,a=0;\epsilon)\,dP(y \mid l,a=0;\epsilon)\}\,dP(l;\epsilon)\\ &+\int\int\{y\,dP(y \mid l,a=1;\epsilon) - y\,dP(y \mid l,a=0;\epsilon)\}\ell'_\epsilon(l;\epsilon)\,dP(l;\epsilon),\end{aligned} \tag{8.19}$$

where we used the fact that $dP'_\epsilon(v \mid w;\epsilon) = \ell'_\epsilon(v \mid w;\epsilon)dP(v \mid w;\epsilon)$. This follows since $\partial\log f(\epsilon)/\partial\epsilon = \{\partial f(\epsilon)/\partial\epsilon\}/f(\epsilon)$ for general functions f by definition of the logarithmic derivative. Recall that when we evaluate the above at $\epsilon = 0$, we have $dP(y \mid l,a;0) = dP(y \mid l,a)$ and $dP(l;0) = dP(l)$.

Now consider the term $P(\varphi S_\epsilon) = \mathbb{E}\{\varphi(Z;\psi,\eta)\ell'_\epsilon(Z;0)\}$, which equals

$$\begin{aligned}&\mathbb{E}\Big[\{m_1(Z;\eta) - m_0(Z;\eta) - \psi\}\{\ell'_\epsilon(Y \mid L,A;0) + \ell'_\epsilon(A \mid L;0) + \ell'_\epsilon(L;0)\}\Big]\\ &= \mathbb{E}\left[\left\{\frac{A}{\pi(L)} - \frac{1-A}{1-\pi(L)}\right\}Y\ell'_\epsilon(Y \mid L,A;0) + \{\mu(L,1) - \mu(L,0)\}\ell'_\epsilon(L;0)\right]\end{aligned}$$

$$= \mathbb{E}\Big[\mathbb{E}\{Y\ell'_\epsilon(Y \mid L, A=1;0) \mid L, A=1\} - \mathbb{E}\{Y\ell'_\epsilon(Y \mid L, A=0;0) \mid L, A=0\}$$
$$+ \{\mu(L,1) - \mu(L,0)\}\ell'_\epsilon(L;0)\Big]$$
$$= \int\int \{y\ell'_\epsilon(y \mid l, a=1;0)\, dP(y \mid l, a=1)$$
$$-y\ell'_\epsilon(y \mid l, a=0;0)\, dP(y \mid l, a=0)\}\, dP(l)$$
$$+\int\int \{y\, dP(y \mid l, a=1) - y\, dP(y \mid l, a=0)\}\ell'_\epsilon(l;0)\, dP(l). \quad (8.20)$$

The first equality follows from iterated expectation and the fact that, by usual properties of score functions, $\mathbb{E}\{\ell'_\epsilon(V \mid W;0) \mid W\} = 0$. The second equality follows from iterated expectation, and the third follows by definition.

Since the last expression for the covariance $P(\varphi S_\epsilon)$ in Eq. (8.20) equals the expression for $\psi'_\epsilon(\epsilon)$ from Eq. (8.19) when evaluated at $\epsilon = 0$, we have shown that φ is in fact the efficient influence function.

3.5 *Full vs. Observed Data Influence Functions*

So far we have introduced the notion of a tangent space and discussed how influence functions φ for regular asymptotically linear estimators can be viewed as elements of a subspace of the Hilbert space $L_2(P)$, namely the orthogonal complement of the nuisance tangent space, i.e., $\varphi \in \mathscr{T}_\eta^\perp$. We also illustrated how to check that a proposed influence function is the efficient influence function. But how does one find the space $\mathscr{T}_\eta^\perp$ in a given problem? In many cases this is a bit of an art: one conjectures the form of $\mathscr{T}_\eta^\perp$ and then checks that the conjectured space satisfies the required properties. For nonparametric models, one can sometimes deduce the form of the efficient influence function from the nonparametric maximum likelihood estimator, assuming discrete data [60]. However, in some settings it can be useful to characterize influence functions with hypothetical "full data" (i.e., had we observed all counterfactuals), and then map these to observed data influence functions [59].

To characterize full-data influence functions in causal inference problems we need to start by presenting causal inference as a missing data problem [53, 59]. Thus far we have supposed that we observe an independent and identically distributed sample of observations $Z \sim P$. In general missing data problems, we conceive of hypothetical full data $\tilde{Z}$, of which the observed data Z is a coarsened version. The problem is that we want to learn about the distribution $\tilde{P}$ of the full data $\tilde{Z}$, but we only get to observe the coarsened version Z of the full data $\tilde{Z}$. In general coarsened data problems, $Z = \Phi(\tilde{Z}, C)$ is a known many-to-one function $\Phi(\cdot)$ of both $\tilde{Z}$ and a coarsening variable C that indicates what portion of $\tilde{Z}$ is observed. In causal inference settings, the coarsening variable generally equals the treatment process so that $C = A$, and

$$\tilde{Z} = \{Z^a : a \in \mathscr{A}\}. \tag{8.21}$$

Thus the full data $\tilde{Z}$ are the potential outcomes under different levels $a \in \mathscr{A}$ of a general treatment process A (here A could be multivariate, e.g., a treatment sequence over multiple timepoints). For a given unit we only get to observe $Z = \Phi(\tilde{Z}, A) = Z^A$, i.e., the potential outcome under the observed treatment process. For instance, in our running example where $Z = (L, A, Y)$ with binary treatment so that $\mathscr{A} = \{0, 1\}$, the full data for a given unit could be represented as

$$\tilde{Z} = \{(L^a, Y^a) : a \in \{0, 1\}\} = (L, Y^0, Y^1). \tag{8.22}$$

Note that the last equality follows since $L^a = L$ if we make the usual assumption that events in the past cannot be affected by the future. In some cases we might also want to include the observed treatment process in the full data, so that in the above example we would have $\tilde{Z} = (L, A, Y^0, Y^1)$. In a longitudinal setting where covariates and a binary treatment are updated at timepoints $t = 1, \ldots, K$ and an outcome is measured at the end of follow-up, we could have

$$\tilde{Z} = \{(L_1, L_2^{a_1}, L_3^{a_1,a_2}, \ldots, L_t^{\overline{a}_{t-1}}, \ldots, L_K^{\overline{a}_{K-1}}, Y^{\overline{a}_K}) : \overline{a}_K \in \{0, 1\}^K\}, \tag{8.23}$$

where $\overline{a}_t = (a_1, \ldots, a_t)$ denotes the past history of a variable through time t. The observed data in this case would be $Z = (L_1, A_1, \ldots, L_t, A_t, \ldots, L_K, A_K, Y)$ for a given unit. Not every causal inference problem fits in the above framework, but when the framework applies it can often be very useful.

Now that we have defined the full data $\tilde{Z}$ and given some examples, we can also define corresponding tangent spaces, influence functions, and parametric submodels, using semiparametric models $\widetilde{\mathscr{P}}$ for the full data just as we did for the observed data previously. The advantage is that it is often more straightforward to derive tangent spaces and influence functions for full data problems (or else results may already be known for common models), and then translate them to observed data, rather than working with observed data directly and using the results from previous subsections. Of course, in order to translate full data influence functions to observed data influence functions, we need identifying assumptions.

Under a coarsening at random assumption [14], results for mapping full data to observed data tangent spaces are given, for example, in [59] and [53]. In general, coarsening at random means $P(Z = z \mid \tilde{Z} = \tilde{z}_1) = P(Z = z \mid \tilde{Z} = \tilde{z}_2)$ whenever $z = \Phi(\tilde{z}_1, a) = \Phi(\tilde{z}_2, a)$ for some $a \in \mathscr{A}$. In many problems [40], this can be equivalently expressed by saying that $P(A = a \mid \tilde{Z} = \tilde{z}_1) = P(A = a \mid \tilde{Z} = \tilde{z}_2)$ only depends on z whenever $z = \Phi(\tilde{z}_1, a) = \Phi(\tilde{z}_2, a)$. Under some conditions, coarsening at random also reduces to a randomization assumption, which says treatment is independent of potential outcomes given the observed past, e.g., $A \perp\!\!\!\perp Y^a \mid L$ in our running example, or $A_t \perp\!\!\!\perp Y^{\overline{a}_K} \mid \overline{L}_t, \overline{A}_{t-1}$ in the above longitudinal example. More details on these issues are given in [40, 59]. Again we point out that this framework does not always apply: sometimes coarsening at random is not equivalent to treatment randomization, or is not the identifying assumption we wish to utilize.

Here we will be content giving a simple example of how to map a full data influence function to the observed data, rather than discussing details in full generality; see [59] and [53] for more general results. Assume coarsening at random holds, and that the treatment assignment process is known. Further suppose the observed data is $Z = (L, A, Y)$ with $A \in \{0, 1\}$ and our goal is to estimate $\mathbb{E}(Y^1 \mid V) = \gamma(V; \psi)$, where $V \subseteq L$ is a subset of the covariates. The full data orthogonal complement of the nuisance tangent space includes functions of the form

$$\tilde{\varphi}_g(Z^*; \psi) = g(V)\{Y^1 - \gamma(V; \psi)\} \tag{8.24}$$

for arbitrary functions g. From Theorem 7.2 in [53], if $\pi(l) = P(A = 1 \mid L = l)$ is bounded away from zero, then the observed data space $\mathcal{T}_\eta^\perp$ comprises functions of the form

$$\frac{A}{\pi(L)}\Big[\tilde{\varphi}_g(Z^*; \psi) + \{1 - \pi(L)\}h(Z)\Big] - (1 - A)h(Z) \tag{8.25}$$

for arbitrary functions h (the simplest estimator would use the above as an estimating function with $h = 0$). Note that functions of the above form only depend on observed data since $Y^1 = Y$ when $A = 1$. This represents an inverse-probability-weighting approach for mapping full data spaces to observed data spaces.

4 Empirical Processes

In the previous section we discussed how to construct influence functions $\varphi(Z; \psi, \eta)$ in semiparametric models. We also discussed how one can use these influence functions to construct estimators $\hat{\psi}$ for ψ, by solving (up to order $o_p(1/\sqrt{n})$) the estimating equation

$$\mathbb{P}_n\{\varphi(Z; \psi, \hat{\eta})\} = 0 \tag{8.26}$$

in ψ, where $\hat{\eta}$ is an estimator of the nuisance function. As in the previous section we let $\mathbb{P}_n = n^{-1}\sum_i \delta_{Z_i}$ denote the empirical measure so that sample averages can be written as $n^{-1}\sum_i f(Z_i) = \int f(z)\, d\mathbb{P}_n = \mathbb{P}_n\{f(Z)\}$. We briefly discussed the asymptotics of the estimators $\hat{\psi}$ given above for the case where $\hat{\eta} \in \mathbb{R}^q$ is a finite-dimensional real-valued parameter, itself estimated from some estimating equation; a standard estimating equation analysis can then be used by simply stacking estimating equations for ψ and η together.

In contrast, in this section we consider how to analyze the asymptotic behavior of $\hat{\psi}$ when the nuisance function η is estimated nonparametrically, in the sense that $\hat{\eta}$ cannot be characterized by a finite-dimensional real-valued parameter. This can be accomplished with tools from empirical process theory. Our discussion in this section comes from work by Andrews [1, 2], Pollard [35, 36], van der

Vaart [64, 65, 67], and Wellner [48, 67], among many others [21, 60]. The field of empirical process theory is vast; we limit our discussion to tools for handling nuisance estimation.

4.1 Motivation and Setup

To motivate our study of empirical processes, consider our running example where the goal is to estimate the average treatment effect $\psi = \mathbb{E}(Y^1 - Y^0)$. Specifically consider the doubly robust estimator for ψ that solves an estimated version of the efficient influence function presented in Sect. 3.4, i.e., the estimator given by $\hat{\psi} = \mathbb{P}_n\{m_1(Z;\hat{\eta}) - m_0(Z;\hat{\eta})\}$ where

$$m_a(Z;\eta) = m_a(Z;\pi,\mu) = \frac{I(A=a)\{Y-\mu(L,a)\}}{a\pi(L)+(1-a)\{1-\pi(L)\}} + \mu(L,a). \tag{8.27}$$

Note that in this case the nuisance function is given by $\eta = (\pi,\mu)$. In observational studies the covariates L are often high-dimensional, and little might be known about the propensity score and outcome regression functions π and μ, in which case it makes sense to use flexible, nonparametric, data-adaptive methods to estimate them. Of course then the asymptotic analysis presented in Sect. 3.2 does not apply, since the estimators used to construct $\hat{\eta} = (\hat{\pi},\hat{\mu})$ will not be described by a single finite-dimensional parameter. Nonetheless under some conditions we can still learn about the asymptotics of $\hat{\psi}$ and obtain valid confidence intervals, using tools from empirical process theory.

Before going further, we need to introduce some notation. Throughout this section we will use $\mathbb{P}\{f(Z)\} = \int f(z)\,d\mathbb{P}$ to denote expectations of $f(Z)$ for a new observation Z (treating the function f as fixed); thus, $\mathbb{P}\{\hat{f}(Z)\}$ is random when $\hat{f}$ is random (e.g., estimated from the sample). Contrast this with the fixed non-random quantity $\mathbb{E}\{\hat{f}(Z)\}$, which averages over randomness in both Z and $\hat{f}$ and thus will not equal $\mathbb{P}\{\hat{f}(Z)\}$ except when $\hat{f} = f$ is fixed and non-random.

Suppose for simplicity that $\hat{\psi} = \mathbb{P}_n\{m(Z;\hat{\eta})\}$ for some m, as in the above example. If we only have $\mathbb{P}_n\{\varphi(Z;\hat{\psi},\hat{\eta})\} = 0$, then we can proceed similarly, with an extra step requiring differentiability of $\mathbb{P}\{\varphi(Z;\psi,\eta)\}$ in ψ, at ψ_0 in a neighborhood of η_0 [64]. Also suppose that $\mathbb{P}\{m(Z;\eta_0)\} = \psi_0$ (alternatively we can define ψ_0 so that this holds by definition). For instance, it is straightforward to check for the doubly robust estimator described above that $\mathbb{P}\{m(Z;\pi_0,\mu)\} = \mathbb{P}\{m(Z;\pi,\mu_0)\} = \psi_0$ where $m = m_1 - m_0$. Then consider the decomposition

$$\begin{aligned}\hat{\psi} - \psi_0 &= \mathbb{P}_n\{m(Z;\hat{\eta})\} - \mathbb{P}\{m(Z;\eta_0)\} \\ &= (\mathbb{P}_n - \mathbb{P})\{m(Z;\hat{\eta})\} + \mathbb{P}\{m(Z;\hat{\eta}) - m(Z;\eta_0)\},\end{aligned} \tag{8.28}$$

where the first line is true by definition, and the second follows by simply adding and subtracting $\mathbb{P}\{m(Z;\hat{\eta})\}$.

We will show that the first term $(\mathbb{P}_n - \mathbb{P})\{m(Z;\hat{\eta})\}$ above can be handled under general conditions with empirical process theory. Specifically, we will discuss conditions under which

$$(\mathbb{P}_n - \mathbb{P})\{m(Z;\hat{\eta})\} = (\mathbb{P}_n - \mathbb{P})\{m(Z;\eta_0)\} + o_p(1/\sqrt{n}), \tag{8.29}$$

where $\hat{\eta}$ converges to η_0, so that $(\mathbb{P}_n - \mathbb{P})\{m(Z;\hat{\eta})\}$ is asymptotically equivalent to its limiting version $(\mathbb{P}_n - \mathbb{P})\{m(Z;\eta_0)\}$ (up to order $o_p(1/\sqrt{n})$) and can be analyzed with a standard central limit theorem. The second term in the decomposition in (8.28) typically requires a case-by-case analysis, but we will give examples shortly. Note that if we have $\mathbb{P}\{m(Z;\hat{\eta}) - m(Z;\eta_0)\} = (\mathbb{P}_n - \mathbb{P})\phi(Z;\eta_0) + o_p(1/\sqrt{n})$ for some finite-variance function ϕ, then

$$\hat{\psi} - \psi_0 = (\mathbb{P}_n - \mathbb{P})\{m(Z;\eta_0) + \phi(Z;\eta_0)\} + o_p(1/\sqrt{n}) \tag{8.30}$$

and thus $\hat{\psi}$ is regular and asymptotically linear with influence function $(m + \phi)$.

4.2 Donsker Classes

From an empirical process perspective, a primary way to control how close the term $(\mathbb{P}_n - \mathbb{P})\{m(Z;\hat{\eta})\}$ is to its limiting version $(\mathbb{P}_n - \mathbb{P})\{m(Z;\eta_0)\}$ (in large samples) is to restrict the complexity of the nuisance function η_0 and its estimator $\hat{\eta}$. If these functions are not too complex, then the terms will not differ by more than $o_p(1/\sqrt{n})$. In this subsection we will discuss characterizing complexity with Donsker classes.

We will start by giving the main result in the context of our example, and will then describe the conditions in detail. Suppose our nuisance estimator $\hat{\eta}$ converges to some limit η_0 in the sense that

$$||m(;\hat{\eta}) - m(;\eta_0)||^2 = \int \{m(z;\hat{\eta}) - m(z;\eta_0)\}^2 dP(z) = o_p(1), \tag{8.31}$$

and suppose the function class $\mathcal{M} = \{m(;\eta) : \eta \in H\}$ is a Donsker class (to be defined shortly), where H is a function class containing the nuisance estimator $\hat{\eta}$. Then the result in (8.29) holds, i.e.,

$$(\mathbb{P}_n - \mathbb{P})\{m(Z;\hat{\eta})\} = (\mathbb{P}_n - \mathbb{P})\{m(Z;\eta_0)\} + o_p(1/\sqrt{n}). \tag{8.32}$$

Thus, asymptotically, nuisance estimation only affects the second term in (8.28).

In order to define a Donsker class, we need to introduce a few concepts first. Throughout this section we use $\mathbb{G}_n = \sqrt{n}(\mathbb{P}_n - \mathbb{P})$ for ease of notation. Let $\mathcal{F}$ denote a class of functions $f : \mathcal{X} \to \mathbb{R}$, and consider the *empirical process*

$$\{\mathbb{G}_n f : f \in \mathcal{F}\}. \tag{8.33}$$

This is a type of *stochastic process* since it is a collection of random variables indexed by a set (the function class $\mathscr{F}$). From one standpoint, given a function f, we can view $\mathbb{G}_n f = \sqrt{n}(\mathbb{P}_n - \mathbb{P})f(Z)$ as a random variable mapping the sample (product) space $\mathscr{Z}^n$ to $\mathbb{R}$. Alternatively, given a sample $(Z_1, \ldots, Z_n)$, we can also view $\mathbb{G}_n f$ as a map from the function class $\mathscr{F}$ to $\mathbb{R}$. Therefore (if these latter maps are bounded) we can view the empirical process as a *random function*, mapping the sample space $\mathscr{Z}^n$ to the space $\ell^\infty(\mathscr{F})$ of bounded functions $h : \mathscr{F} \to \mathbb{R}$ with $\sup_{f \in \mathscr{F}} |h(f)| = ||h||_{\mathscr{F}} < \infty$.

The above discussion of the empirical process $\{\mathbb{G}_n f : f \in \mathscr{F}\}$ was all for a fixed sample size n. Now consider a sequence of empirical processes $\{\mathbb{G}_n f : f \in \mathscr{F}\}_{n \geq 1}$. We say this sequence *converges in distribution* to element $\mathbb{G}$ (equivalently, converges weakly to $\mathbb{G}$) in the space $\ell^\infty(\mathscr{F})$, denoted $\mathbb{G}_n \rightsquigarrow \mathbb{G}$, if

$$\mathbb{E}^* h(\mathbb{G}_n) \to \mathbb{E} h(\mathbb{G}) \tag{8.34}$$

for all continuous bounded functions $h : \ell^\infty(\mathscr{F}) \to \mathbb{R}$, where $\mathbb{E}^*$ denotes outer expectation. (Outer expectation is a measure-theoretic subtlety that we will largely sidestep here; roughly, $\mathbb{E}^*$ can be viewed as a generalization of expectation that accounts for the fact that $h(\mathbb{G}_n)$ may not be measurable). Thus we have a notion of convergence for empirical processes viewed as random functions. Finally, we say a generic measurable random element $\mathbb{G}$ is *tight* if for all $\epsilon > 0$ there is a compact set S for which $P(\mathbb{G} \in S) > 1 - \epsilon$, i.e., if the element $\mathbb{G}$ stays in a compact set with high probability.

We are now ready to define a Donsker class. A function class $\mathscr{F}$ is called a *Donsker class* if the sequence of empirical processes $\{\mathbb{G}_n f : f \in \mathscr{F}\}_{n \geq 1}$ converges in distribution to some tight limit $\mathbb{G}$ (in fact this limit must be a zero-mean Gaussian process $\mathbb{G}_P$, known as a P-Brownian bridge).

The Donsker property, along with the continuous mapping theorem, allows us to obtain results like that given in (8.29). Specifically, suppose $\hat{f} \in \mathscr{F}$ for a Donsker class $\mathscr{F}$, and suppose $\hat{f}$ converges to f_0 in the sense that $||\hat{f} - f_0|| = o_p(1)$, where $||f||^2 = Pf^2$ denotes the $L_2(P)$ norm as before. Then (as in Lemma 19.24 of [64]) we can apply the continuous mapping theorem to $(\mathbb{G}_n, \hat{f}) \rightsquigarrow (\mathbb{G}_P, f_0)$ with function $h(z, f) = z(f) - z(f_0)$ to obtain that

$$\mathbb{G}_n \hat{f} = \mathbb{G}_n f_0 + o_p(1). \tag{8.35}$$

Thus $(\mathbb{P}_n - \mathbb{P})\hat{f} = n^{-1/2}\mathbb{G}_n \hat{f}$ is asymptotically equivalent to $(\mathbb{P}_n - \mathbb{P})f_0$, up to $o_p(1/\sqrt{n})$ error.

In our setting, where $\hat{\psi} = \mathbb{P}_n\{m(Z; \hat{\eta})\}$, it is often more natural to put Donsker conditions on the estimated nuisance functions themselves, i.e., to assume that $\hat{\eta} \in H$ for a Donsker class H, rather than to put conditions on the transformed function class $\mathscr{M} = \{m(; \eta) : \eta \in H\}$. Fortunately, "nice enough" transformations of Donsker function classes will also be Donsker. Specifically, suppose the function classes $\mathscr{F}$ and $\mathscr{F}_j$ are Donsker; then, as discussed in Sect. 2.10 of [67], as in [1, 64], the following transformations of $\mathscr{F}$ and $\mathscr{F}_j$ are also Donsker:

1. *Subsets:* $\mathscr{G} \subset \mathscr{F}$
2. *Unions:* $\mathscr{G} = \mathscr{F}_1 \cup \mathscr{F}_2$
3. *Closures:* $\mathscr{G} = \{g : f_m \to g$ pointwise and in L_2, for $f_m \in \mathscr{F}\}$
4. *Convex combinations:* $\mathscr{G} = \{g : g = \sum_i w_i f_i$ for $f_i \in \mathscr{F}, \sum_i |w_i| \leq 1\}$
5. *Lipschitz transformations:* $\mathscr{G} = \{g : g = \phi(f_1, \ldots, f_k)$ for $f_j \in \mathscr{F}_j\}$ if ϕ satisfies $|\phi(f_1, \ldots, f_k)(x) - \phi(f_1', \ldots, f_k')(x)|^2 \leq \sum_j (f_j - f_j')(x)^2$ for all f_j, f_j', and x, and if $\sup_{f \in \mathscr{F}_j} |Pf| < \infty$ and $\int \phi(f_1, \ldots, f_k)(x)^2 dx < \infty$.

The convex combination result suggests using ensemble methods that use weighted combinations of estimators, e.g., Super Learner [58, 60, 62]. The Lipschitz transformation result given above is particularly useful. It means, for example, that the following function classes are Donsker [1, 64, 67]:

1. *Minimums:* $\mathscr{G} = \{g : g = \min(f_1, f_2)$ for $f_j \in \mathscr{F}_j\}$
2. *Maximums:* $\mathscr{G} = \{g : g = \max(f_1, f_2)$ for $f_j \in \mathscr{F}_j\}$
3. *Sums:* $\mathscr{G} = \{g : g = f_1 + f_2$ for $f_j \in \mathscr{F}_j\}$
4. *Products:* $\mathscr{G} = \{g : g = f_1 f_2$ for $f_j \in \mathscr{F}_j\}$ if $\mathscr{F}_j$ are uniformly bounded
5. *Ratios:* $\mathscr{G} = \{g : g = 1/f$ for $f \in \mathscr{F}\}$ if $f \geq \delta > 0$ for all $f \in \mathscr{F}$

Repeated use of stability results like those above often allows one to conclude Donsker properties for the class $\mathscr{M} = \{m(; \eta) : \eta \in H\}$ based on Donsker assumptions about the class H.

For example, consider the doubly robust estimator $\hat{\psi} = \mathbb{P}_n\{m_1(Z; \hat{\eta}) - m_0(Z; \hat{\eta})\}$ given in (8.27). If $\hat{\pi}$ and $\hat{\mu}$ take values in Donsker classes $\mathscr{F}_\pi$ and $\mathscr{F}_\mu$, respectively, then $m_a(Z; \hat{\eta})$ does as well (provided that π is bounded away from zero and one for all $\pi \in \mathscr{F}_\pi$). This follows from Lipschitz results 3 and 5 for sums and ratios above.

4.3 Examples of Donsker Classes

To this point we have seen that, if we assume the estimated nuisance functions $\hat{\eta}$ are contained in Donsker function classes, we can use a standard central limit theorem to analyze $(\mathbb{P}_n - \mathbb{P})m(Z; \hat{\eta})$ since it is asymptotically equivalent to $(\mathbb{P}_n - \mathbb{P})m(Z; \eta_0)$ up to order $o_p(1/\sqrt{n})$. We have defined Donsker classes and shown how they can be combined and modified to produce new Donsker classes, but we have yet to give any specific examples of such classes. For the prior results to be useful over and above more standard parametric techniques, we need Donsker classes to be able to capture sufficiently flexible functions. Luckily, this is in fact the case, as we will discuss in this subsection using specific examples.

First we will simply provide a short list of function classes that are Donsker, and then we will briefly discuss how one typically shows that a particular class is Donsker (using bracketing and covering numbers). Results showing that certain classes are Donsker are somewhat scattered across the literature, but examples and nice overviews are given by [64, 67], for example. Among many other kinds of classes, the following simple classes of functions are Donsker classes [13, 64, 67]:

1. *Indicator functions:* $\mathscr{F} = \{f : f(x) = I(x < t), t \in \mathbb{R}\}$
2. *Vapnik–Cervonenkis (VC) classes*
3. *Bounded monotone functions*
4. *Lipschitz parametric functions:* $\mathscr{F} = \{f : f(x) = f(x;\theta), \theta \in \Theta \subset \mathbb{R}^q\}$ with $|f(x;\theta_1) - f(x;\theta_2)| \leq b(x)||\theta_1 - \theta_2||$ for some b with $\int |b(x)|^r dP(x) < \infty$
5. *Smooth functions:* $\mathscr{F} = \{f : \sup_x |\frac{\partial^\alpha f(x_1,\ldots,x_q)}{\partial^{\alpha_1} x_1 \ldots \partial^{\alpha_q} x_q}| < B < \infty$, with $\alpha > q/2\}$
6. *Sobolev classes:* $\{f : \sup_x |f(x)| \leq 1, f^{(k-1)}\text{absolutelycts.}, \int |f^{(k)}(x)|^2\, dx \leq 1\}$
7. *Uniform sectional variation:* $\{f : \sup_{x_1} ||f(x_1,\cdot)||_{tv} \leq B_1, \sup_{x_2} ||f(\cdot,x_2)||_{tv} \leq B_2\}$ where $B_1, B_2 < \infty$ and $||\cdot||_{tv}$ denotes the total variation norm.

Thus we see that Donsker classes include usual parametric classes, but many other classes as well, including infinite-dimensional classes that only require certain smoothness or boundedness. Many other function classes can also be shown to be Donsker. For example, any appropriate combination or transformation of the above classes as discussed in the previous subsection will also be Donsker.

Showing that a function class is Donsker is often accomplished using bracketing or covering numbers [64, 67], which are measures of the size of a class $\mathscr{F}$. These measures also provide simple sufficient conditions for a function class being Donsker. An ϵ-bracket (in $L_2(P)$) is defined as all functions f bracketed by functions $[l, u]$ (i.e., $l \leq f \leq u$) satisfying $\int \{u(z) - l(z)\}^2 dP(z) < \epsilon^2$. The *bracketing number* of a class $\mathscr{F}$ is the smallest number of ϵ-brackets needed to cover $\mathscr{F}$, and is denoted by $N_B(\epsilon, \mathscr{F})$. Similarly, the *covering number* of a class $\mathscr{F}$ (with envelope F, i.e., $\sup_{\mathscr{F}} |f| \leq F$) is the smallest number of $L_2(Q)$ balls of radius ϵ needed to cover $\mathscr{F}$, and is denoted by $N_C(\epsilon, \mathscr{F})$. Then the class $\mathscr{F}$ is Donsker if either

$$\int_0^1 \sqrt{\log N_B(\epsilon, \mathscr{F})}\, d\epsilon < \infty, \quad \text{or} \quad \int_0^1 \sqrt{\log \sup_Q N_C(\epsilon \sqrt{QF^2}, \mathscr{F})}\, d\epsilon < \infty. \tag{8.36}$$

4.4 Average Treatment Effect Example

Now we return to analyze the asymptotic behavior of the doubly robust estimator of the average treatment effect $\psi = E(Y^1 - Y^0)$ from Sect. 3.4, which is given by $\hat{\psi} = \mathbb{P}_n\{m(Z;\hat{\eta})\} = \mathbb{P}_n\{m_1(Z;\hat{\eta}) - m_0(Z;\hat{\eta})\}$ with

$$m_a(Z;\eta) = m_a(Z;\pi,\mu) = \frac{I(A=a)\{Y - \mu(L,a)\}}{a\pi(L) + (1-a)\{1-\pi(L)\}} + \mu(L,a). \tag{8.37}$$

Throughout we assume the identification assumptions from Sect. 2.2, or else suppose we are estimating the observed data quantity $\mathbb{E}\{\mu(L,1) - \mu(L,0)\}$ under the positivity assumption. Suppose the estimator $\hat{\eta} = (\hat{\pi}, \hat{\mu})$ converges to some $\overline{\eta} = (\overline{\pi}, \overline{\mu})$ in the sense that $||\hat{\eta} - \overline{\eta}|| = o_p(1)$, where either $\overline{\pi} = \pi_0$ or $\overline{\mu} = \mu_0$ (or both) correspond to the true nuisance function. Thus at least one nuisance estimator needs to converge to the correct function, but one can be misspecified. Then $\mathbb{P}\{m(Z; \overline{\eta})\} = \mathbb{P}\{m(Z; \eta_0)\} = \psi_0$, from the easy-to-check fact that $\mathbb{P}\{m(Z; \pi_0, \mu)\} = \mathbb{P}\{m(Z; \pi, \mu_0)\}$ for any $\overline{\pi}$ and $\overline{\mu}$. Thus as in Sect. 4.1 we can write

$$\hat{\psi} - \psi_0 = (\mathbb{P}_n - \mathbb{P})m(Z; \hat{\eta}) + \mathbb{P}\{m(Z; \hat{\eta}) - m(Z; \overline{\eta})\}. \tag{8.38}$$

As discussed in Sect. 4.2, if the estimators $\hat{\pi}$ and $\hat{\mu}$ take values in Donsker classes, then $m_a(Z; \hat{\eta})$ does as well (as long as functions in the class containing $\hat{\pi}$ are uniformly bounded away from zero and one). Therefore the result in (8.29) applies, and we have

$$\hat{\psi} - \psi_0 = (\mathbb{P}_n - \mathbb{P})m(Z; \overline{\eta}) + \mathbb{P}\{m(Z; \hat{\eta}) - m(Z; \overline{\eta})\} + o_p(1/\sqrt{n}). \tag{8.39}$$

Now it remains to analyze $\mathbb{P}\{m(Z; \hat{\eta}) - m(Z; \overline{\eta})\}$. By iterated expectation this term equals

$$\sum_{a \in \{0,1\}} \mathbb{P}\left[\frac{\pi_0(L) - \hat{\pi}(L)}{a\hat{\pi}(L) + (1-a)\{1 - \hat{\pi}(L)\}}\{\mu_0(L,a) - \hat{\mu}(L,a)\}\right]. \tag{8.40}$$

Therefore, by the fact that $\hat{\pi}$ is bounded away from zero and one, along with the Cauchy–Schwarz inequality ($P(fg) \leq ||f||\ ||g||$), we have that (up to a multiplicative constant) $|\mathbb{P}\{m(Z; \hat{\eta}) - m(Z; \overline{\eta})\}|$ is bounded above by

$$\sum_{a \in \{0,1\}} ||\pi_0(L) - \hat{\pi}(L)||\ ||\mu_0(L,a) - \hat{\mu}(L,a)||. \tag{8.41}$$

Thus, for example, if $\hat{\pi}$ is based on a correctly specified parametric model, so that $||\hat{\pi} - \pi_0|| = O_p(n^{-1/2})$, then we only need $\hat{\mu}$ to be consistent, $||\hat{\mu} - \mu_0|| = o_p(1)$, to make the product term $\mathbb{P}\{m(Z; \hat{\eta}) - m(Z; \overline{\eta})\} = o_p(1/\sqrt{n})$ asymptotically negligible. Then the doubly robust estimator satisfies $\hat{\psi} - \psi_0 = (\mathbb{P}_n - \mathbb{P})m(Z; \eta_0) + o_p(1/\sqrt{n})$ and it is efficient with influence function $\varphi(Z; \psi, \eta) = m(Z; \eta) - \psi$. Thus if we know the treatment mechanism, the outcome model can be very flexible.

Another way to achieve efficiency is if we have both $||\hat{\pi} - \pi_0|| = o_p(n^{-1/4})$ and $||\hat{\mu} - \mu_0|| = o_p(n^{-1/4})$, so that the product term is $o_p(1/\sqrt{n})$ and asymptotically negligible. This of course occurs if both $\hat{\pi}$ and $\hat{\mu}$ are based on correctly specified models, but it can also hold even for estimators that are very flexible and not based on parametric models. However, completely nonparametric (e.g., kernel or nearest-neighbor) estimators are typically not an option in this setting since they will generally converge at rates slower than $n^{-1/4}$; exceptions include cases where there are very few covariates or very strong smoothness assumptions. Explicit conditions ensuring given convergence rates for kernel estimators are described, for example, in [27]. Thus some modeling is in general required to attain $n^{-1/4}$ rates, but luckily numerous semiparametric models yield estimators that can satisfy this condition. In particular, faster than $n^{-1/4}$ rates are possible with single index models, generalized additive models, and partially linear models (see, for example, [17] for a review of such models, which typically yield estimators with $n^{-2/5}$ rates), as well as regularized estimators such as the Lasso [5, 6]. Cross-validation-based weighted combinations of such estimators (e.g., Super Learner) can also satisfy this rate condition if one of the candidate estimators does [58].

Inference after nonparametric estimation of η in truly doubly robust settings where one arbitrary nuisance estimator can be misspecified is more complicated. If one of the estimators $\hat{\pi}$ or $\hat{\mu}$ is misspecified so that either $||\hat{\pi} - \pi_0|| = O_p(1)$ or $||\hat{\mu} - \mu_0|| = O_p(1)$, then obtaining root-$n$ rate inference for standard estimators will typically require knowledge of which estimator is correctly specified, as well as that the correctly specified estimator is based on a parametric model. More sophisticated estimators that weaken this requirement are discussed in the next section (e.g., [56]).

5 Extensions and Future Directions

In this section we briefly describe some future directions and extensions to semiparametric causal inference beyond the theory we have presented in this review. A number of authors have worked to extend semiparametric causal inference to, for example, settings involving non-standard sampling, estimation and inference under yet weaker conditions on the nuisance estimators, and complex non-regular or non-smooth parameters.

Throughout this review we presumed access to an independent and identically distributed sample from the distribution P of interest; however, many studies use alternative sampling schemes. For example, authors have developed results for semiparametric causal inference in case control studies [45, 50, 54, 70, 71] and matched cohort studies [19, 63]. There has also been progress made for causal inference in studies using network data with possible interference [18, 29, 52, 55]. Much more work is needed in settings related to both study designs with non-standard sampling and network data with interference. The latter should be a growing concern as data from, e.g., social networks becomes more commonplace.

In Sect. 4 we showed that semiparametric estimators can have appealing asymptotic behavior, including standard root-n rates of convergence and straightforward confidence intervals, even when using flexible nonparametric estimates of nuisance functions. However, as noted in Sect. 4.4, this can require a delicate balance in settings where one does not want to rely on parametric models, and also wants to be agnostic about whether the treatment or outcome process is correctly estimated. Efforts to weaken the conditions needed on the nuisance estimation have been made using approaches based on higher-order estimation [8, 12, 56], which were inspired by work by Robins et al. [42, 44, 66] that focused on minimax estimation in settings where root-n rates of convergence are not possible. Further, Donsker-type regularity conditions (though not rate conditions) can be weakened via cross-validation approaches, proposed, for example, by Zheng and van der Laan [72].

We also supposed in this review that our target parameter was a low-dimensional Euclidean parameter $\psi \in \mathbb{R}^p$ that admitted regular asymptotically linear estimators. However, in some settings these conditions fail to hold. As mentioned above, Robins et al. [42, 44, 66] considered semiparametric minimax estimation in settings where the parameter of interest is Euclidean, but root-n rates of convergence cannot be attained due to high-dimensional covariates. Estimation of functional effect parameters was considered by Diaz and van der Laan [11], Kennedy et al. [20] in the context of continuous treatment effects; in such settings the target parameter is a non-pathwise differentiable curve, and root-n rates of convergence are again not possible. Inference for a non-regular parameter in an optimal treatment regime setting was considered by Luedtke and van der Laan [22]; in this case, non-regularity does not preclude the existence of root-n rate inference.

Numerous other authors have also made important contributions extending semiparametric causal inference to novel settings; unfortunately, we cannot list all of them here. In addition, much important work is left to be done, both in the areas mentioned above and in many other interesting settings.

Acknowledgements Edward Kennedy acknowledges support from NIH grant R01-DK090385, and thanks Jason Roy and Bret Zeldow for very helpful comments and discussion.

References

1. Andrews, D.W.K.: Empirical process methods in econometrics. Handb. Econ. **4**, 2247–2294 (1994)
2. Andrews, D.W.K.: Asymptotics for semiparametric econometric models via stochastic equicontinuity. Econometrica **62**, 43–72 (1994)
3. Angrist, J.D., Imbens, G.W., Rubin, D.B.: Identification of causal effects using instrumental variables. J. Am. Stat. Assoc. **91**, 444–455 (1996)
4. Begun, J.M., Hall, W.J., Huang, W.M., Wellner, J.A.: Information and asymptotic efficiency in parametric-nonparametric models. Ann. Stat. **11**, 432–452 (1983)
5. Belloni, A., Chernozhukov, V., Hansen, C.: Inference on treatment effects after selection among high-dimensional controls. Rev. Econ. Stud. **81**, 608–650 (2014)
6. Belloni, A., Chernozhukov, V., Kato, K.: Uniform post-selection inference for least absolute deviation regression and other Z-estimation problems. Biometrika **102**, 77–94 (2015)
7. Bickel, P.J., Klaassen, C.A.J., Ritov, Y., Wellner, J.A.: Efficient and Adaptive Estimation for Semiparametric Models. Springer, New York (1993)
8. Carone, M., Diaz, I., van der Laan, M.J.: Higher-order targeted minimum loss-based estimation. U.C. Berkeley Division of Biostatistics Working Paper Series, vol. 331, pp. 1–39 (2015)
9. Chakraborty, B., Moodie, E.E.M.: Statistical Methods for Dynamic Treatment Regimes. Springer, New York (2013)
10. Dawid, P.A.: Causal inference without counterfactuals. J. Am. Stat. Assoc. **95**, 407–424 (2000)
11. Diaz, I., van der Laan, M.J.: Targeted data adaptive estimation of the causal dose - response curve. J. Causal Inf. **1**, 171–192 (2013)

12. Diaz, I., Carone, M., van der Laan, M.J.: Second order inference for the mean of a variable missing at random. U.C. Berkeley Division of Biostatistics Working Paper Series, vol. 337, pp. 1–22 (2015)
13. Gill, R.D., van der Laan, M.J., Wellner, J.A.: Inefficient estimators of the bivariate survival function for three models. Ann. Inst. Henri Poincare. **31**, 545–597 (1995)
14. Gill, R.D., van der Laan, M.J., Robins, J.M.: Coarsening at random: characterizations, conjectures, counter-examples. In: Proceedings of the First Seattle Symposium in Biostatistics, pp. 255–294. Springer, New York (1997)
15. Hahn, J.: On the role of the propensity score in efficient semiparametric estimation of average treatment effects. Econometrica **66**, 315–333 (1998)
16. Hernan, M.A., Robins, J.M.: Instruments for causal inference: an epidemiologist's dream? Epidemiology **17**, 360–372 (2006)
17. Horowitz, J.L.: Semiparametric and Nonparametric Methods in Econometrics. Springer, New York (2009)
18. Hudgens, M.G., Halloran, M.E.: Toward causal inference with interference. J. Am. Stat. Assoc. **103**, 832–842 (2012)
19. Kennedy, E.H., Sjolander, A., Small, D.S.: Semiparametric causal inference in matched cohort studies. Biometrika **102**, 739–746 (2015)
20. Kennedy, E.H., Ma, Z., McHugh, M.D., Small, D.S.: Nonparametric methods for doubly robust estimation of continuous treatment effects. arXiv preprint, arXiv:1507.00747 (2015)
21. Kosorok, M.R.: Introduction to Empirical Processes and Semiparametric Inference. Springer, New York (2007)
22. Luedtke, A.R., van der Laan, M.J.: Statistical inference for the mean outcome under a possibly non-unique optimal treatment strategy. U.C. Berkeley Division of Biostatistics Working Paper Series, vol. 332, pp. 1–37 (2014)
23. Manski, C.F.: Partial Identification of Probability Distributions. Springer, New York (2003)
24. Murphy, S.A.: Optimal dynamic treatment regimes. J. R. Stat. Soc. B **65**, 331–355 (2003)
25. Neugebauer, R., van der Laan, M.J.: Nonparametric causal effects based on marginal structural models. J. Stat. Plan. Infer. **137**, 419–434 (2007)
26. Newey, W.K.: The asymptotic variance of semiparametric estimators. Econometrica **62**, 1349–1382 (1994)
27. Newey, W.K., McFadden, D.: Large sample estimation and hypothesis testing. Handb. Econ. **4**, 2111–2245 (1994)
28. Neyman, J.: On the application of probability theory to agricultural experiments: essay on principles. Excerpts reprinted (1990) in English (D. Dabrowska and T. Speed, trans.) Stat. Sci. **5**, 463–472 (1923)
29. Ogburn, E.L., VanderWeele, T.J.: Causal diagrams for interference. Stat. Sci. **29**, 559–578 (2014)
30. Pearl, J.: Causal diagrams for empirical research. Biometrika **82**, 669–688 (1995)
31. Pearl, J.: Causality. Cambridge University Press, Cambridge (2009)
32. Petersen, M.L., Porter, K.E., Gruber, S., Wang, Y., van der Laan, M.J.: Diagnosing and responding to violations in the positivity assumption. Stat. Methods Med. Res. **21**, 31–54 (2010)
33. Pfanzagl, J.: Contributions to a General Asymptotic Statistical Theory. Springer, New York (1982)
34. Pfanzagl, J.: Estimation in Semiparametric Models. Springer, New York (1990)
35. Pollard, D.: Convergence of stochastic processes. Springer, New York (1984)
36. Pollard, D.: Empirical processes: theory and applications. In: NSF-CBMS Regional Conference Series in Probability and Statistics. Institute of Mathematical Statistics and the American Statistical Association (1990)
37. Robins, J.M.: A new approach to causal inference in mortality studies with a sustained exposure period - application to control of the healthy worker survivor effect. Math. Mod. **7**, 1393–1512 (1986)

38. Robins, J.M., Rotnitzky, A., Zhao, L.P.: Estimation of regression coefficients when some regressors are not always observed. J. Am. Stat. Assoc. **89**, 846–866 (1994)
39. Robins, J.M., Rotnitzky, A., Zhao, L.P.: Analysis of semiparametric regression models for repeated outcomes in the presence of missing data. J. Am. Stat. Assoc. **90**, 106–121 (1995)
40. Robins, J.M., Rotnitzky, A., Scharfstein, D.O.: Sensitivity analysis for selection bias and unmeasured confounding in missing data and causal inference models. In: Statistical Models in Epidemiology, the Environment, and Clinical Trials, pp. 1–94. Springer, New York (1999)
41. Robins, J.M., Hernan, M.A., Brumback, B.: Marginal structural models and causal inference in epidemiology. Epidemiology **11**, 550–560 (2000)
42. Robins, J.M., Li, L., Tchetgen, E., van der Vaart, A.W.: Higher order influence functions and minimax estimation of nonlinear functionals. In: Probability and Statistics: Essays in Honor of David A. Freedman, pp. 335–421. Beachwood, Ohio, USA, Institute of Mathematical Statistics (2008)
43. Robins, J.M., Hernan, M.A.: Estimation of the causal effects of time-varying exposures. In: Fitzmaurice, G., Davidian, M., Verbeke, G., Molenberghs, G. (eds.) Longitudinal Data Analysis, pp. 553–600. Chapman & Hall, London (2009)
44. Robins, J.M., Li, L., Tchetgen, E., van der Vaart, A.W.: Quadratic semiparametric von mises calculus. Metrika **69**, 227–247 (2009)
45. Rose, S., van der Laan, M.J.: A double robust approach to causal effects in case-control studies. Am. J. Epid. **179**, 662–669 (2014)
46. Rubin, D.B.: Estimating causal effects of treatments in randomized and nonrandomized studies. J. Educ. Psychol. **66**, 688–701 (1974)
47. Rubin, D.B.: Bayesian inference for causal effects: the role of randomization. Ann. Stat. **6**, 34–58 (1978)
48. Shorack, G.R., Wellner, J.A.: Empirical Processes with Applications to Statistics. Wiley, New York (1986)
49. Stefanski, L.A., Boos, D.D.: The calculus of M-estimation. Am. Stat. **56**, 29–38 (2002)
50. Tchetgen, E., Rotnitzky, A.: Double-robust estimation of an exposure-outcome odds ratio adjusting for confounding in cohort and case-control studies. Stat. Med. **30**, 335–347 (2011)
51. Tchetgen, E., Shpitser, I.: Semiparametric theory for causal mediation analysis: efficiency bounds, multiple robustness and sensitivity analysis. Ann. Stat. **40**, 1816–1845 (2012)
52. Tchetgen, E., VanderWeele, T.J.: On causal inference in the presence of interference. Stat. Methods Med. Res. **21**, 55–75 (2012)
53. Tsiatis, A.A.: Semiparametric Theory and Missing Data. Springer, New York (2006)
54. van der Laan, M.J.: Estimation based on case-control designs with known prevalence probability. Int. J. Biostat. **4** (2008). Article 17
55. van der Laan, M.J.: Causal inference for a population of causally connected units. J. Causal Inf. **2**, 13–74 (2014)
56. van der Laan, M.J.: Targeted estimation of nuisance parameters to obtain valid statistical inference. Int. J. Biostat. **10**, 29–57 (2014)
57. van der Laan, M.J.: Targeted learning: From MLE to TMLE. In: Lin, X., Genest, C., Banks, D.L., et al. (eds.) Past, Present, and Future of Statistical Science, pp. 465–480. Chapman & Hall, London (2014)
58. van der Laan, M.J., Dudoit, S.: Unified cross-validation methodology for selection among estimators and a general cross-validated adaptive epsilon-net estimator: finite sample oracle inequalities and examples. U.C. Berkeley Division of Biostatistics Working Paper Series. vol. 130, pp. 1–103 (2003)
59. van der Laan, M.J., Robins, J.M.: Unified Methods for Censored Longitudinal Data and Causality. Springer, New York (2003)
60. van der Laan, M.J., Rose, S.: Targeted Learning: Causal Inference for Observational and Experimental Data. Springer, New York (2011)
61. van der Laan, M.J., Rubin, D.: Targeted maximum likelihood learning. Int. J. Biostat. **2**, 1–38 (2006)

62. van der Laan, M.J., Polley, E.C., Hubbard, A.E.: Super learner. Stat. Appl. Genet. Mol. **6**, 1–21 (2007)
63. van der Laan, M.J., Petersen, M., Zheng, W.: Estimating the effect of a community-based intervention with two communities. J. Causal Inf. **1**, 83–106 (2013)
64. van der Vaart, A.W.: Asymptotic Statistics. Cambridge University Press, Cambridge (2000)
65. van der Vaart, A.W.: Part III: Semiparametric Statistics. In: Bernard, P. (ed.) Lectures on Probability Theory and Statistics, pp. 331–457. Springer, New York (2002)
66. van der Vaart, A.W.: Higher order tangent spaces and influence functions. Stat. Sci. **29**, 679–686 (2014)
67. van der Vaart, A.W., Wellner, J.A.: Weak Convergence and Empirical Processes. Springer, New York (1996)
68. VanderWeele, T.J.: Concerning the consistency assumption in causal inference. Epidemiology **20**, 880–883 (2009)
69. VanderWeele, T.J.: Explanation in Causal Inference: Methods for Mediation and Interaction. Oxford University Press, Oxford (2015)
70. VanderWeele, T.J., Vansteelandt, S.: A weighting approach to causal effects and additive interaction in case-control studies: marginal structural linear odds models. Am. J. Epidemiol. **174**, 1197–1203 (2011)
71. VanderWeele, T.J., Vansteelandt, S.: Invited commentary: some advantages of the relative excess risk due to interaction (RERI) - towards better estimators of additive interaction. Am. J. Epidemiol. **179**, 670–671 (2014)
72. Zheng, W., van der Laan, M.J.: Asymptotic theory for cross-validated targeted maximum likelihood estimation. U.C. Berkeley Division of Biostatistics Working Paper Series, vol. 273, pp. 1–58 (2010)

Chapter 9
Structural Nested Models for Cluster-Randomized Trials

Shanjun Helian, Babette A. Brumback, Matthew C. Freeman, and Richard Rheingans

Abstract In clinical trials and epidemiologic studies, adherence to the assigned components is not always perfect. In this book chapter, we are interested in estimating the causal effect of cluster-level adherence on an individual-level outcome. Two different methodologies will be provided, based on ordinary and weighted structural nested models (SNMs). We also applied the jackknife to construct confidence intervals. The computation is straightforward with application of instrumental variables software, and the programming schemes are developed for both ordinary and weighted structural nested models. Simulation studies under ordinary structural nested models with different link functions (loglinear SNM, logistic SNM, and linear SNM) were conducted to validate our methods. We then applied the methods to a school-based water, sanitation, and hygiene study to estimate the causal effect of increased adherence to intervention components on student absenteeism. The results calculated from these two methodologies are quite close.

1 Introduction

Estimating the causal effect of treatment or exposure on subjects' outcomes is the main purpose of many clinical trials and epidemiologic studies. It is common to use instrumental variables to adjust for unmeasured confounding when estimating

S. Helian • B.A. Brumback (✉)
Department of Biostatistics, University of Florida, 2004 Mowry Road, Gainesville, FL 32611-7450, USA
e-mail: hlsj2012@ufl.edu; brumback@ufl.edu

M.C. Freeman
Departments of Environmental Health, Epidemiology, and Global Health, Rollins School of Public Health, Emory University, 1518 Clifton Road NE, Atlanta, GA 30322, USA
e-mail: mcfreem@emory.edu

R. Rheingans
Chair, Department of Sustainable Development, Appalachian State University, 287 Rivers St, Boone, NC 28608, USA
e-mail: rheingansrd@appstate.edu

H. He et al. (eds.), *Statistical Causal Inferences and Their Applications in Public Health Research*, ICSA Book Series in Statistics, DOI 10.1007/978-3-319-41259-7_9

the causal effect of adherence in randomized studies. Researchers tend to use the instrumental variables within one of three frameworks: regression with an endogenous variable [3, 5, 6, 17, 21, 22, 31], principal stratification [1, 2, 7–9, 11, 13–15, 19, 20, 26], or structural nested models [4, 12, 16, 18, 23, 24, 27–30]. In this book chapter, we will focus on using structural nested models to estimate the effect of cluster-level adherence on individual level outcomes.

Structural nested models (SNMs) were introduced by Robins [23, 24] to address non-compliance in clinical trials. Vansteelandt and Goetghebeur [28] further developed a double-logistic structural mean model to estimate the effect of observed exposure on the success of treatment in a randomized trial with non-compliance. Korhonen et al. [18] developed and applied SNMs for time-to-event outcomes. Hernan and Robins [12] provided an accessible introduction to instrumental variables embedded within structural nested models, and the models have proven to be useful for adjusting estimated causal effects of adherence for unmeasured confounding. Vansteelandt et al. [30] offered a review of the use of SNMs with binary outcomes, and pointed out that in some instances, the estimating equation has no solution.

In this book chapter, we will focus on using structural nested models to estimate the causal effect of cluster-level adherence on an individual-level outcome. Two different methodologies will be provided, based on ordinary and weighted structural nested models [4]. With the ordinary SNMs, we will adjust for the individual-level confounders by including the individual-level covariates into our structural nested model. In the weighted SNMs, we will adjust for the individual-level confounders by weighting the sampled data as in Brumback et al. [4]. For both methodologies, we will develop an easily programmed iterative algorithm for solving the estimating equations, using Newton's method. To construct confidence intervals, we will consider a sandwich estimator, the bootstrap, and the jackknife for complex survey data. Furthermore, for each methodology, we will apply and compare three different structural nested modeling approaches to estimating causal relative risks based on a linear, loglinear, and logistic SNM. We will investigate the performance of both methodologies using simulated data sets, and then apply them to the school-based water, sanitation, and hygiene (WASH) study data. Both methodologies work quite well, but if the initial setup values are too far away from real ones when applying iterative algorithm or when the SNM does not fit the data, either the iterative estimation algorithm may not converge or there may be no solution to the estimating equation.

2 Motivating Example

As described by Brumback et al. [4], the school-based WASH study in Nyanza is designed to affect several outcomes, including pupil absence. The study area is divided into three geographical strata—Nyando/Kisumu East, Rachuonyo, and Suba Districts. The WASH intervention randomized public primary schools nested in three geographical strata to one of three study arms: water treatment and hygiene (hand-washing) promotion (WH), additional sanitation improvement that included latrine construction (WH+S), or control. Freeman et al. [10] showed that there is no overall effect of the intervention on absence. However, among schools in two of the geographical areas (Rachuonyo and Suba Districts) not affected by post-election violence, the school-based WASH components can improve school attendance, particularly for girls. In this book chapter, as in Freeman et al. [10], we will focus on these two geographical strata, and girls only. Also, because pupils were selected into the study with unequal probabilities, sampling weights need to be incorporated into the analysis.

In the WASH study, adherence at schools to the assigned components (W, H, or S) is far from perfect, but the randomized assignment can be used as an instrumental variable. For the purpose of analysis, we dichotomized the measure of adherence for each of the three components as adequate or not, then categorized it into three levels: an inadequate degree of water treatment, hygiene promotion, and sanitation improvement, an adequate degree of exactly one of these three components, or an adequate degree of two or more of these components.

In this book chapter, we hypothesize that increased adherence to intervention components would reduce absenteeism. In Sect. 6, we will use ordinary structural nested models to analyze the school-based WASH intervention study, and compare the results provided by Brumback et al. [4] based on weighted structural nested models.

3 Estimands

In this section, we will introduce the concept of potential outcomes and provide notations we are going to use in this chapter. We choose the causal relative risk as the estimand of interest in the study. Some sophisticated assumptions are required under different approaches. The causal relative risks are defined for ordinary and weighted structural nested models, respectively.

3.1 Potential Outcomes

Potential outcomes are possible values of an individual's measurement of interest when competing treatments are received. Suppose all potential outcomes are well defined. We let Z_i denote the randomized treatment arm, let A_i denote the adherence level of cluster i, and let X_{ij} denote the individual-level covariates for individual j in cluster i. Define $Y_{ij}(a, z)$ as the potential outcome for individual j in cluster i who had been assigned to treatment arm $Z_i = z$ with subsequent adherence $A_i = a$. We assume that $Y_{ij}(a, z)$ does not depend on z. We also assume $Y_{ij}(a) = Y_{ij}(A_i) = Y_{ij}$ when A_i is observed to equal a.

3.2 Estimands

For the ordinary SNM approach, the estimand of interest is the causal relative risk, as a factor of a

$$\mathrm{RR}(a) = \frac{E[Y_{ij}(a)|A_i = a]}{E[Y_{ij}(0)|A_i = a]}. \tag{9.1}$$

For the weighted SNM approach, the estimand of interest is

$$\mathrm{RR}^{W_1}(a) = \frac{E^{W_1}[Y_{ij}(a)|A_i = a]}{E^{W_1}[Y_{ij}(0)|A_i = a]}, \tag{9.2}$$

where $Y_{ij}(a)$ as the potential outcome for individual j in cluster i who had adherence $A_i = a$ and W_1 is the weight adjustment. Define $P^p(V)$ as the probability that V equals its observed value based on the distribution of the population data. Let $W_{ij1} = P^p(Z_i)/P^p(Z_i|X_{ij})$. We define $P^{W_1}(Y_{ij}(0), Z_i, X_{ij}) \equiv P^p(Y_{ij}(0), Z_i, X_{ij})W_{ij1}$.

3.3 Assumptions

For the ordinary SNM approach, we first posit a model representing the effect of individual-level covariates X_{ij} on $Y_{ij}(0)$ as

$$E[Y_{ij}(0)|X_{ij}] = f(X_{ij}; \gamma), \tag{9.3}$$

where $f(X_{ij}; \gamma)$ is a simple linear model.

Besides requiring that the potential outcomes are well defined at baseline and that $Y_{ij}(A) = Y_{ij}$, the methodology requires two assumptions. They are

Assumption 1 given $X_{ij} = x$, the population distribution of $Y_{ij}(0)$ does not depend on Z_i; that is, $P(Y_{ij}(0)|Z_i, X_{ij} = x) = P(Y_{ij}(0)|X_{ij} = x)$.

Assumption 2 $h\{E(Y_{ij}(a)|A_i = a, X_{ij} = x, Z_i)\} = h\{E(Y_{ij}(0)|A_i = a, X_{ij} = x, Z_i)\} + f(a_v, x; \xi)$, where $f(0, x; \xi) = 0$, a_v is defined as a dummy vector indicates the adherence level, and $h(\cdot)$ is a canonical link corresponding to a generalized linear model, such as $h(p) = p$, $h(p) = \log(p)$, or $h(p) = \log(p/1-p)$.

For the weighted structural nested model approach developed by Brumback et al. [4], besides requiring that the potential outcomes are well defined at baseline and that $Y_{ij}(A) = Y_{ij}$, we require two additional assumptions.

Assumption 3 conditional on X_{ij}, the population distribution of $Y_{ij}(0)$ does not depend on Z_i; that is, $P^p(Y_{ij}(0)|Z_i, X_{ij}) = P^p(Y_{ij}(0)|X_{ij})$.

By Assumption 3, $P^{W_1}(Y_{ij}(0), Z_i, X_{ij}) = P^p(Y_{ij}(0), X_{ij})P^p(Z_i)$. This weighted distribution reflects the distribution of the population data we would have observed if we could have randomized schools so that the distribution of X_{ij} were the same at each level of Z_i (e.g., by paired matching or frequency matching [25] of schools); note that for this distribution, $Y_{ij}(0) \perp Z_i$. Thus, Assumption 3 implies that $E^{W_1}(Y_{ij}(0)|Z_i) = E^{W_1}(Y_{ij}(0)) = E^p(Y_{ij}(0))$, where $E^{W_1}(V|C)$ is the conditional expectation of V given C with respect to the weighted distribution $P^{W_1}(V|C)$, and $E^p(V)$ is the expectation of V with respect to the population distribution $P^p(V)$. We further assume that

Assumption 4 $h\{E^{W_1}(Y_{ij}(a)|A_i = a, Z_i)\} = h\{E^{W_1}(Y_{ij}(0)|A_i = a, Z_i)\} + a_v\xi$, where $h(\cdot)$ is a canonical link corresponding to a generalized linear model, such as $h(p) = p$, $h(p) = \log(p)$, or $h(p) = \log(p/(1-p))$.

4 Estimation

In this section, the estimation methodology based on ordinary structural nested models with different link functions (linear SNM, loglinear SNM, and logistic SNM) is provided, and the weighted structural nested models developed by Brumback et al. [4] are reviewed. Different approaches to constructing confidence intervals are compared and discussed. Computing and programming schemes are provided for both ordinary and weighted structural nested models.

4.1 *Estimation Using an Ordinary Structural Nested Model*

For the ordinary SNM approach, suppose that for each level of X_{ij} we could have randomized all clusters in the population and observed both cluster-level adherence and individual-level outcomes, so that X_{ij}, Z_i, A_i, and the potential outcomes $Y_{ij}(a)$ for all a in all levels of x are defined for each individuals in the population. Based on Assumption 1, we have that conditional on X_{ij}, $Y_{ij}(0)$ does not depend on Z_i. We further let $f(X_{ij}; \gamma) = X_{vij}\gamma_1 + \gamma_0$, where X_{vij} is defined as a vector-valued function of X_{ij} (perhaps denoting dummy variables, e.g. when X_{ij} is a multinomial random variable).

Let W_{ij2} be the inverse probability that individual j from cluster i was selected into the study. Let $\mu(A_i, Z_i, X_{ij}; \eta)$ be a parametric model for $E(Y_{ij}|A_i, Z_i, X_{ij})$ with parameter η. When A_i and X_{ij} are multinomial random variables, one could use the model $\mu(A_i, Z_i, X_{ij}; \eta) = g(A_{vi}\eta_1 + Z_{vi}\eta_2 + A_i{*}Z_i\eta_3 + X_{vij}\eta_4)$, where A_{vi} and Z_{vi} are defined as vector functions of A_i and Z_i (perhaps denoting dummy variables, e.g. when A_i and Z_i are multinomial random variables), and $A_i{*}Z_i$ represents a multidimensional interaction. Under Assumption 2, letting $f(a_v, x; \xi) = a_v x\xi$ and assuming that $\mu(A_i, Z_i, X_{ij}; \eta)$ is correctly specified, we can consistently estimate

(ξ, η) by solving the estimating equations

$$\sum_i \sum_j W_{ij2}(A_{vi}, Z_{vi}, A_{vi}{*}Z_{vi}, X_{vij})^T [Y_{ij} - \mu(A_i, Z_i, X_{ij}; \eta)] = 0$$

$$\sum_i \sum_j W_{ij2}(Z_{vi}, X_{vij})^T \{h^{-1}[h(\mu(A_i, Z_i, X_{ij}; \eta)) - A_{vi}X_{ij}\xi] - X_{vij}\gamma - \gamma_0\} = 0 \tag{9.4}$$

for (ξ, η, γ). The first estimating equation at (9.4) is unbiased assuming $\mu(A_i, Z_i, X_{ij}; \eta) = g(A_{vi}\eta_1 + Z_{vi}\eta_2 + A_i{*}Z_i\eta_3 + X_{vij}\eta_4)$ is correctly specified. The second equation at (9.4) is unbiased because

$$\begin{aligned} &E\{h^{-1}[h(\mu(A_i, Z_i, X_{ij}; \eta)) - A_{vi}X_{ij}\xi] - X_{vij}\gamma_1 - \gamma_0 | Z_i, X_{ij}\} \\ &= E\{h^{-1}[h\{E(Y_{ij}(0)|A_i = a, X_{ij} = c, Z_i)\}] - X_{vij}\gamma_1 - \gamma_0 | Z_i, X_{ij}\} \\ &= E[E(Y_{ij}(0)|A_i, Z_i, X_{ij}) - X_{ij}\gamma_1 - \gamma_0 | Z_i, X_{ij}] \\ &= E[Y_{ij}(0)|Z_i, X_{ij}] - X_{vij}\gamma_1 - \gamma_0 \\ &= E[Y_{ij}(0)|X_{ij}] - X_{vij}\gamma_1 - \gamma_0 = 0. \end{aligned} \tag{9.5}$$

If we use a generalized linear model for $\mu(A_i, Z_i, X_{ij}; \eta)$ with a canonical link function $g^{-1}(\cdot)$, the first estimating equation at (9.4) can be solved by using weighted GLM software (e.g., PROC GLM in SAS). If we furthermore let $h(\cdot) = g^{-1}(\cdot)$ and $D_i \equiv (A_{vi}, Z_{vi}, A_i{\cdot}Z_i, X_{vij})$, then substituting $\hat{\eta}$ for η into the second estimating equation at (9.4), it reduces to

$$\sum_i \sum_j W_{ij2}(Z_{vi}, X_{vij})^T [g(D_i\hat{\eta} - A_{vi}X_{ij}\xi) - X_{vij}\gamma - \gamma_0] = 0, \tag{9.6}$$

which can be solved iteratively using Newton's method by linearizing $g(D_i\hat{\eta} - A_{vi}X_{ij}\xi)$ about a current estimate of ξ, then solve the second estimating equation at (9.4) by using weighted instrumental software (e.g., PROC SYSLIN in SAS).

For example, define $g(x) = \exp(x)$, and let ξ^t denote the current estimate of ξ. We have

$$g'(x) = g(x) = \exp(x). \tag{9.7}$$

Let $f(\xi) = g(D_i\hat{\eta} - A_{vi}X_{ij}\xi)$, then

$$f'(\xi) = -A_{vi}X_{ij}g'(D_i\hat{\eta} - A_{vi}X_{ij}\xi) = -A_{vi}X_{ij}g(D_i\hat{\eta} - A_{vi}X_{ij}\xi). \tag{9.8}$$

By the Delta method, we have

$$f(\xi) - f(\xi^t) \approx (\xi - \xi^t) f'(\xi^t) = (\xi - \xi^t) f(\xi^t). \tag{9.9}$$

Let $A^*_{vi} = A_{vi} g(D_i \hat{\eta} - A_{vi} X_{ij} \xi)$; then $f(\xi) - f(\xi^t) = -A^*_{vi} X_{ij} (\xi - \xi^t)$, where $f(\xi^t) = g(D_i \hat{\eta} - A_{vi} X_{ij} \xi^t)$. Also, letting $Y^*_{ij} = f(\xi^t) + A^*_{vi} X_{ij} \xi^t$, the second estimating equation at (9.4) simplifies to

$$\sum_i \sum_j W_{ij2}(Z_{vi}, X_{vij})^T (Y^*_{ij} - A^*_{vi} X_{ij} \xi - X_{vij} \gamma - \gamma_0) = 0, \tag{9.10}$$

which can be solved iteratively using weighted instrumental variables software (e.g., PROC SYSLIN in SAS) with Y^*_{ij} as the outcome, $A^*_{vi} X_{ij}$ and X_{vij} as the endogenous regressor, and Z_{vi} and X_{vij} as the instrument variables.

The linearization can also be applied to logit link function. We have

$$g(x) = \frac{\exp(x)}{1 + \exp(x)}$$

$$g'(x) = \frac{\exp(x)}{(1 + \exp(x))^2} = g(x)[1 - g(x)]. \tag{9.11}$$

Let $f(\xi) = g(D_i \hat{\eta} - A_{vi} X_{ij} \xi)$, then

$$\begin{aligned} f'(\xi) &= -A_{vi} X_{ij} g'(D_i \hat{\eta} - A_{vi} X_{ij} \xi) \\ &= -A_{vi} X_{ij} g(D_i \hat{\eta} - A_{vi} X_{ij} \xi)[1 - g(D_i \hat{\eta} - A_{vi} X_{ij} \xi)]. \end{aligned} \tag{9.12}$$

Let $A^*_{vi} = A_{vi} g(D_i \hat{\eta} - A_{vi} X_{ij} \xi)[1 - g(D_i \hat{\eta} - A_{vi} X_{ij} \xi)]$ and $Y^*_{ij} = f(\xi^t) + A^*_{vi} X_{ij} \xi^t$. Again, the second estimating equation at (9.4) simplifies to

$$\sum_i \sum_j W_{ij2}(Z_{vi}, X_{vij})^T (Y^*_{ij} - A^*_{vi} X_{ij} \xi - X_{vij} \gamma - \gamma_0) = 0, \tag{9.13}$$

which can be solved iteratively using weighted instrumental variables software (e.g., PROC SYSLIN in SAS) with Y^*_{ij} as the outcome, $A_{vi} X_{ij}$ and X_{vij} as the endogenous regressor, and Z_{vi} and X_{vij} as the instruments.

If we let $h(p) = p$ in Assumption 2, then the estimating equations at (9.4) can be solved by using weighted linear regression, and the second estimating equation at (9.4) becomes

$$\sum_i \sum_j W_{ij2}(Z_{vi}, X_{vij})^T (D_i \hat{\eta} - A_{vi} X_{ij} \xi - X_{vij} \gamma - \gamma_0) = 0, \tag{9.14}$$

which can be solved without iteration using weighted instrumental variables software (e.g., PROC SYSLIN in SAS) with $D_i \hat{\eta}$ as the outcome, $A_{vi} X_{ij}$ and X_{vij} as the endogenous regressor, and Z_{vi} and X_{vij} as the instruments.

Assumption 2 states that the distribution of potential outcomes $Y_{ij}(a)$ in the population satisfies an ordinary generalized structural nested model. Then we can estimate $E[Y_{ij}(0)|A_i = a]$ and $E[Y_{ij}(a)|A_i = a]$ via

$$\hat{E}[Y_{ij}(0)|A_i = a] = \sum_i \sum_j W_{ij2}[g(D_i\eta - ac\xi; \hat{\eta}, \hat{\xi})]I(A_i = a)$$

$$\hat{E}[Y_{ij}(a)|A_i = a] = \sum_i \sum_j W_{ij2}[g(D_i\eta; \hat{\eta})]I(A_i = a), \quad (9.15)$$

where $\hat{\xi}$ is the estimator solved by the estimating equations at (9.4), and $I(A_i = a)$ is an indicator function taking the value 1 when $A_i = a$, and otherwise equaling 0.

Then the Causal relative risk is estimated as

$$\hat{\mathrm{RR}}(a) = \frac{\hat{E}[Y_{ij}(a)|A_i = a]}{\hat{E}[Y_{ij}(0)|A_i = a]}. \quad (9.16)$$

4.2 *Estimation Using a Weighted Structural Nested Model*

We next review the weighted structural nested modeling approach presented in Brumback et al. [4]. Let $W_{ij} = W_{ij1}W_{ij2}$, where $W_{ij1} = P^p(Z_i)/P^p(Z_i|X_{ij})$ as we discussed in Sect. 3. Let $\mu(A_i, Z_i; \eta)$ be a parametric model for $E^{W_1}(Y_{ij}|A_i, Z_i)$ with parameter η. When A_i is multinomial random variables, one could use the saturated model $\mu(A_i, Z_i, ; \eta) = g(A_{vi}\eta_1 + Z_{vi}\eta_2 + A_i{*}Z_i\eta_3)$, where A_{vi}, Z_{vi}, and $A_i{*}Z_i$ have been defined in the previous section. Let $D_i \equiv (A_{vi}, Z_{vi}, A_i{*}Z_i)^T$. Under Assumption 3 and 4, define $E^{W_1}(Y_{ij}(0)) \equiv \alpha$ and assume $\mu(A_i, Z_i; \eta)$ is correctly specified, we can consistently estimate (ξ, η) by solving the estimating equations

$$\sum_i \sum_j W_{ij}{D_i}^T[Y_{ij} - \mu(A_i, Z_i; \eta)] = 0$$

$$\sum_i \sum_j W_{ij}{Z_{vi}}^T\{h^{-1}[h(\mu(A_i, Z_i; \eta)) - A_{vi}\xi] - \alpha\} = 0 \quad (9.17)$$

for (ξ, η). The first estimating equation at (9.17) is unbiased conditional on A_i and Z_i provided $\mu(A_i, Z_i; \eta)$ is correctly specified; if one uses a saturated model, that is automatic. The second estimating equation at (9.17) is unbiased because

$$\begin{aligned} &E^{W_1}\{h^{-1}[h(\mu(A_i, Z_i; \eta)) - A_{vi}\xi] - \alpha\} \\ &= E^{W_1}\{h^{-1}[h\{E(Y_{ij}(0)|A_i = a, Z_i)\}] - \alpha\} \\ &= E^{W_1}[E(Y_{ij}(0)|A_i = a, Z_i) - \alpha] \\ &= E^{W_1}[Y_{ij}(0)|Z_i] - \alpha \\ &= E^{W_1}[Y_{ij}(0)] - \alpha = 0. \end{aligned} \quad (9.18)$$

Similarly, if we use a generalized linear model for $\mu(A_i, Z_i; \eta)$ with a canonical link function $g^{-1}(\cdot)$, the first estimating equation at (9.17) can be solved using weighted GLM software (e.g., PROC GLM in SAS). If we further assume that $h(\cdot) = g^{-1}(\cdot)$, then substituting $\hat{\eta}$ for η into the second estimating equation at (9.17), the equation reduces to

$$\sum_i \sum_j W_{ij} Z_{vi}{}^T [g(D_i\hat{\eta} - A_{vi}\xi) - \alpha] = 0, \quad (9.19)$$

which can be solved iteratively using Newton's method by linearizing $g(D_i\hat{\eta} - A_{vi}\xi)$ about a current estimate of ξ, then solve the second estimating equation at (9.17) by using weighted instrumental software (e.g., PROC SYSLIN in SAS).

Similarly as the linearization procedure in the ordinary SNM approach, the second estimating equation at (9.17) can be simplified as

$$\sum_i \sum_j W_{ij} Z_{vi}{}^T (Y^*_{ij} - A^*_{vi}\xi - \alpha) = 0, \quad (9.20)$$

where Y^*_{ij} is the outcome, A^*_{vi} is the endogenous regressor, and Z_{vi} is the instrument variable.

Assumption 2 states that the distribution of potential outcomes $Y_{ij}(a)$ in the population satisfies a weighted generalized structural nested mean model. Then we can estimate

$$\hat{E}^{W_1}[Y_{ij}(0)|A_i = a] = \sum_i \sum_j W_{ij} g(D_i\eta - a\hat{\xi}; \hat{\eta}) I(A_i = a)$$

$$\hat{E}^{W_1}[Y_{ij}(a)|A_i = a] = \sum_i \sum_j W_{ij} g(D_i\eta; \hat{\eta}) I(A_i = a), \quad (9.21)$$

where $\hat{\xi}$ is the estimator solved by the estimating equations at (9.17).

Then the Causal relative risk is estimated as

$$\hat{RR}^{W_1}(a) = \frac{\hat{E}^{W_1}[Y_{ij}(a)|A_i = a]}{\hat{E}^{W_1}[Y_{ij}(0)|A_i = a]}. \quad (9.22)$$

4.3 *Constructing Confidence Intervals*

As discussed in Brumback et al. [4], to construct confidence intervals, one could use the sandwich estimator of variance. The estimating equations at (9.4) and (9.17) have the form $U(\theta) = \sum_{h=1}^{H} \sum_{c=1}^{C_h} U_{hc}(\theta) = 0$, where θ is a vector of parameters, c indexes primary sampling units (PSU, e.g. the schools), and h indexes the primary strata. $U_{hc}(\theta)$ is a sum of weighted estimating equations, with the weighted components each having an expected value of zero unconditionally, but not conditionally upon stratum h. The sandwich estimator of variance for the $\hat{\theta}$ which solves $U(\theta) = 0$ has the form

$$\widehat{\text{var}}(\hat{\theta}) = \{\nabla U(\hat{\theta})\}^{-1} V(\hat{\theta}) \{\nabla U(\hat{\theta})^T\}^{-1}, \tag{9.23}$$

where $\nabla U(\hat{\theta})$ is the gradient of $U(\theta)$ with respect to θ, and

$$V(\hat{\theta}) = \sum_{h=1}^{H} \{C_h/(C_h - 1)\} \sum_{c=1}^{C_h} \{U_{hc}(\hat{\theta}) - U_{h\cdot}(\hat{\theta})\}\{U_{hc}(\hat{\theta}) - U_{h\cdot}(\hat{\theta})\}^T, \tag{9.24}$$

where $U_{h\cdot}(\hat{\theta}) = (1/C_h) \sum_{c=1}^{C_h} U_{hc}(\hat{\theta})$. By the law of large numbers and central limit theorem, $\hat{\theta}$ approximately follows multivariate normal distribution with mean θ and variance $\widehat{\text{var}}(\hat{\theta})$.

However, the sandwich estimator of variance is difficult to program. An easier way to estimate $\widehat{\text{var}}(\hat{\theta})$ is to use the bootstrap or jackknife for complex survey data. Let $\hat{\theta}^b$ be an estimate of θ based on the data from the bth bootstrap sample, then the estimator of variance is

$$\widehat{\text{var}}_B(\hat{\theta}) = \{1/(B-1)\} \sum_{b=1}^{B} \left[\hat{\theta}^b - \left\{ (1/B) \sum_{b=1}^{B} \hat{\theta}^b \right\} \right]^2, \tag{9.25}$$

where B is the total number of bootstrap samples, and $\hat{\theta}$ approximately follows normal distribution.

Unfortunately, if we use the bootstrap, sometimes the estimating equations have no solution. Thus, we will use the jackknife to estimate variance. Let $\hat{\theta}^{hc}$ be an estimate of θ based on deleting the cth PSU within stratum h. The jackknife estimator of variance we used is

$$\widehat{\text{var}}_J(\hat{\theta}) = \sum_{h=1}^{H} \{(C_h - 1)/C_h\} \sum_{c=1}^{C_h} (\hat{\theta}^{hc} - \hat{\theta})^2. \tag{9.26}$$

For estimating confidence intervals for functions $\varphi(\theta)$ of θ, such as relative risks, we use the normal approximation to the log of $\varphi(\theta)$.

5 Simulation Study

For the weighted SNM approach, Brumback et al. [4] provide a simulation study. For the ordinary SNM approach, we conducted three additional sets of simulations, the first based on a loglinear SNM, with $h(p) = \log(p)$ in Assumption 2; the

Table 9.1 Distribution of X_{ij} given Z_i

	$Z_i = 0$	$Z_i = 1$
$P(X_{ij} = 1 \mid Z_i)$	1/2	1/3
$P(X_{ij} = 2 \mid Z_i)$	1/2	2/3

Table 9.2 Joint distribution of X_{ij} and Z_i

	$Z_i = 0$	$Z_i = 1$
$P(X_{ij} = 1, Z_i)$	1/6	1/9
$P(X_{ij} = 2, Z_i)$	1/6	2/9

second based on a logistic SNM, with $h(p) = \log(p/(1-p))$ in Assumption 2; and the third based on a linear SNM, with $h(p) = p$ in Assumption 2. Constructing simulations for the ordinary SNM approach is more difficult than for the weighted SNM approach, due to needing to satisfy more modeling assumptions. For the simulation, we let $Z_i = 0, 1$ with equal probability. Then generate the distribution of X_{ij} conditional on Z_i as listed in Table 9.1, where Z_i, A_i, and X_{ij} all have two levels.

Based on the distribution of X_{ij} given Z_i, we can calculate the joint distribution of X_{ij} and Z_i as listed in Table 9.2.

For the loglinear SNM, we let $\eta_0 = -1.5$, $\eta_1 = 0.05$, $\eta_2 = 0.1$, $\eta_3 = 0.05$ and $\eta_4 = 0.1$. For the logistic SNM, we let $\eta_0 = -1.25$, $\eta_1 = 0.05$, $\eta_2 = 0.1$, $\eta_3 = 0.05$ and $\eta_4 = 0.1$. For the linear SNM, we let $\eta_0 = 0.55$, $\eta_1 = -0.05$, $\eta_2 = -0.08$, $\eta_3 = -0.05$ and $\eta_4 = -0.04$. First, we generated $Y_{ij}(a)$ according to $P(Y_{ij}(a) = 1 \mid A_i, Z_i, X_{ij})$ based on $\mu(A_i, Z_i, X_{ij}; \eta)$. We then generated $Y_{ij}(0)$ according to $P(Y_{ij}(0) = 1 \mid A_i, Z_i, X_{ij})$ based on Assumption 2, which for the loglinear and logistic SNMs, we let $\xi = 0.25$, and for the linear SNM, we let $\xi = 0.08$. The distribution of $Y_{ij}(0)$ and $Y_{ij}(a)$ given A_i, Z_i, and X_{ij} are listed in Table 9.3. The distribution of $P(Y_{ij}(a) = 1 \mid A_i = a, X_{ij}, Z_i)$ is given as $P^1(Y_{ij}(a) = 1 \mid A_i = a, X_{ij}, Z_i)$ for the logistic SNM, as $P^2(Y_{ij}(a) = 1 \mid A_i = a, X_{ij}, Z_i)$ for the loglinear SNM and as $P^3(Y_{ij}(a) = 1 \mid A_i = a, X_{ij}, Z_i)$ for the linear SNM.

Now, to satisfy the Assumption 1 that $Y_{ij}(0) \perp Z_i \mid X_{ij}$, we let $\gamma_0 = 0.217$ and $\gamma_1 = 0.015$ for the loglinear SNM; let $\gamma_0 = 0.218$ and $\gamma_1 = 0.011$ for the logistic SNM; let $\gamma_0 = 0.442$ and $\gamma_1 = -0.107$ for the linear SNM. Take $P(A_i = 0 \mid Z_i = 0, X_{ij} = 1)$ and $P(A_i = 1 \mid Z_i = 0, X_{ij} = 1)$ as an example, then the distribution of A_i given Z_i and X_{ij} can be solved by the equations below:

$$P(A_i = 0 \mid Z_i = 0, X_{ij} = 1) + P(A_i = 1 \mid Z_i = 0, X_{ij} = 1) = 1$$

$$\sum_{a=0}^{1} P[Y_{ij}(0) = 1 \mid A_i = a, X_{ij} = 1, Z_i = 0] \times P(A_i = a \mid Z_i = 0, X_{ij} = 1)$$
$$= P[Y_{ij}(0) = 1 \mid Z_i = 0, X_{ij} = 1], \tag{9.27}$$

Table 9.3 Distribution of $Y_{ij}(0)$ and $Y_{ij}(a)$ given A_i, Z_i, X_{ij}

	$Z_i = 0$	$Z_i = 1$
$P^1(Y_{ij}(0) = 1 \mid A_i = 0, X_{ij} = 1, Z_i)$	0.2231	0.2466
$P^2(Y_{ij}(0) = 1 \mid A_i = 0, X_{ij} = 1, Z_i)$	0.2227	0.2405
$P^3(Y_{ij}(0) = 1 \mid A_i = 0, X_{ij} = 1, Z_i)$	0.55	0.47
$P^1(Y_{ij}(0) = 1 \mid A_i = 1, X_{ij} = 1, Z_i)$	0.1827	0.2122
$P^1(Y_{ij}(1) = 1 \mid A_i = 1, X_{ij} = 1, Z_i)$	0.2346	0.2725
$P^2(Y_{ij}(0) = 1 \mid A_i = 1, X_{ij} = 1, Z_i)$	0.1900	0.2142
$P^2(Y_{ij}(1) = 1 \mid A_i = 1, X_{ij} = 1, Z_i)$	0.2315	0.2592
$P^3(Y_{ij}(0) = 1 \mid A_i = 1, X_{ij} = 1, Z_i)$	0.42	0.29
$P^3(Y_{ij}(1) = 1 \mid A_i = 1, X_{ij} = 1, Z_i)$	0.50	0.37
$P^1(Y_{ij}(0) = 1 \mid A_i = 0, X_{ij} = 2, Z_i)$	0.2466	0.2725
$P^2(Y_{ij}(0) = 1 \mid A_i = 0, X_{ij} = 2, Z_i)$	0.2405	0.2592
$P^3(Y_{ij}(0) = 1 \mid A_i = 0, X_{ij} = 2, Z_i)$	0.51	0.43
$P^1(Y_{ij}(0) = 1 \mid A_i = 1, X_{ij} = 2, Z_i)$	0.1572	0.1827
$P^1(Y_{ij}(1) = 1 \mid A_i = 1, X_{ij} = 2, Z_i)$	0.2592	0.3012
$P^2(Y_{ij}(0) = 1 \mid A_i = 1, X_{ij} = 2, Z_i)$	0.1680	0.1900
$P^2(Y_{ij}(1) = 1 \mid A_i = 1, X_{ij} = 2, Z_i)$	0.2497	0.2789
$P^3(Y_{ij}(0) = 1 \mid A_i = 1, X_{ij} = 2, Z_i)$	0.30	0.17
$P^3(Y_{ij}(1) = 1 \mid A_i = 1, X_{ij} = 2, Z_i)$	0.46	0.33

where from Eq. (9.3) with $f(X_{ij};\gamma) = X_{vij}\gamma_1 + \gamma_0$ and Assumption 1, we have

$$P[Y_{ij}(0) = 1 \mid Z_i = 0, X_{ij} = 1] = P[Y_{ij}(0) = 1 \mid Z_i = 1, X_{ij} = 1] = \gamma_0$$
$$P[Y_{ij}(0) = 1 \mid Z_i = 0, X_{ij} = 2] = P[Y_{ij}(0) = 1 \mid Z_i = 1, X_{ij} = 2] = \gamma_0 + \gamma_1. \tag{9.28}$$

The distribution of A_i given Z_i and X_{ij} is shown in Table 9.4, where $P^1(A_i = a \mid X_{ij}, Z_i)$ and $P^1(Y_{ij}(0) = 1 \mid Z_i, X_{ij})$ correspond to the loglinear SNM, $P^2(A_i = a \mid X_{ij}, Z_i)$ and $P^2(Y_{ij}(0) = 1 \mid Z_i, X_{ij})$ correspond to the logistic SNM, and $P^3(A_i = a \mid X_{ij}, Z_i)$ and $P^3(Y_{ij}(0) = 1 \mid Z_i, X_{ij})$ correspond to the linear SNM.

Based on the above distributions, we calculate $P(A_i = a \mid Z_i = z) = \sum_{c=1}^{2} P(A_i = a \mid Z_i = z, X_{ij} = c)P(X_{ij} = c \mid Z_i = z)$ and $P(A_i = a) = \sum_{z=0}^{1} P(A_i = a \mid Z_i = z)P(Z_i = z)$. The distribution of A_i given Z_i is listed in Table 9.5, where $P^1(A_i = a \mid Z_i)$ is for the loglinear SNM, $P^2(A_i = a \mid Z_i)$ is for the logistic SNM, and $P^3(A_i = a \mid Z_i)$ is for the linear SNM.

In the loglinear simulated model, we would have $P(A_i = 0) = \sum_{z=0}^{1} P(A_i = 0 \mid Z_i = z)P(Z_i = z) = 0.6273$ and $P(A_i = 1) = \sum_{z=0}^{1} P(A_i = 1 \mid Z_i = z)P(Z_i = z) = 0.3727$. Similarly, in the logistic simulated model, we would have $P(A_i = 0) = 0.6365$ and $P(A_i = 1) = 0.3635$; and in the linear simulated model, we would have $P(A_i = 0) = 0.4363$ and $P(A_i = 1) = 0.5637$. Then we can calculate the joint distribution of X_{ij} and Z_i given A_i as listed in Table 9.6, where $P^1(X_{ij} = c, Z_i \mid A_i = a)$ is for the loglinear SNM, $P^2(X_{ij} = c, Z_i \mid A_i = a)$ is for the logistic SNM, and $P^3(X_{ij} = c, Z_i \mid A_i = a)$ is for the linear SNM.

Table 9.4 Distribution of A_i given Z_i and X_{ij}

	$X_{ij} = 1$	$X_{ij} = 2$
$P^1(A_i = 0 \mid Z_i = 0, X_{ij})$	0.8484	0.8366
$P^1(A_i = 1 \mid Z_i = 0, X_{ij})$	0.1516	0.1634
$P^1(A_i = 0 \mid Z_i = 1, X_{ij})$	0.1384	0.5489
$P^1(A_i = 1 \mid Z_i = 1, X_{ij})$	0.8617	0.4511
$P^1(Y_{ij}(0) = 1 \mid Z_i, X_{ij})$	0.217	0.232
$P^2(A_i = 0 \mid Z_i = 0, X_{ij})$	0.8563	0.8415
$P^2(A_i = 1 \mid Z_i = 0, X_{ij})$	0.1437	0.1585
$P^2(A_i = 0 \mid Z_i = 1, X_{ij})$	0.1457	0.5634
$P^2(A_i = 1 \mid Z_i = 1, X_{ij})$	0.8543	0.4366
$P^2(Y_{ij}(0) = 1 \mid Z_i, X_{ij})$	0.218	0.229
$P^3(A_i = 0 \mid Z_i = 0, X_{ij})$	0.1692	0.1667
$P^3(A_i = 1 \mid Z_i = 0, X_{ij})$	0.8308	0.8333
$P^3(A_i = 0 \mid Z_i = 1, X_{ij})$	0.8444	0.6346
$P^3(A_i = 1 \mid Z_i = 1, X_{ij})$	0.1556	0.3654
$P^3(Y_{ij}(0) = 1 \mid Z_i, X_{ij})$	0.442	0.335

Table 9.5 Distribution of A_i given Z_i

	$Z_i = 0$	$Z_i = 1$
$P^1(A_i = 0 \mid Z_i)$	0.8425	0.4120
$P^1(A_i = 1 \mid Z_i)$	0.1575	0.5880
$P^2(A_i = 0 \mid Z_i)$	0.8489	0.4241
$P^2(A_i = 1 \mid Z_i)$	0.1511	0.5759
$P^3(A_i = 0 \mid Z_i)$	0.1679	0.7046
$P^3(A_i = 1 \mid Z_i)$	0.8321	0.2954

Table 9.6 Joint distribution of X_{ij} and Z_i given A_i

	$Z_i = 0$	$Z_i = 1$
$P^1(X_{ij} = 1, Z_i \mid A_i = 0)$	0.3381	0.0368
$P^1(X_{ij} = 2, Z_i \mid A_i = 0)$	0.3334	0.2917
$P^1(X_{ij} = 1, Z_i \mid A_i = 1)$	0.1017	0.3853
$P^1(X_{ij} = 2, Z_i \mid A_i = 1)$	0.1096	0.4035
$P^2(X_{ij} = 1, Z_i \mid A_i = 0)$	0.3363	0.0381
$P^2(X_{ij} = 2, Z_i \mid A_i = 0)$	0.3305	0.2950
$P^2(X_{ij} = 1, Z_i \mid A_i = 1)$	0.0989	0.3917
$P^2(X_{ij} = 2, Z_i \mid A_i = 1)$	0.1090	0.4004
$P^3(X_{ij} = 1, Z_i \mid A_i = 0)$	0.0970	0.3226
$P^3(X_{ij} = 2, Z_i \mid A_i = 0)$	0.0955	0.4849
$P^3(X_{ij} = 1, Z_i \mid A_i = 1)$	0.3684	0.0460
$P^3(X_{ij} = 2, Z_i \mid A_i = 1)$	0.3696	0.2160

Table 9.7 Loglinear simulation results

Simulation	γ_0	stderr	TRUE	Pvalue	γ_1	stderr	TRUE	Pvalue
100	0.2166	0.0009	0.2170	0.6577	0.0165	0.0009	0.0150	0.0987
500	0.2172	0.0004	0.2170	0.6173	0.0149	0.0003	0.0150	0.7390
1000	0.2169	0.0003	0.2170	0.7390	0.0154	0.0003	0.0150	0.1827
Simulation	ξ	stderr	TRUE	Pvalue	log(RR(1))	stderr	TRUE	Pvalue
100	0.2492	0.0063	0.2500	0.8992	0.3844	0.0088	0.3781	0.4757
50	0.2506	0.0026	0.2500	0.8176	0.3806	0.0057	0.3781	0.5323
1000	0.2509	0.0019	0.2500	0.6358	0.3709	0.0042	0.3781	0.1434

Table 9.8 Logistic simulation results

Simulation	γ_0	stderr	TRUE	Pvalue	γ_1	stderr	TRUE	Pvalue
100	0.2185	0.0009	0.2180	0.5755	0.0111	0.0008	0.0110	0.9008
500	0.2179	0.0004	0.2180	0.7950	0.0113	0.0003	0.0110	0.3178
1000	0.2181	0.0003	0.2180	0.7390	0.0108	0.0003	0.0110	0.5051
Simulation	ξ	stderr	TRUE	Pvalue	log(RR(1))	stderr	TRUE	Pvalue
100	0.2545	0.0083	0.2500	0.5889	0.2960	0.0100	0.2896	0.5237
500	0.2489	0.0035	0.2500	0.7554	0.2970	0.0044	0.2896	0.0932
1000	0.2528	0.0025	0.2500	0.2630	0.2933	0.0030	0.2896	0.2177

Based on the above distributions, we calculate $E(Y_{ij}(0)|A_i = a)$ as

$$E(Y_{ij}(0)|A_i = a) = P(Y_{ij}(0) = 1|A_i = a)$$
$$= \sum_{z=0}^{1} \sum_{c=1}^{2} P(Y_{ij}(0) = 1|A_i = a, Z_i = z, X_{ij} = c) P(Z_i = z, X_{ij} = c|A_i = a). \tag{9.29}$$

To check our results, we simulated a data set with 10,000 observations for 100, 500, and 1000 repetitions. Assume all parameter estimates (e.g., ξ and γ) and $\log(\hat{RR}(1))$ approximately follow normal distribution. We can use one sample t-test to check the bias. The simulation results of all three SNMs are listed in Tables 9.7, 9.8, and 9.9. From the results we can see that generally, our estimating procedure performs well.

To study the performance of the jackknife method, we simulated 500 data sets with 500 observations for each SNM approach and computed the confidence intervals with jackknife variance estimators. The coverage of 95 % confidence intervals for γ_0, γ_1, ξ, and causal relative risk RR(1) for logistic SNM, loglinear SNM, and linear SNM are listed in Table 9.10. From the results, we can conclude that the jackknife performs well.

Table 9.9 Linear simulation results

Simulation	γ_0	stderr	TRUE	Pvalue	γ_1	stderr	TRUE	Pvalue
100	0.4424	0.0009	0.4420	0.6541	−0.1086	0.0014	−0.1070	0.3559
500	0.4415	0.0005	0.4420	0.3178	−0.1078	0.0006	−0.1070	0.1830
1000	0.4424	0.0003	0.4420	0.1324	−0.1070	0.0004	−0.1070	1.0000
Simulation	ξ	stderr	TRUE	Pvalue	log(RR(1))	stderr	TRUE	Pvalue
100	0.0812	0.0013	0.0800	0.3582	0.3280	0.0057	0.3376	0.0953
500	0.0807	0.0005	0.0800	0.1621	0.3392	0.0026	0.3376	0.5386
1000	0.0798	0.0004	0.0800	0.6172	0.3385	0.0018	0.3376	0.6172

Table 9.10 Jackknife simulation results

Approaches	γ_0 (%)	γ_1 (%)	ξ (%)	RR(1) (%)
logistic	95.0 ± 1.9	95.0 ± 1.9	96.4 ± 1.6	93.0 ± 2.2
loglinear	95.2 ± 1.9	97.4 ± 1.4	96.6 ± 1.6	95.4 ± 1.8
linear	95.4 ± 1.8	95.2 ± 1.9	94.6 ± 2.0	96.0 ± 1.7

6 Analysis of the WASH Intervention

For the school-based WASH analysis, we let A_i be a three-level ordinal variable, $a = 0$, 1, or 2. Define $a = 0$ as the reference level, representing inadequate degrees of water treatment, hygiene promotion, and sanitation improvement; $a = 1$ as an adequate degree of exactly one of those three components; and $a = 2$ as an adequate degree of two or more of those components. Since we are interested in the effect of adherence on pupils school absence, we let Y_{ij} indicate that outcome. W_{ij2} is the inverse probability of an individual being selected into the study. Let Z_i denote randomization level (Control, WH, or WH+S). Let C_{ij} denote grade level as individual-level covariate.

In the ordinary SNMs, we used linear, logistic, and loglinear SNMs to analyze the effect of intervention adherence on absenteeism for the school-based WASH trial. Table 9.11 summarizes the estimates and 95 % confidence intervals based on the jackknife estimators of variance. The analysis of the WASH study based on the weighted structural nested models can be found in Brumback et al. [4]. However, some of those results are incorrect due to accidentally using the bootstrap formula, rather than the jackknife formula, to combine the jackknife estimates to compute the CI. The corrected results are listed in Table 9.12.

Comparing the results in Tables 9.11 and 9.12, we can see that both ordinary and weighted methodologies generated similar results. All three approaches (linear, loglinear, and logistic) generated similar point estimates, and the results support the hypothesis that increased adherence to intervention components would reduce absenteeism. The relative risk is closer to one for the $A = 2$ group than for the $A = 1$ group, but this is due to the estimate of the risk of absence had been set to 0 being higher in the $a = 1$ group. For example, for the logistic ordinary SNM, the estimate

Table 9.11 Estimated relative risks and 95 % confidence intervals with ordinary SNM

Approach	RR(1)	RR(2)
Linear	0.44 (0.23, 0.83)	0.64 (0.40, 1.03)
Logistic	0.38 (0.17, 0.85)	0.67 (0.39, 1.14)
Loglinear	0.42 (0.17, 1.02)	0.67 (0.40, 1.14)

Table 9.12 Estimated relative risks and 95 % confidence intervals with weighted SNM

Approach	$RR^{W_1}(1)$	$RR^{W_1}(2)$
Linear	0.45 (0.24, 0.86)	0.66 (0.41, 1.08)
Logistic	0.41 (0.19, 0.89)	0.69 (0.40, 1.19)
Loglinear	0.40 (0.15, 1.03)	0.72 (0.39, 1.31)

for $A = 1$ was 0.51, whereas that for $A = 2$ was 0.27. Therefore more reduction in risk of absenteeism was possible in the schools with $A_i = 1$. However, when using Newton's method to solve the estimating equations at (9.4) with the loglinear or logistic ordinary generalized SNM approaches, our initial values for ξ and γ have to be close enough to the true ones. Otherwise, the iterative algorithm we use to solve the estimating equations at (9.4) may fail to converge for some jackknife samples.

7 Discussion

In this book chapter, we presented two methods based on structural nested models for the analysis of multi-armed cluster-randomized trials with unequal probabilities of sampling individuals. With the weighted structural nested model, we used individual-level covariates X_{ij} to construct weights in order to adjust for individual-level confounding. With the ordinary structural nested model, we included the individual covariates into our structural nested model. Software and programming schemes were provided for both weighted and ordinary structural nested models assuming different link functions (linear SNM, loglinear SNM, and logistic SNM). We also applied our methods to analyze the effect of adherence in the school-based WASH study. The computation is straightforward with the application of instrumental variables software (e.g., SAS PROC SYSLIN). With nonlinear link functions (e.g., loglinear SNM and logistic SNM), we can solve the estimating equations at (9.4) and (9.17) iteratively using Newton's method to linearize the canonical link functions. However, with the ordinary structural nested models, when we used Newton's method to solve the estimating equations, the starting values have to be close to the true ones, otherwise, we may have convergence issues. To construct confidence intervals, we discussed three different methodologies, sandwich estimator, bootstrap, and jackknife. However, the sandwich estimator of variance is difficult to program, and the bootstrap may fail to generate solutions of the estimating equations. Thus, we turned to the jackknife variance estimator.

To verify our methodologies, we conducted a simulation study for the ordinary structural nested model. A simulation study for the weighted structural nested model can be found in Brumback et al. [4]. Generally, the simulation results supported our methods, and jackknife variance estimations performed well. However, when using the ordinary SNM with nonlinear link functions (e.g., loglinear and logistic) to analyze data sets, the initial values have to be close enough to the real ones in order to avoid convergence issues.

References

1. Albert, J.M.: Estimating efficacy in clinical trials with clustered binary responses. Stat. Med. **21**, 649–661 (2002)
2. Angrist, J.D., Imbens, G.W., Rubin, D.B.: Identification of causal effects using instrumental variables. J. Am. Stat. Assoc. **91**, 444–455 (1996)
3. Bhattacharya, J., Goldman, D., McCaffrey, D.: Estimating probit models with self-selected treatments. Stat. Med. **25**, 389–413 (2006)
4. Brumback, B.A., He, Z.L., Prasad, M., Freeman, M.C., Rheingans, R.: Using structural-nested models to estimate the effect of cluster-level adherence on individual-level out-comes with a three-armed cluster-randomized trial. Stat. Med. **33**, 1490–1502 (2014)
5. Burgess, S., Collaboration, C.C.G.: Identifying the odds ratio estimated by a two-stage instrumental variable analysis with a logistic regression model. Stat. Med. **32**, 4726–4747 (2013)
6. Burgess, S., Thompson, S.G.: Improving bias and coverage in instrumental variable analysis with weak instruments for continuous and binary outcomes. Stat. Med. **31**, 1582–1600 (2012)
7. Cai, B., Small, D.S., Ten Have, T.R.: Two-stage instrumental variable methods for estimating the causal odds ratio: analysis of bias. Stat. Med. **30**, 1809–1824 (2011)
8. Cheng, J., Small, D.S.: Bounds on causal effects in three-arm trials with non-compliance. J. R. Stat. Soc. Ser. B Stat Methodol. **68**, 815–836 (2006)
9. Frangakis, C.E., Rubin, D.B.: Principal stratification in causal inference. Biometrics **58**, 21–29 (2002)
10. Freeman, M.C., Greene, L.E., Dreibelbis, R., Saboori, S., Muga, R., Brumback, B., Rheingans, R.: Assessing the impact of a school-based water treatment, hygiene and sanitation program on pupil absence in Nyanza province, Kenya: a cluster-randomized trial. Tropical Med. Int. Health **17**, 380–391 (2012)
11. Greenland, S.: An introduction to instrumental variables for epidemiologists. Int. J. Epidemiol. **29**, 1102–1102 (2000)
12. Hernan, M.A., Robins, J.M.: Instruments for causal inference - an epidemiologist's dream? Epidemiology **17**, 360–372 (2006)
13. Jo, B., Asparouhov, T., Muthen, B.O.: Intention-to-treat analysis in cluster randomized trials with noncompliance. Stat. Med. **27**, 5565–5577 (2008)
14. Jo, B., Asparouhov, T., Muthen, B.O., Ialongo, N.S., Brown, C.H.: Cluster randomized trials with treatment noncompliance. Psychol. Methods **13**, 1–18 (2008)
15. Jo, B., Stuart, E.A.: On the use of propensity scores in principal causal effect estimation. Stat. Med. **28**, 2857–2875 (2009)
16. Joffe, M.M., Brensinger, C.: Weighting in instrumental variables and G-estimation. Stat. Med. **22**, 1285–1303 (2003)
17. Johnston, K.M., Gustafson, P., Levy, A.R., Grootendorst, P.: Use of instrumental variables in the analysis of generalized linear models in the presence of unmeasured confounding with applications to epidemiological research. Stat. Med. **27**, 1539–1556 (2008)
18. Korhonen, P.A., Laird, N.M., Palmgren, J.: Correcting for non-compliance in randomized trials: an application to the ATBC study. Stat. Med. **18**, 2879–2897 (1999)

19. Long, Q., Little, R.J.A., Lin, X.H.: Estimating causal effects in trials involving multitreatment arms subject to non-compliance: a Bayesian framework. J. R. Stat. Soc.: Ser. C: Appl. Stat. **59**, 513–531 (2010)
20. Ma, Y., Roy, J., Marcus, B.: Causal models for randomized trials with two active treatments and continuous compliance. Stat. Med. **30**, 2349–2362 (2011)
21. Nagelkerke, N., Fidler, V., Bernsen, R., Borgdorff, M.: Estimating treatment effects in randomized clinical trials in the presence of non-compliance. Stat. Med. **19**, 1849–1864 (2000)
22. Rassen, J.A., Schneeweiss, S., Glynn, R.J., Mittleman, M.A., Brookhart, M.A.: Instrumental variable analysis for estimation of treatment effects with dichotomous outcomes. Am. J. Epidemiol. **169**, 273–284 (2009)
23. Robins, J.M.: Correcting for noncompliance in randomized trials using structural nested mean models. Commun. Stat. Theory Methods **23**, 2379–2412 (1994)
24. Robins, J.M.: Correction for non-compliance in equivalence trials. Stat. Med. **17**, 269–302 (1998)
25. Rothman, K.J., Greenland, S., Lash, T.L.: Modern Epidemiology, 3rd edn. Wolters Kluwer Health/Lippincott Williams and Wilkins, Philadelphia (2008)
26. Small, D.S., Ten Have, T.R., Joffe, M.M., Cheng, J.: Random effects logistic models for analyzing efficacy of a longitudinal randomized treatment with non-adherence. Stat. Med. **25**, 1981–2007 (2006)
27. Ten Have, T.R., Joffe, M., Cary, M.: Causal logistic models for non-compliance under randomized treatment with univariate binary response. Stat. Med. **22**, 1255–1283 (2003)
28. Vansteelandt, S., Goetghebeur, E.: Causal inference with generalized structural mean models. J. R. Stat. Soc. Ser. B Stat. Methodol. **65**, 817–835 (2003)
29. Vansteelandt, S., Goetghebeur, E.: Sense and sensitivity when correcting for observed exposures in randomized clinical trials. Stat. Med. **24**, 191–210 (2005)
30. Vansteelandt, S., Bowden, J., Babanezhad, M., Goetghebeur, E.: On instrumental variables estimation of causal odds ratios. Stat. Med. **26**, 403–422 (2011)
31. Wooldridge, J.M.: Econometric Analysis of Cross Section and Panel Data. MIT Press, Cambridge, MA (2002)

Chapter 10
Causal Models for Randomized Trials with Continuous Compliance

Yan Ma and Jason Roy

Abstract In behavioral medicine trials, such as smoking cessation trials, 2 or more active treatments are often compared. Noncompliance by some subjects with their assigned treatment poses a challenge to the data analyst. In particular, the joint distribution of the observed and counterfactual compliance variables cannot be identified, without imposing strong assumptions. However, due to randomization, each marginal compliance distribution can be identified. These marginal distributions impose bounds on the joint distribution. Our approach is to use a copula model to link the two marginal distributions, up to a sensitivity parameter. We then take a principal stratification approach to estimate causal effects. We develop this approach when compliance is either binary (yes/no) or continuous (dose).

1 Introduction and Motivating Example

We consider the situation where subjects were randomized to one of two active treatments, and compliance with each treatment was measured on a continuous scale. Examples of continuous measures of compliance include the duration of compliance and the proportion of assigned treatment actually received.

The motivating example for this research was a smoking cessation clinical trial. The Commit to Quit (CTQ) trials [1–3] comprise two longitudinal follow-up studies of supervised exercise to promote smoking cessation. One arm included cognitive-behavioral smoking cessation therapy (CBT) augmented by an individualized, supervised exercise program. In the control arm, CBT was augmented by a wellness education program that included lectures, films, handouts, and discussions covering issues such as healthy eating and prevention of cardiovascular disease. Interest is in

Y. Ma (✉)
Department of Epidemiology and Biostatistics, The George Washington University, Washington, DC, USA
e-mail: yanma@gwu.edu

J. Roy
Center for Clinical Epidemiology and Biostatistics, University of Pennsylvania, Philadelphia, PA, USA

H. He et al. (eds.), *Statistical Causal Inferences and Their Applications in Public Health Research*, ICSA Book Series in Statistics, DOI 10.1007/978-3-319-41259-7_10

the comparison between standard therapy augmented by wellness education and standard therapy augmented by an exercise regimen. However, many subjects only attended some of the exercise or wellness classes.

One approach that has been proposed for inferring causal effects from trials with non-compliance is structural mean models [4–6]. A structural model must be specified that relates the mean of the outcome at the observed compliance level with the mean of the potential outcome at some reference level. For example, the reference outcome could be the treatment-free outcome. These methods have primarily been used for placebo-controlled trials. An appealing aspect of that approach is causal effects can be inferred for the entire population at various compliance levels. However, for two-arm behavioral intervention trials it is difficult to conceive of fixing compliance levels for the entire population. Whatever incentives would have to be used to entice non-compliers with exercise into perfect compliance might also affect the outcome directly. Further, interest is in comparing the two interventions in their current form. If the interventions were implemented in practice, which one would yield better results overall (the intention-to-treat effect)? What is the extent to which the causal effect depends on the compliance level with each arm? We propose answering these questions using principal stratification [7].

The principal stratification approach stratifies the population based on what each subject's compliance status would be under assignment to each of the two treatments. Thus, only compliance levels that could have been observed in the particular trial are considered. A difficulty with this approach is it stratifies on the pair of potential compliance variables, only one of which is observed. Thus, assumptions about the joint compliance distribution are necessary.

Efron and Feldman [8] assumed the percentiles of the two potential compliance variables in a placebo-controlled trial would be the same for each subject. Jin and Rubin [9] considered the same situation, but instead proposed a weaker assumption—side-effect monotonicity. Essentially, the assumption is that side-effects would be greater for the drug compared to placebo, and therefore compliance would not be smaller in the placebo arm. However, in behavioral intervention trials, it is not difficult to imagine scenarios where different patients have different intervention preferences. We therefore deal with the problem from a different perspective, and explicitly model the joint distribution of compliance. The joint distribution of potential values of compliance is modeled by linking the two marginal distributions using a copula model [10]. This is similar to the approach taken by Bartolucci and Grilli [11]. One difference between the approaches is we use a Gaussian copula and they use a Plackett copula [12]. In a simulation study, we explore the sensitivity to the choice of copula [13]. Another difference is they estimate the copula association parameter, which cannot be identified non-parametrically, using a profile likelihood approach. Instead, we vary the copula parameter as part of a sensitivity analysis. In addition, we propose to estimate causal effects not just at particular points, but within regions of the compliance distributions. In that way, our approach is similar to the work of Dominici et al.[14]. A key difference is that our approach allows the outcome to be from any exponential family distribution. We also propose some computational simplifications for maximum likelihood estimation.

For the CTQ trial, we measure compliance by the proportion of assigned classes that were actually attended. We then consider the effect of treatment assignment among interesting subpopulations. Examples could include: the subpopulation that would be perfectly compliant with either treatment; the subpopulation that would be highly compliant with either treatment; the subpopulation that would be highly compliant with wellness but not exercise.

The remainder of the chapter is organized as follows. In Sect. 2 we introduce the notation, assumptions, and models. In Sect. 3 we describe our estimation procedure. The example is presented in Sect. 4. Finally, we conclude with a discussion in Sect. 5.

2 The Structural Principal Effects Model

2.1 Notation and Assumptions

We consider experimental trials with two active treatments. Let $R \in \{0, 1\}$ denote a randomization indicator, where $R = 1$ indicates randomization to the new treatment (e.g., supervised exercise plus CBT), and $R = 0$ indicates randomization to standard therapy (e.g., wellness sessions plus CBT). Let A_r denote compliance with *assigned* treatment under assignment r. We assume A_r is continuous and possibly bounded (e.g., the proportion of assigned treatment actually taken). Similarly, define Y_r to be the outcome under assignment r. Each person has two potential compliance levels, A_0 and A_1, that characterize compliance under either treatment assignment; however only $A = RA_1 + (1 - R)A_0$ is observed. Similarly, each subject has two potential outcomes, Y_0 and Y_1, with $Y = RY_1 + (1 - R)Y_0$ observed.

We make three standard assumptions for the development of analytic methods: (1) the stable unit treatment value assumption (SUTVA), which is the assumption that the value of the potential outcomes and potential compliance variables for subject i only depend on the treatment assigned to subject i, not on the treatment assigned to other subjects; (2) randomization ($R \perp\!\!\!\perp \{Y_0, Y_1, A_0, A_1\}$); and (3) the exclusion restriction. SUTVA essentially states that there is no interference between subjects. Randomization requires that treatment assignment was unrelated to potential outcomes. The exclusion restriction is the assumption that treatment assignment affects the outcome entirely through its affect on treatment received; knowledge of treatment assignment alone will not affect the outcome. These assumptions have been described in detail elsewhere [15]. We make one additional assumption that subjects in group $R = r$ do not have access to the treatment assigned in arm $R = 1 - r$, for $r = 0, 1$. This assumption has been referred to as the "treatment access restriction" [16] or "strong treatment access monotonicity" [9]. Without the treatment access restriction assumption, we would have four potential compliance levels rather than two: compliance to treatment j if assigned to treatment k, for $j, k \in \{0, 1\}^2$. The assumption holds for the CTQ study, as subjects were not allowed to attend sessions to which they were not assigned.

Other identifying assumptions could be made as well. For example, a side-effect monotonicity assumption [9], such as $A_1 \leq A_0$ for every subject, could be specified. This states that every subject would be less compliant with treatment 1 than they would have been with treatment 0. The assumption is plausible in many placebo-controlled trials, but is unrealistic in many behavioral intervention trials (and therefore will not be used in our analysis).

2.2 *Causal Effects*

Our interest is in the intention to treat (ITT) effects in subpopulations that have similar compliance behavior. Let $\mu(r, a_0, a_1) = \mathrm{E}(Y_r | A_0 = a_0, A_1 = a_1) = \mathrm{E}(Y | R = r, A_0 = a_0, A_1 = a_1)$. We are interested in comparing quantities $\mu(1, a_0, a_1)$ and $\mu(0, a_0, a_1)$. Such comparisons have been called principal effects [7]. For example, if compliance was a proportion of assigned dose actually received, investigators might be interested in the comparison $\mu(1, 1, 1)$ and $\mu(0, 1, 1)$, which is the causal effect of assignment to new treatment compared to standard treatment, among perfect compliers.

Even if both A_0 and A_1 were observed for all subjects, we would not be able to identify $\mu(r, a_0, a_1)$ non-parametrically. Notice that this differs from the binary (yes/no) compliance case, where the principal effects could be identified if the potential compliance variables were known. With continuous measures of compliance, additional structure is needed to identify $\mu(r, a_0, a_1)$. One possibility would be just estimate the causal effects within regions of the compliance space. Alternatively, fully parametric (e.g., linear model) or semi-parametric (e.g., smoothing spline) could be specified.

In order to identify the causal parameters, we propose the use of a structural model for the principal effects. We assume the data Y_i given A_{i0} and A_{i1} are from an exponential family with distribution

$$f(y_i) = \exp[\{y_i \eta_i - \psi(\eta_i)\}/(m_i \omega) + h(y_i, \omega)]$$

where $\mathrm{E}(Y_i) = g^{-1}(\eta_i) = \psi'(\eta_i)$, η_i is the linear predictor, $\psi(\cdot)$ is a known function, ω is a scale parameter, and m_i is the prior weight. This family includes normal ($\psi(x) = x^2/2$), binomial ($\psi(x) = \log(1 + e^x)$), and Poisson ($\psi(x) = e^x$) distributions, among others. We then model $\mathrm{E}(Y_i)$ as a function of the potential compliance variables and regression coefficients β. We denote this by

$$\mu(r, a_0, a_1; \beta) = g^{-1}\{\eta(r, a_0, a_1; \beta)\}, \tag{10.1}$$

where $g()$ is a link function and η is a linear predictor. For example, if Y is binary, one might specify a logistic model $\mu(r, a_0, a_1; \beta) = \mathrm{logit}^{-1}(\beta_0 + \beta_1 a_0 + \beta_2 a_1 + \beta_3 r a_0 + \beta_4 r a_1)$. By the exclusion restriction, the model should be specified so that

$\mu(1, 0, 0; \beta) = \mu(0, 0, 0; \beta)$. Baseline covariates X could potentially be included in the model by including, for example, $\beta_x^T x$ as an additional term in the linear predictor.

2.3 Compliance Distributions

We next consider the joint distribution of the potential compliance variables, A_0 and A_1. Because only one of A_0 and A_1 is observed for each subject, there is an identifiability problem. The identifiability problem would persist, even if compliance was binary [16, 17]. For the continuous case the problem is even more pronounced.

Due to the randomization assumption, we can identify the two marginal distributions $f(a_0) = f(a|r = 0)$ and $f(a_1) = f(a|r = 1)$. Our strategy therefore is to first specify the two marginal models, which are identifiable. We will then link the two marginal distributions using a copula model, which will include an association parameter that cannot be identified.

For the marginal models, we assume that $f(a_r)$ follows a parametric distribution. For example, if A_r is continuous and bounded in the interval $(0, 1)$, as would be the case if compliance is a proportion, then a Beta distribution might be appropriate. One could include covariates via a beta regression model [18], with $E(A_r)=\text{logit}^{-1}(X^T\gamma_r)$ and scale parameter ϕ_r. Alternatively, if compliance is an unbounded positive scalar (a duration), then a gamma distribution might be appropriate. Standard diagnostic techniques can be used to check the adequacy of the assumptions.

To link the two marginal distributions, which are identifiable, to the joint distribution, we propose the use of a Gaussian copula model with correlation ρ [10]. To make notation easier to follow, we will use the following shorthand notation for the marginal compliance cumulative distribution functions (CDFs), $F_{A_0}(a_0) = F_{A_0}(a_0; \gamma_0, \phi_0)$ and $F_{A_1}(a_1) = F_{A_1}(a_1; \gamma_1, \phi_1)$. We will use similar notation for the joint CDF. The joint distribution is then

$$F_{A_0,A_1}(a_0, a_1) = \Phi_2\left[\Phi_1^{-1}\{F_{A_0}(a_0)\}, \Phi_1^{-1}\{F_{A_1}(a_1)\}\right]$$

where Φ_1 is the univariate standard normal CDF and Φ_2 is the bivariate normal CDF with mean $(0, 0)^T$, variance $(1, 1)^T$ and correlation ρ. Essentially, this implies that the joint distribution $f(z_0, z_1)$ is bivariate normal with correlation ρ, where $z_0 = \Phi_1^{-1}\{F_{A_0}(a_0)\}$ and $z_1 = \Phi_1^{-1}\{F_{A_1}(a_1)\}$. Therefore, the joint distribution $f(a_0, a_1)$ is

$$\begin{aligned} f(a_0, a_1) &= f(z_0, z_1)\det(J) \\ &= \frac{1}{2\pi\sqrt{1-\rho^2}} \cdot \exp\left\{-\frac{z_0^2 - 2\rho z_0 z_1 + z_1^2}{2(1-\rho^2)}\right\} \cdot \det(J), \end{aligned}$$

where J is the Jacobian and $f()$ is a density function.

There is no information in the data about the parameter ρ, as it represents the association between two variables that are never observed simultaneously. We will therefore treat ρ as known and vary it as part of a sensitivity analysis.

Special Case Assuming $\rho = 1$ in the marginal models for A_0 and A_1 would be equivalent to the "equipercentile equating of compliances" assumption of Efron and Feldman [8] and discussed in Jin and Rubin [9].

3 The Likelihood and Inference Methods

3.1 Likelihood

Without loss of generality, suppose the first n_0 subjects are in group $R = 0$ and the next n_1 subjects are in group $R = 1$ ($n = n_0 + n_1$). The likelihood function involves integrating out missing data from the complete data likelihood. The likelihood contribution for a subject in group $R = 0$ can be written

$$L_{i0}(\beta, \gamma_0, \gamma_1, \phi_0, \phi_1; \rho) = \int f(y_{i0}|a_{i0}, a_{i1}; \beta) f(a_{i0}, a_{i1}; \gamma_0, \gamma_1, \phi_0, \phi_1, \rho) da_{i1} \tag{10.2}$$

where the distributions $f(y_{i0}|a_{i0}, a_{i1}; \beta)$ and $f(a_{i0}, a_{i1}; \gamma_0, \gamma_1, \phi_0, \phi_1, \rho)$ were defined previously. Define $L_{i1}(\beta, \gamma_0, \gamma_1, \phi_0, \phi_1; \rho)$ similarly for subjects in arm $R = 1$, except there A_{i0} is integrated out of the likelihood. The loglikelihood is therefore $\log L(\beta, \gamma_0, \gamma_1, \phi_0, \phi_1; \rho) = \sum_{i=1}^{n_0} \log L_{i0}(\beta, \gamma_0, \gamma_1, \phi_0, \phi_1; \rho) + \sum_{i=n_0+1}^{n} \log L_{i1}(\beta, \gamma_0, \gamma_1, \phi_0, \phi_1; \rho)$.

3.2 Estimation: Two-Stage Approach

Maximizing the full likelihood can be computationally intensive, due to numerical integration and optimization procedures. We propose a two-stage estimation approach as an alternative, which is potentially faster and more stable than full MLE. In the first stage, the parameters from the marginal compliance distributions are estimated using only the compliance data. For example, γ_0 and ϕ_0 are estimated by maximizing the likelihood corresponding to the distribution $f(a_0; \gamma_0, \phi_0) = f(a|R = 0; \gamma_0, \phi_0)$. Similar calculations are carried out for the $R = 1$ group. In the second stage, we find the values of β that maximize $\log L(\beta, \hat{\gamma}_0, \hat{\gamma}_1, \hat{\phi}_0, \hat{\phi}_1; \rho)$, where the estimated values of the compliance parameters from stage one are plugged in to the likelihood.

The right-hand side of (10.2) will typically not have closed form. We therefore look to approximate the integral, before proceeding with estimation. To approximate the integration, we first use a transformation, in order to take advantage of the fact that $[\Phi_1^{-1}\{F_{A_{i0}}(a_{i0})\}, \Phi_1^{-1}\{F_{A_{i1}}(a_{i1})\}]$ follows a bivariate normal distribution. The right-hand side of (10.2) can be written as

$$\int_{-\infty}^{\infty} f\left[y_{i0}\middle|a_{i0}, a_{i1} = F_{A_{i1}}^{-1}\{\Phi_1(z_{i1})\}; \beta\right] f(a_{i0}) f(z_{i1}|z_{i0}; \rho) dz_{i1} \tag{10.3}$$

where $z_{i0} = \Phi_1^{-1}\{F_{A_{i0}}(a_{i0})\}$. The distribution of $[z_{i0}]$ is standard normal and $[z_{i1}|z_{i0}; \rho] \sim N(\rho z_{i0}, 1-\rho^2)$. We can then approximate (10.3) using a Gauss–Hermite quadrature as follows:

$$\widetilde{L}_{i0}(\beta, \hat{\gamma}_0, \hat{\gamma}_1, \hat{\phi}_0, \hat{\phi}_1; \rho) = \sum_{j=1}^{J} f\left[y_{i0}\middle|a_{i0}, a_{i1} = F_{A_{i1}}^{-1}\{\Phi_1(z_{i1}^j)\}; \beta\right] f(a_{i0}) w^j$$

where $(z_{i1}^j - \rho z_{i0})/\sqrt{1-\rho^2}$ is the jth of J nodes from a standard normal distribution and w^j is the corresponding weight. Typically, $J = 10$ points provides sufficient accuracy. We define $\widetilde{L}_{i1}(\beta, \hat{\gamma}_0, \hat{\gamma}_1, \hat{\phi}_0, \hat{\phi}_1; \rho)$ analogously. Our approximation of the loglikelihood is

$$\tilde{l}(\beta, \hat{\gamma}_0, \hat{\gamma}_1, \hat{\phi}_0, \hat{\phi}_1; \rho) = \sum_{i=1}^{n_0} \log \widetilde{L}_{i0}(\beta, \hat{\gamma}_0, \hat{\gamma}_1, \hat{\phi}_0, \hat{\phi}_1; \rho) + \sum_{i=n_0+1}^{n} \log \widetilde{L}_{i1}(\beta, \hat{\gamma}_0, \hat{\gamma}_1, \hat{\phi}_0, \hat{\phi}_1; \rho).$$

We estimate the parameters β by maximizing $\tilde{l}(\beta, \hat{\gamma}_0, \hat{\gamma}_1, \hat{\phi}_0, \hat{\phi}_1; \rho)$. Variances are estimated by inverting the observed information matrix, which is approximated using numerical differentiation.

Intuitively, there should be little information in the outcome data about the compliance distribution parameters. Therefore, we do not anticipate much loss in efficiency with the two-stage approach. The gain in computational efficiency comes from estimating fewer parameters in the likelihood that requires integration. The performance of the two-stage estimator is explored in a simulation study [13].

3.3 Estimation of Effects in Compliance Regions

The ITT effect for a particular combination of the compliance variables might be of limited interest, primarily because very few subjects would have potential compliance equal to those two values. Therefore, researchers might also be interested in causal effects within certain regions defined by a range of values of the two compliance variables. Consider the situation where A_0 and A_1 represent the proportion of assigned treatment actually taken (so that 0 represents non-compliance and 1 represents perfect compliance). Suppose we would like to estimate $E(Y_1 - Y_0|A_0, A_1 \in \Theta)$, where Θ is some region of $[0, 1]^2$. Consider the following examples. The region $\Theta = \{A_0, A_1 \in [0.7, 1], |A_1 - A_0| < 0.2\}$ includes subjects who would be at least 70 % compliant with either treatment, and whose compliance level would not differ by more than 20 % between the two arms. This region would be of interest if investigators wanted to know the ITT effect among highly compliant subjects. The region $\Theta = \{A_0 > 0.7, A_1 < 0.3\}$ would include subjects who would be highly compliant with treatment $Z = 0$ but poorly compliant with treatment $Z = 1$. In the CTQ example, this would include subjects that seemed to prefer wellness to exercise. Finally, one could consider $\Theta = \{A_1 > 0.8\}$. In our example, this would include subjects who would be at least 80 % compliant with exercise, regardless of how compliant they would be with wellness. One could imagine many regions Θ that might be of interest. Once the model parameters are estimated, causal effects within a region Θ can then be estimated in a separate step. We next provide the computational details for a specific example.

In general, the causal effect is

$$E(Y_1 - Y_0|a_0, a_1 \in \Theta) = \lim_{n\to\infty} \frac{1}{n}\sum_i E(Y_{i1} - Y_{i0}|a_{i0}, a_{i1} \in \Theta)$$
$$\approx \frac{1}{n}\sum_i E(Y_{i1} - Y_{i0}|a_{i0}, a_{i1} \in \Theta),$$

where the approximation holds for large n, and

$$E(Y_{i1} - Y_{i0}|a_{i0}, a_{i1} \in \Theta) = \frac{\int_\Theta E(Y_{i1} - Y_{i0}|a_{i0}, a_{i1}) f(a_{i1}, a_{i0}) da_{i1} da_{i0}}{\int_\Theta f(a_{i1}, a_{i0}) da_{i1} da_{i0}}.$$

Suppose we would like to estimate $E(Y_1 - Y_0|A_0, A_1 \in \Theta)$ where $\Theta = \{A_0, A_1 \in [0.7, 1], |A_1 - A_0| < 0.2\}$. This group might be of interest because it represents a group of people who would take a similar dose of either treatment, and would receive most of their assigned treatment. We can approximate the double integrals in the numerator and denominator using Gauss–Hermite quadrature. For example, the numerator can be written as follows:

$$\int_\Theta E(Y_{i1} - Y_{i0}|a_{i0}, a_{i1}) f(a_{i0}, a_{i1}) da_{i1} da_{i0}$$
$$= \int_{0.7}^{1} \int_{\max(0.7, a_{i1}-0.2)}^{\min(1, a_{i1}+0.2)} \Delta(a_{i0}, a_{i1}; \hat{\beta}) f(a_{i0}, a_{i1}) da_{i0} da_{i1},$$

where $\Delta(a_{i0}, a_{i1}; \hat{\beta}) = \mu(1, a_{i0}, a_{i1}; \hat{\beta}) - \mu(0, a_{i0}, a_{i1}; \hat{\beta})$. After a transformation of variables, the above expression can be written

$$\int_{l_1}^{u_1} \int_{l_0(z_{i1})}^{u_0(z_{i1})} \Delta\left[F_{A_{i0}}^{-1}\{\Phi_1(z_{i0})\}, F_{A_{i1}}^{-1}\{\Phi_1(z_{i1})\}; \hat{\beta}\right] f(z_{i0}, z_{i1}) dz_{i0} dz_{i1} \qquad (10.4)$$

where $l_1 = \Phi_1^{-1}\{F_{A_{i1}}(0.7)\}$, $u_1 = \Phi_1^{-1}\{F_{A_{i1}}(1)\}$,

$$l_0(z_{i1}) = \max\left[\Phi_1^{-1}\left\{F_{A_{i0}}(0.7)\right\},\ \Phi_1^{-1}\left\{F_{A_{i0}}\left(F_{A_{i1}}^{-1}(\Phi_1(z_{i1})) - 0.2\right)\right\}\right]$$

and

$$u_0(z_{i1}) = \min\left[\Phi_1^{-1}\left\{F_{A_{i0}}(1)\right\},\ \Phi_1^{-1}\left\{F_{A_{i0}}\left(F_{A_{i1}}^{-1}(\Phi_1(z_{i1})) + 0.2\right)\right\}\right].$$

In the above expressions, parameters are replaced by their MLEs, e.g., $F_{A_{i0}}(0.7) = F_{A_{i0}}(0.7; \hat{\gamma}_0, \hat{\phi}_0)$.

The joint distribution $f(z_{i0}, z_{i1})$ is bivariate normal with mean 0, variance 1 and correlation ρ. This joint distribution can also be written as $f(z_{i0}|z_{i1})f(z_{i1})$, where $[z_{i0}|z_{i1}] \sim N(\rho z_{i1}, 1 - \rho^2)$ and $[z_{i1}] \sim N(0, 1)$. Thus, we can apply a Gaussian quadrature to approximate the integral in (10.4). In particular, (10.4) can be approximated by

$$\sum_{k=1}^{K}\sum_{j=1}^{J} \Delta\left[F_{A_{i0}}^{-1}\left\{\Phi_1(z_{i0}^j)\right\}, F_{A_{i1}}^{-1}\left\{\Phi_1(z_{i1}^k)\right\}; \hat{\beta}\right] I\left[z_{i0}^j \in \{l_0(z_{i1}^k), u_0(z_{i1}^k)\}, z_{i1}^k \in (l_1, u_1)\right] w_0^j w_1^k,$$

where $I()$ is the indicator function, z_{i1}^k and $(z_{i0}^j - \rho z_{i1}^k)/\sqrt{1-\rho^2}$ are nodes from a standard normal distribution, and w_1^k and w_0^j are weights. The totals J and K are selected to ensure an adequate (e.g., 10) number of valid nodes.

4 Example: Commit to Quit Trial

4.1 Data

The CTQ study [1] was a randomized controlled trial designed to assess the efficacy of supervised vigorous exercise as an adjuvant to cognitive behavioral therapy (CBT) for promotion of smoking cessation among women. The study enrolled and assigned 134 women to receive CBT plus vigorous exercise (the new treatment) and 147 to receive CBT plus a wellness education program (the control treatment). CBT represents the standard of care for smoking cessation; the wellness education was added to the control arm to equalize staff contact time between the two arms. The CBT program was administered to all women in group format weekly over the course of 12 weeks. The exercise program was supervised, and individually tailored to each woman based on achieving a target heart rate. Women in the control arm participated in a program of supervised lectures, films, and discussions. Both the wellness and exercise interventions were held three times per week. None of the women in the control arm had access to the supervised exercise program, and none in the exercise group had access to wellness classes.

Cessation status was evaluated weekly, assessed by self-report and verified by carbon monoxide (<8 ppm) and saliva cotinine (<10 ng/mL), over the course of 12 weeks. To be considered *abstinent*, an individual needed to submit to testing and meet both the carbon monoxide and saliva cotinine criteria.

The target quit date was week 5 following randomization. The primary outcome of the study was continuous abstinence during the 8 weeks after the quit date. By definition, an individual who was not present for scheduled testing at one or more occasions could not be counted as continuously abstinent. The compliance variable was defined during the pre-quit date period (weeks 1–4). Specifically, we defined A_r as the proportion of the 12 classes that were attended, where $A_r \in [0, 1]$, $r = 1$ for the exercise training and $r = 0$ for the wellness training.

4.2 Model Specification

We next describe the specific models that were fitted to the CTQ data.

Compliance Model Because A_r was a proportion (approximately continuous, bounded between 0 and 1), we specified beta distributions for the compliance variables A_r. To assess the fit of the models, we plotted the empirical probability density function (PDF) and the model-based PDF. Specifically, the empirical PDF was obtained as a histogram of A_0 and A_1, using ten bins of size 0.1 each. The model-based PDF was based on the assumption that A_r follows a beta distribution with parameters γ_r and α_r estimated by maximizing the likelihood. This model-based estimated PDF was smoothed over the distribution of a_r. The empirical and model-based PDFs are plotted in Fig. 10.1 for $\rho = 0.1$ (the figure looks the same for other values of ρ, as ρ affects the joint distribution, but not the marginal). The marginal PDFs that were estimated from the model appear to capture the key features of the data. For example, in the wellness arm, the function appears to decrease initially, then increase as A_0 approaches 1. The distribution of A_1 appears to be an increasing function.

Principal Effects To model the causal effects $\mu(r, a_0, a_1)$ in Sect. 2.2, we assumed the distribution $f(Y_r|A_0, A_1; \beta)$ was Bernoulli with

$$P(Y_r = 1|a_0, a_1) = \texttt{logit}^{-1}(\beta_{0r} + \beta_{1r}a_0 + \beta_{2r}a_1),$$

for $r = 0, 1$. Recall that Y_r is the indicator that a subject assigned treatment r would abstain from smoking during the final 8 weeks of the trial. The exclusion restriction implies that randomization should have no impact among non-compliers (those with $A_0 = A_1 = 0$), and thus $\beta_{00} = \beta_{01}$. We believe this assumption is plausible for the CTQ study. However, as part of a sensitivity analysis we allow the two intercepts to vary. We define δ as the ratio of risks among subjects who would be non-compliant with either intervention, i.e.,

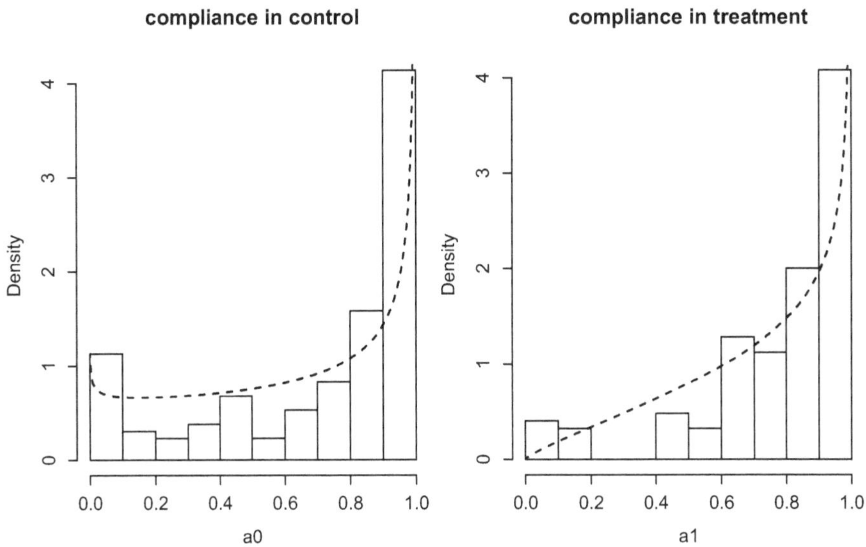

Fig. 10.1 Empirical PDF (*histogram*) and model-based estimated PDF (*dashed line*) of compliance from the Commit to Quit trial

$$\delta = \frac{P(Y_1 = 1|A_0 = 0, A_1 = 0)}{P(Y_0 = 1|A_0 = 0, A_1 = 0)}.$$

We can write β_{01} as a function of β_{00} and δ: $\beta_{01} = \log\left(\delta/(1 + e^{-\beta_{00}} - \delta)\right)$. We propose to estimate β_{00} and fix the value of δ, which will determine the value of β_{01}. The exclusion restriction implies $\delta = 1$. We also consider δ equal to 1.2 and 1/1.2. For example, $\delta = 1.2$ implies that, among subjects that would not comply with either intervention, the risk of the outcome is 20 % greater if randomized to $Z = 1$ (exercise).

The model also relies on the assumption that the effects of A_0 and A_1 are linear and additive on the logit scale. Non-linear terms or interactions could also be specified, but we found that inference was relatively unaffected by these added complexities.

We fitted the models at values of ρ equal to 0.1, 0.5, and 0.9. These values represent three values within the range of plausible values of ρ. Recall that ρ is the correlation between transformed values of A_0 and A_1. Independence between A_0 and A_1 would occur if $\rho = 0$. We believe it is unlikely that A_0 and A_1 are independent, as, for example, a subject might miss a visit for personal reasons that are unrelated to the treatment itself. We also believe that negative correlation is unlikely. We therefore focus on positive values of ρ, while acknowledging that we cannot rule out 0 or negative values.

4.3 Results

Estimation In order to obtain the estimates of the parameters, we maximized the likelihood in two stages using the approach described in Sect. 3.2. Shown in Table 10.1 are the estimates of the parameters from wellness/exercise compliance models, and from the principal effects model. The results are only displayed for one value of ρ ($\rho = 0.1$), as the compliance parameter estimates have no dependence on ρ, and the principal effects model parameters are only of secondary interest.

To get an idea of what affect ρ has on the joint distribution of the compliance variables, we simulated 500 values from marginal Beta distributions, using the parameter values from Table 10.1, at ρ equal to 0.1 and 0.9. The results are displayed in Fig. 10.2. The plot demonstrates that there is less information in the data about causal effects near the diagonal when ρ is small. When ρ is close to 1, there

Table 10.1 Maximum likelihood estimates (standard error) from Commit to Quit trial when $\rho = 0.1$ and $\delta = 1$

Wellness compliance model	
γ_0	0.60 (0.10)
α_0	−0.28 (0.10)
Exercise compliance model	
γ_1	1.02 (0.11)
α_1	−0.75 (0.11)
Principal effects model	
β_{00}	−7.66 (10.60)
β_{10}	6.87 (6.14)
β_{20}	−0.24 (20.62)
β_{11}	0.54 (15.17)
β_{21}	6.97 (2.51)

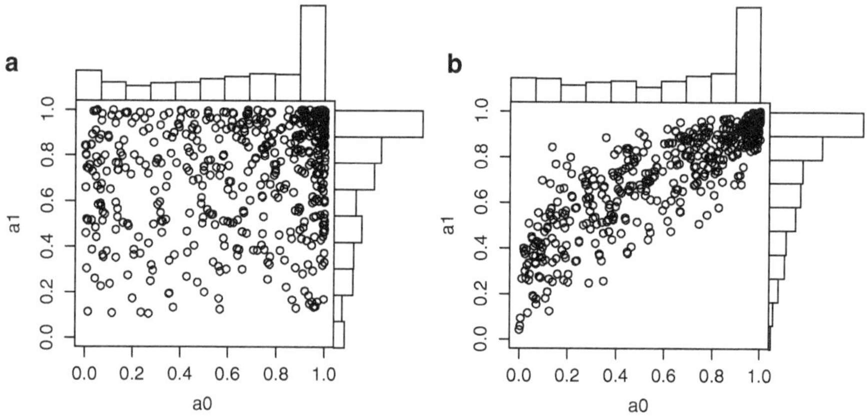

Fig. 10.2 Plots of 500 simulated values of a_0 and a_1 from a copula model with Beta marginal distributions estimated from the Commit to Quit trial and correlation. (**a**) $\rho = 0.1$ and (**b**) $\rho = 0.9$

are regions in the graph with little to no data. For example, inference about the population of people who would be highly compliant with exercise and poorly compliant if assigned to wellness would be based on extrapolation if ρ is 0.9.

Causal Effects We next consider parameters from the causal model. First, we focus on principal effects in subpopulations that would have the same compliance in either treatment, i.e., $\Delta_1 = \mu(1, a, a) - \mu(0, a, a)$ where $a_0 = a_1 = a$. We estimated the effects by plugging the MLEs of β into the following formula

$$\texttt{logit}^{-1}\{\beta_{01} + (\beta_{11} + \beta_{12})a\} - \texttt{logit}^{-1}\{\beta_{00} + (\beta_{01} + \beta_{02})a\}$$

In Table 10.2 we present the estimated principal effects $\widehat{\Delta}_1$ and their standard errors at compliance levels ≥ 0.7 when $\rho = 0.1$, 0.5, and 0.9 and $\delta = 0.83$, 1, and 1.2. For all values of ρ, the causal effects increased as the compliance level

Table 10.2 Estimated causal effects from Commit to Quit trial

$\delta = 1$	$\rho = 0.1$	$\rho = 0.5$	$\rho = 0.9$
Compliance level	$\widehat{\Delta}_1$ (se)		
0.7	0.038 (0.05)	0.062 (0.033)	0.06(0.03)
0.8	0.075 (0.14)	0.13(0.093)	0.11(0.05)
0.9	0.13 (0.29)	0.25 (0.23)	0.19(0.11)
1.0	0.20 (0.36)	0.38 (0.39)	0.28(0.18)
Compliance region	$\widehat{\Delta}_2$ (se)		
Θ	0.11 (0.33)	0.24 (0.26)	0.20 (0.10)
$\delta = 1.2$	$\rho = 0.1$	$\rho = 0.5$	$\rho = 0.9$
Compliance level	$\widehat{\Delta}_1$ (se)		
0.7	0.037 (0.045)	0.064 (0.031)	0.059(0.033)
0.8	0.076 (0.15)	0.13(0.091)	0.11(0.049)
0.9	0.13 (0.32)	0.25 (0.23)	0.19(0.11)
1.0	0.19 (0.38)	0.38 (0.40)	0.28 (0.19)
Compliance region	$\widehat{\Delta}_2$ (se)		
Θ	0.12 (0.36)	0.24 (0.26)	0.20 (0.10)
$\delta = 1/1.2$	$\rho = 0.1$	$\rho = 0.5$	$\rho = 0.9$
Compliance level	$\widehat{\Delta}_1$ (se)		
0.7	0.033 (0.05)	0.06 (0.034)	0.054(0.032)
0.8	0.076 (0.16)	0.13(0.095)	0.11(0.051)
0.9	0.13 (0.32)	0.24 (0.23)	0.19(0.11)
1.0	0.19 (0.38)	0.38 (0.40)	0.28(0.18)
Compliance region	$\widehat{\Delta}_2$ (se)		
Θ	0.12 (0.36)	0.24 (0.26)	0.20 (0.095)

Reported are $\widehat{\Delta}_1$ and $\widehat{\Delta}_2$ when $\rho = 0.1$, 0.5 and 0.9, and $\delta = 0.83$, 1, and 1.2, where $\Delta_1 = \{\mu(1, a_0, a_1) - \mu(0, a_0, a_1) | a_0 = a_1\}$, $\Delta_2 = \{\mu(1, a_0, a_1) - \mu(0, a_0, a_1) | a_0, a_1 \in \Theta\}$ and $\Theta = \{A_0, A_1 \in [0.7, 1], |A_1 - A_0| < 0.2\}$

increased. There were no prominent causal effects (effects are less than 0.01) for the compliance levels below 0.6 (results not displayed). Estimated causal effects were greater than 0.1 for compliance levels 0.9 or above. While the point estimates suggest a benefit from exercise when compliance is high, the evidence was only strong (estimate about twice as large as SE) when $\rho = 0.9$. Causal effect estimation at the compliance levels that we focused on was insensitive to variations in δ within the range of values that we considered (0.83–1.2).

Causal Effects in Compliance Regions Finally, we consider estimation of $\Delta_2 = E(Y_1 - Y_0 | a_0, a_1 \in \Theta)$ where $\Theta = \{a_0, a_1 \in [0.7, 1], |a_1 - a_0| < 0.2\}$, as discussed in Sect. 3.3. This is the estimated effect of being randomized to exercise compared to wellness, among the group of people who would attend a similar (not differ by more than 20 %) number of classes in either arm, and would attend most (at least 70 %) of their assigned classes. The estimates and SEs are given in Table 10.2. For this group of subjects, those in the exercise arm appeared more likely to quit than those in the wellness arm. For 0.9, the estimated difference in quit rates between treatment arms was about 0.20, with p-values of about 0.05. Thus, for the group that would be highly compliant with either arm, there was moderate evidence of a benefit of exercise.

5 Discussion

We have developed methods that are designed to estimate the principal effects in clinical trials, such as smoking cessation trials, in which subjects have access to only one of two active treatments and the compliance variable is continuous (and possibly bounded). The joint distribution of the observed and counterfactual compliance is specified by linking the two marginal compliance distributions utilizing a Gaussian copula with a sensitivity correlation parameter. At each of the value of the correlation parameter, we obtain the MLEs of all parameters and estimate the causal effects at a particular combination of the compliance variables or within certain compliance regions in subpopulations that have similar compliance behavior. In the smoking cessation analysis, the exercise arm appeared to have lower quit rates among subjects that would be highly compliant with either intervention.

The two-stage ML approach is relatively easy to implement. However, we found that for small sample sizes the optimization algorithm can be unstable. A fully Bayesian approach is a viable alternative that could potentially resolve some of the convergence problems by helping to identify parameters using informative priors.

Our approach relies heavily on the structural form of outcome model. In particular, we specify a parametric model for $\mu(r, a_0, a_1)$—the mean of the potential outcome given the potential compliance variables. In principle, it would not be difficult to extend our approach to the semiparametric setting, where, for example, $\mu(r, a_0, a_1)$ could be modeled using bivariate smoothing via penalized splines. However, because one of the compliance variables is always missing, too much flexibility in the model can lead to identifiability problems.

Another related issue is that when ρ is close to 1, a_0 and a_1 contain about the same information. In that case, it might be sufficient to just include one of the compliance variables in the model, which should improve computational stability. This suggests that the form of $\mu(r, a_0, a_1)$ could depend on ρ. These issues, among others, are in need of additional research.

References

1. Marcus, B.H., Albrecht, A.E., King, T.K., Parisi, A.F., Pinto, B.M., Roberts, M., Niaura, R., Abrams, D.B.: The efficacy of exercise as an aid for smoking cessation in women. Arch. Intern. Med. **159**, 1229–1234 (1999)
2. Marcus, B.H., Lewis, B.A., King, T.K., Albrecht, A.E., Hogan, J., Bock, B., Parisi, A.F., Abrams, D.B.: Rationale, design and baseline data for commit to quit II: an evaluation of the efficacy of moderate-intensity physical activity as an aid to smoking cessation in women. Prev. Med. **36**, 479–492 (2003)
3. Marcus, B.H., Lewis, B.A., Hogan, J., King, T.K., Albrecht, A.E., Bock, B., Parisi, A.F., Niaura, R., Abrams, D.B.: The efficacy of moderate-intensity exercise as an aid for smoking cessation in women: a randomized controlled trial. Nicotine Tob. Res. **7**, 871–80 (2005)
4. Robins, J.M.: Correcting for non-compliance in randomized trials using structural nested mean models. Commun. Stat. Theory Methods **23**, 2379—2412 (1994)
5. Robins, J.M., Rotnitzky, A.: Estimation of treatment effects in randomised trials with non-compliance and a dichotomous outcome using structural mean models. Biometrika **91**, 763—783 (2005)
6. Vansteelandt, S., Goetghebeur, E.: Causal inference with generalized structural mean models. J. R. Stat. Soc. Ser. B. **65**, 817–835 (2003)
7. Frangakis, C., Rubin, D.: Principal stratification in causal inference. Biometrics **58**, 21–29 (2002)
8. Efron, B., Feldman, D.: Compliance as an explanatory variable in clinical trials. J. Am. Stat. Assoc. **86**, 9–17 (1991)
9. Jin, H., Rubin, D.B.: Principal stratification for causal inference with extended partial compliance. J. Am. Stat. Assoc. **103**, 101–111 (2008)
10. Frees, E.W., Valdez, E.A.: Understanding relationships using copulas. N. Am. Actuarial J. **2**, 1–25 (1998)
11. Bartolucci, F., Grilli, L.: Modeling partial compliance through copulas in a principal stratification framework. J. Am. Stat. Assoc. **106**, 469–479 (2011)
12. Plackett, R.L.: A class of bivariate distributions. J. Am. Stat. Assoc. **60**, 516–522 (1965)
13. Ma, Y., Roy, J., Marcus, B.: Causal models for randomized trials with two active treatments and continuous compliance. Stat. Med. **30**, 2349–2362 (2011)
14. Dominici, F., Zeger, S.L., Parmigiani, G., Katz, J., Christian, P.: Estimating percentile-specific effects in counterfactual models: a case study of micronutrient supplementation, birth weight, and infant mortality. J. R. Stat. Soc. C **55**, 261–280 (2006)
15. Angrist, J., Imbens, G., Rubin, D.: Identification of causal effects using instrumental variables. J. Am. Stat. Assoc. **91**, 444–455 (1996)
16. Roy, J., Hogan, J.W., Marcus, B.W.: Principal stratification with predictors of compliance for randomized trials with 2 active treatments. Biostatistics **9**, 277–289 (2008)
17. Cheng, J., Small, D.S.: Bounds on causal effects in three-arm trials with non-compliance. J. R. Stat. Soc. B **68**, 815–836 (2006)
18. Ferrari, S.L.P., Cribari-Neto, F.: Beta regression for modeling rates and proportions. J. Appl. Stat. **31**, 799–815 (2004)

Chapter 11
Causal Ensembles for Evaluating the Effect of Delayed Switch to Second-Line Antiretroviral Regimens

Li Li and Brent A. Johnson

Abstract Transitioning from a failing antiretroviral regimen to a new regimen is a critical period in managing treatments to suppress HIV-1 RNA because it can have lasting effects on the durability of disease and likelihood of developing resistant mutations. Evaluating the timing of a switch to the subsequent therapy is difficult because patients are not randomly assigned to switch failing regimens at designed time points. Li et al. (J. Am. Stat. Assoc. 107:542–554, 2012) proposed and applied doubly robust semi-parametric methods to evaluate the effect of early versus late regimen switch in a two-stage design setting. These semi-parametric estimators are consistent if a parametric treatment model is correctly specified and achieve optimal performance if a parametric outcome model is also correctly specified. Here, we propose a new non-parametric estimator of the same causal estimand using an ensemble-type statistical learner. Compared to earlier estimators, the proposed estimator requires fewer model assumptions and can easily accommodate a large number of potential confounders. We illustrate the methods through simulation studies and application to data from the AIDS Clinical Trials Group Study A5095.

1 Introduction

In current clinical practice, HIV-1 infected patients are treated through a sequence of combined antiretroviral therapies (cART). Although HIV-1 is a viral agent and causes acquired immunodeficiency syndrome (AIDS), modern treatment successes and the lack of a cure suggest similarities in treatment of HIV-1 infection and a chronic disease. The primary goal of cART is to reduce viremia below a limit of

L. Li
Eli Lilly and Company, Lilly Corporate Center, Indianapolis, IN 46285, USA
e-mail: Li_Li_X1@Lilly.com

B.A. Johnson (✉)
Department of Biostatistics and Computational Biology, University of Rochester, 601 Elmwood Avenue, Box 630, Rochester, NY 14642, USA
e-mail: brent_johnson@urmc.rochester.edu

H. He et al. (eds.), *Statistical Causal Inferences and Their Applications in Public Health Research*, ICSA Book Series in Statistics, DOI 10.1007/978-3-319-41259-7_11

detection, but providers also make treatment decisions to help patients manage adverse side effects and opportunistic infections. For a variety of reasons, including co-morbidities, genetic mutations, and poor adherence, patients eventually fail their current cART and move to the next-in-line cART. Then, similar to individuals that live with chronic diseases, HIV-1 infected individuals transition from treatment regimen to regimen, as necessary, until all treatment options have been exhausted or death.

Despite all that the scientific community has learned about treating HIV-1 infection and AIDS over the last four decades, there is still much that is unknown. In particular, there are many open questions about the timing of treatment decisions that leads to better patient outcomes given a patient's medical and treatment history. Collecting scientific evidence to identify better treatment decisions is difficult because of patient heterogeneity and because designing and enrolling patients in randomized controlled trials to investigate these scientific questions is challenging [25]. In the absence of controlled clinical trials, investigators may conduct secondary analyses of existing databases in an attempt to address the same questions. However, in secondary analyses of observational data, there is often confounding between treatment and outcome and this issue must be addressed statistically. For HIV-1 infected patients who move failing cARTs to new cARTs, the reasons for switching cARTs can be extremely important and are expected to be related to clinical outcome. In addition to controlling for confounding, there may be other features of the database that are tangential to the scientific question of interest but must nevertheless be addressed by the data analyst. Therefore, a fair and objective evaluation of treatment decisions in a sequence of cARTs is challenging for several reasons but important for public health and the infected individual's quality of life.

Our methods are motivated by data from the AIDS Clinical Trials Group (ACTG) Study A5095, a controlled clinical trial designed to compare two efavirenz-containing regimens and a triple nucleoside regimen [15]. After patients failed their initial cARTs, patients were allowed to switch to second-line cARTs and then followed per study protocol. We are interested in assessing whether it is clinically beneficial to delay switching to second-line cART post-virologic failure or whether patients ought to switch as soon as possible. Because switching from a failing initial cART may depend on the initial cART, Li et al. [20] argued that it may be prudent to analyze the ACTG A5095 data as a two-stage design problem [21, 31], where "two-stage" refers to treatment assignment/decision at two points in time. Here, in our early versus late two-stage framework, patients are randomly assigned to one of two treatment arms at the first stage. Then, if a patient fails the initial cART, the patient decides to switch cART immediately or delay switch at the second stage. The two levels for each of initial cART and delay versus immediate switch comprise the four treatment combinations. A key feature of the statistical framework for two-stage methods is the introduction of a Bernoulli indicator for eligibility to second-stage randomization [21] and makes evident the connection to intent-to-treat inference in a randomized controlled trial.

In the methods below, we propose and investigate a non-parametric extension of our earlier work in Li et al. (2012) for evaluating early versus late switch from a failing cART. The estimators in Li et al. (2012) are semi-parametric insofar as they rely on parametric regression models of treatment assignment and outcome. The proposed estimator differs fundamentally from our earlier work because it is non-parametric and based on regression trees and boosting. We illustrate the method in an analysis of the data from ACTG A5095 and evaluate it through simulation studies.

2 Methods

Because of the close connection between missing data problems and causal inference [28], it will be instructive to develop methods in both contexts. Methods for estimating $E(Y)$ when some outcomes Y are missing are described in Sect. 2.1 while methods for two-stage designs are described in Sect. 2.2.

2.1 Missing Data

Without loss of generality, assume that the full data are $\{(X_1, Y_1), \ldots, (X_n, Y_n)\}$, i.i.d. pairs from the distribution of (X_1, Y_1), and the scientific interest is to estimate $\mu = E(Y_1)$. However, some outcomes Y are missing and the missingness mechanism depends on X but not on Y. The observed data are

$$O \equiv \{(X_1, \gamma_1 Y_1, \gamma_1), \ldots, (X_n, \gamma_n Y_n, \gamma_n)\},$$

i.i.d. random triples from the distribution of $(X_1, \gamma_1 Y_1, \gamma_1)$, where $\gamma_i = 0$ if Y_i is missing and $\gamma_i Y_i$ is the product of γ_i and Y_i for $i = 1, \ldots, n$. The conditional probability that Y_i is missing given X_i is denoted by $\pi(X_i) = P(\gamma_i = 1|X_i)$ and the conditional expectation of Y_i given X_i is $f(X_i) = E(Y_i|\gamma_i = 1, X_i)$. Our working assumption is that $\pi(X_i) = P(\gamma_i = 1|Y_i, X_i)$, also known as the missing at random assumption.

Under the missing at random assumption, two estimators of μ are the inverse probability weighted estimator and outcome regression estimator,

$$\hat{\mu}_{\text{IPW}} = \frac{1}{n}\sum_{i=1}^{n} \frac{\delta_i Y_i}{\hat{\pi}(X_i)}, \qquad \hat{\mu}_{\text{OR}} = \frac{1}{n}\sum_{i=1}^{n} \hat{f}(X_i),$$

where $\hat{\pi}(\cdot)$ and $\hat{f}(\cdot)$ are consistent estimators for $\pi(\cdot)$ and $f(\cdot)$. In many applications, $\pi(X)$ and $f(X)$ are assumed to be simple parametric functions of the covariates X; for example, $\pi(X)$ and $f(X)$ are fitted quantities from a logistic regression model for

$\pi(X)$ and linear model for $f(X)$. Then, one can show that $\hat{\mu}_{\mathrm{IPW}}$ and $\hat{\mu}_{\mathrm{OR}}$ are unbiased for μ if $\pi(X)$ and $f(X)$ are correctly specified. Using semiparametric theory, Robins et al. [26] proposed a doubly-robust estimator which would be consistent for μ if either $\pi(X)$ or $f(X)$ was correctly specified. Also, the doubly-robust estimator will be semi-parametric efficient if both $\pi(X)$ and $f(X)$ are correctly specified. However, assuming that $\pi(X)$ is correctly specified, authors subsequently showed that a consistent, doubly robust estimators can be rather imprecise when $f(X)$ is misspecified even if it is semi-parametric efficient when $f(X)$ is correctly modeled. To overcome this shortcoming, Robins et al. [27] proposed an estimator that aims to minimize the variance of the doubly robust estimating function when $f(X)$ is misspecified; however, the resulting estimator is no longer doubly robust. Tan [32] proposed a constrained maximum likelihood estimator that is doubly robust, semi-parametric efficient when $\pi(X)$ and $f(X)$ are correctly modeled, and minimum variance when $\pi(X)$ is correctly modeled but $f(X)$ may be misspecified. All three of these estimators may be expressed in the form of

$$\hat{\mu}_{\bullet} = \hat{\mu}_{\mathrm{IPW}} - k_{\bullet}\left[\frac{1}{n}\sum_{i=1}^{n}\left\{\frac{\gamma_i - \hat{\pi}(X_i)}{\hat{\pi}(X_i)}\right\}\hat{f}(X_i)\right]. \tag{11.1}$$

We define $\hat{\mu}_{\mathrm{AIPW}}$ in (11.1) as the doubly robust, augmented inverse probability weighted (AIPW) estimator with $k_{\mathrm{AIPW}} = 1$, $\hat{\mu}_{\mathrm{RRZ}}$ and $\hat{\mu}_{\mathrm{Tan}}$ are defined by k_{RRZ} and k_{Tan} given in [33, p.563]. For the family of semi-parametric estimators defined through (11.1), $\pi(X)$ and $f(X)$ are parametric functions of X through known link functions [22].

In the interest of robustness, we investigate non-parametric extensions of $\hat{\mu}_{\mathrm{OR}}$. Note that the outcome regression estimator may be written

$$\hat{\mu}_{\mathrm{OR}} = \frac{1}{n}\sum_{i=1}^{n}\left\{\gamma_i\hat{f}(X_i) + (1-\gamma_i)\hat{f}(X_i)\right\} = \frac{1}{n}\sum_{i=1}^{n}\left\{\gamma_i Y_i + (1-\gamma_i)\hat{f}(X_i)\right\}. \tag{11.2}$$

So, a completely non-parametric estimator for μ can be defined through (11.2) with a non-parametric regression estimator $\hat{f}(\cdot)$ for $f(\cdot)$. When there is only one covariate X, one can simply use the Nadaraya-Watson [23] estimator [30]; for example, see [6]. However, as the dimension of X increases, the curse of dimensionality precludes any simple extension of kernel regression and one must typically impose more structure on the data to propose practical, interpretable solutions. Some nonparametric regression methods include local polynomial regression [8], generalized additive models [1, 17], and smoothing splines [14]. See [30] for a review of these methods. Alternatively, one can use statistical methods that have little or no interpretability and this is the approach we take here.

In this note, we use blackbox statistical learners to construct a prediction model $\hat{f}(\cdot)$ and subsequently define our estimator $\hat{\mu}_{\mathrm{OR}}$ in (11.2). Blackbox method is a generic umbrella term used to describe classification algorithms in artificial intelligence, computer science, engineering, machine learning, mathematics, and statistics that aim to separate a vector of class labels ($n \times 1$) using a data input matrix X ($n \times p$). Authors noted that, in most if not all cases, the same algorithms could be used to construct predictions for a continuous response Y by simply modifying the loss function. There are many overlapping names and methods associated with these algorithms including ensemble methods, aggregation, bagging, and boosting. There exists a massive literature on these topics and we refer the interested reader elsewhere for a review (e.g., [18]).

Our estimator of μ is based on boosted regression trees (e.g., [4]), one of many blackbox statistical learners that scales up easily to handle large dimensional covariates X under mild restrictions on the data [9, 10] and is tersely outlined in the Appendix. There is some empirical evidence to suggest that boosting offers improvements in the misclassification rates compared to bagging [2] but such a comparison is beyond the scope of this note. In short, our estimator is constructed in the following three steps. First, the ensemble prediction $\hat{f}(\cdot)$ is built with the complete data $\{(X_i, \gamma_i Y_i, \gamma_i) : \gamma_i = 1\}$. Second, predictions are computed for the observations with missing outcomes where $\gamma_i = 0$, finally, the estimator is defined

$$\hat{\mu}_{\mathrm{New}} = \frac{1}{n} \sum_{i=1}^{n} \left\{ \gamma_i Y_i + (1 - \gamma_i) \hat{f}(X_i) \right\}.$$

We use the `blackboost` function from the `mboost` package in `R`, fivefold cross-validation to determine the stopping iteration, and all other default settings. We adopt the nonparametric bootstrap to estimate $\mathrm{var}(\hat{\mu}_{\mathrm{New}})$ as outlined in Sect. 2.3.

2.2 Two-Stage Designs

Following the framework in [21], let Y^*_{ab} be the potential outcome if a patient followed a treatment policy ($A = a, B = b$), where A and B are the first- and second-stage treatment random variables, respectively. In our problem, A is a binary random variable and represents the initial cART in ACTG A5095: $A = 0$ denotes the triple nucleoside regimen and $A = 1$ represents the combined efavirenz-containing regimens. The second-stage treatment B is also binary and denotes switching early or late to second-line regimen after confirmed virologic failure on the initial regimen. Based on discussions with our collaborators, we define an early switch to second-line regimen as less than 8 weeks after confirmed virologic failure. Hence, the four potential outcomes for this simple design are (Y^*_{ab}, $a, b = 0, 1$) and the goal is to estimate $E(Y^*_{ab})$. If we can derive a consistent estimator for $E(Y^*_{ab})$, then we can extend those statistics to draw inference on the expected difference in potential outcomes $E(Y^*_{a1} - Y^*_{a0})$ to assess the value in switching to second-line regimen within 8 weeks of virologic failure on the initial regimen. This is a summary of the introductory arguments given in [20].

If the observed data consisted of i.i.d. copies of (Y, A, B, X), where treatment assignment to (a, b) depended on X only, then, under suitable regularity conditions, unbiased estimators for $E(Y^*_{ab})$ could be derived using usual arguments from causal inference. However, the observed data are more complex. A challenge in the analysis of data from two-stage designs is the possibility that not all patients fail their first-line regimen and, hence, do not switch to second-line regimen. Lunceford et al. [21] refer to this random variable as an indicator of eligibility to second-stage treatment assignment and include it as part of the definition of treatment policy; see also, [20, p.543]. Define Δ as the binary random variable indicating whether patients fail the initial regimen and are therefore eligible to switch to second-line regimen. Scientifically, we have no priori interest in Δ. In an intent-to-treat analysis of data from a randomized clinical trial where we knew (A, B) at baseline, the indicator Δ would be ignored. In observational data, we do not know (A, B) for a randomly selected patient from the population. Instead, we observe (A, B) when $\Delta = 1$ and only observe A when $\Delta = 0$. Nevertheless, the intent-to-treat estimand is still the parameter of interest. This makes the analysis of data from two-stage designs different and interesting.

In our framework, the observed data are

$$O \equiv \{(X_1, A_1, \Delta_1, \Delta_1 B_1, Y_1), \ldots, (X_n, A_n, \Delta_n, \Delta_n B_n, Y_n)\},$$

assumed to be i.i.d. copies from the distribution of $(X_1, A_1, \Delta_1, \Delta_1 B_1, Y_1)$. Under standard assumptions in causal inference and the identifiability condition,

$$Y^*_{a0} = Y^*_{a1}, \qquad \text{if } \Delta = 0, \text{ for } a = 0, 1,$$

Lunceford et al. [21] show that

$$\hat{\mu}^{ab}_{\mathrm{IPW}} = \frac{1}{n}\sum_{i=1}^{n} Y_i I(A_i = a)\left\{(1 - \Delta_i) + \frac{\Delta_i I(B_i = b)}{P(B_1 = b|A_i = a, \Delta_i = 1, X_i)}\right\}$$

is an unbiased estimator for $E(Y^*_{ab})$. Along the lines described in Sect. 2.1, Li et al. [20] showed that a general family of augmented inverse probability weighted estimators for $E(Y^*_{a1})$ in a two-stage design is

$$\hat{\mu}^{a1}_{\bullet} = \hat{\mu}^{a1}_{\mathrm{IPW}} - k_{\bullet}\left[\frac{1}{n}\sum_{i=1}^{n}\left\{\Delta_i\left(\frac{B_i - \pi_a(X_i)}{\pi_a(X_i)}\right) f_a(X_i)\right\}\right], \tag{11.3}$$

where $\pi_a(X_i) = P(B_i = 1|A_i = a, \Delta_i = 1, X_i)$ and $f_{a1}(X_i) = E(Y_i|A_i = a, \Delta_i = 1, B_i = 1, X_i)$. Setting $k_{\bullet} = 1$ leads to the AIPW estimator, k_{RRZ} and k_{Tan} lead to estimators in the spirit of Robins et al. [27] and Tan [32], respectively. See [20, p.547] for details of this approach. Estimators for $E(Y^*_{a0})$ are defined analogously.

To construct the new estimator for $E(Y^*_{a1})$ that accommodates blackbox prediction tools, we must extend the definition of outcome regression estimator in

Sect. 2.1 and consider the role of the eligibility random variable Δ. Using standard arguments, one can show that the outcome regression estimator for $E(Y^*_{a1})$ is

$$\hat{\mu}^{a1}_{\text{New}} = \frac{1}{n}\sum_{i=1}^{n} I(A_i = a)\left[(1-\Delta_i)Y_i + \Delta_i\left\{B_iY_i + (1-B_i)\hat{f}_{a1}(X_i)\right\}\right]. \quad (11.4)$$

The corresponding estimator for $E(Y^*_{a0})$ is $\hat{\mu}^{a0}_{\text{New}}$ given by the expression in (11.4) except that B_i and $(1-B_i)$ in curly brackets are reversed and $f_{a0}(X_i) = E(Y|A_i = a, \Delta_i = 1, B_i = 0, X_i)$ replaces $f_{a1}(X_i)$. As in Sect. 2.1, the estimator $\hat{\mu}^{a1}_{\text{New}}$ is computed in three steps: build the blackbox $\hat{f}_{a1}(\cdot)$ using the complete data, calculate the predictions for subjects where $B_i = 0$, and compute the estimate. The stopping iteration in the blackbox algorithm is chosen via fivefold cross-validation and minimizing a L_2 loss function.

2.3 *Variance Estimate*

To estimate $\text{var}(\hat{\mu}_{\text{New}})$ for missing data or $\text{var}(\hat{\mu}^{ab}_{\text{New}})$ in the two-stage design, we used a nonparametric bootstrap procedure. More details were in [7]. Because missing data are imputed in our estimators, we follow the recommendation of Shao and Sitter [29] who argue that "…the bootstrap data set should also be imputed in the same way as the original data set was imputed." This is in contrast to some authors who have suggested that the imputed data should be regarded as truth, which clearly underestimates the sampling variance. Therefore, in the case of $\text{var}(\hat{\mu}_{\text{New}})$, we (1) draw a simple random sample of size n with replacement from the original data set, then (2) compute the bootstrap estimate $\hat{\mu}^*_{\text{New,b}}$ using the bth resampled dataset. We repeat the process B times and then take the sample variance of $\{\hat{\mu}^*_{\text{New,1}}, \ldots, \hat{\mu}^*_{\text{New,B}}\}$. We estimate $\text{var}(\hat{\mu}^{ab}_{\text{New}})$ in the two-stage design using an identical resampling plan.

3 Analysis of ACTG A5095 Data

The goal of our analysis is to compare clinical endpoints for patients who switch early versus late from a failing efavirenz-containing combined antiretroviral therapy (cART). The question of when to switch from a failing cART has been discussed and debated in the HIV & AIDS literature for more than 15 years although some recent research has suggested a preference for switching "early" from a failing regimen (e.g., [19, 20, 24]). The data analysis below expands on an earlier analysis performed by our research team [20]. Here, we present new comparisons of the mean clinical endpoints using non-parametric ensemble methods discussed in Sect. 2.2 that make weaker modeling assumptions than the semi-parametric methods in [20].

The data are taken from the AIDS Clinical Trials Group (ACTG) Study A5095. Briefly, ACTG A5095 was a randomized, multi-center clinical trial designed to compare three antiretroviral regimens in HIV-infected, antiretroviral therapy-naive patients with HIV-RNA levels ≤400 copies/mL. The goal of the study was to suppress and maintain HIV-1 RNA levels <200 copies/mL and the study was designed to last 96 weeks. After 32 weeks of follow-up, 82 of 382 patients (21 %) in the triple NRTI group versus 85 of 765 patients (11 %) in the combined efavirenz group experienced virologic failure; hence, the triple nucleoside reverse-transciptase inhibitor (NRTI) regimen appeared inferior when compared to the combined efavirenz-containing regimens. All study patients who failed on the initial cART had the opportunity to switch regimens in favor of another regimens, subject to study protocol restrictions and recommendations. Although the initial regimen was randomized independent of patient characteristics, the switch to second-line regimen was left to patient and provider discretion. Thus, the comparison of early versus late regimen change is subject to confounding. Additional study details are provided in the primary sources [15, 16].

Using the same length-adjusted area under the curve (AUC) outcomes presented in [20], we compared doubly robust and optimal semi-parametric estimates to non-parametric causal ensembles proposed in Sect. 2.2. In addition to these statistically justified procedures, we also presented naive estimates that simply took the sample average of outcomes conditional on those patients that failed the initial regimen. All methods used six potential confounding variables: age, height, weight, baseline CD4 cell counts, baseline CD8, and time to first failure. The parameter estimates and their standard errors are given in Table 11.1. As reported in [20], we found that

Table 11.1 Estimates of mean potential outcomes $E(Y^*_{ab})$ for $n = 744$ patients switching less than (early) or greater than (late) 8-weeks after confirmed virologic failure on an efavirenz-containing regimen

		HIV-1 RNA[a]		Detection limit[b]		CD4[c]	
Method	Switch	Est. (SE)	T	Est. (SE)	T	Est. (SE)	T
Naive	Early	2.600 (0.181)	0.513	0.592 (0.054)	0.800	2.436 (0.055)	0.458
	Late	2.685 (0.068)		0.546 (0.023)		2.466 (0.026)	
IPW	Early	1.835 (0.041)	4.970	0.837 (0.030)	2.720	2.21 (0.093)	0.369
	Late	1.914 (0.032)		0.787 (0.011)		2.564 (0.015)	
AIPW	Early	1.848 (0.048)	2.325	0.829 (0.033)	1.614	2.593 (0.035)	0.764
	Late	1.915 (0.033)		0.787 (0.011)		2.563 (0.015)	
RRZ	Early	1.833 (0.043)	4.218	0.828 (0.01)	19.860	2.600 (0.015)	9.800
	Late	1.914 (0.033)		0.787 (0.011)		2.561 (0.015)	
Tan	Early	1.835 (0.040)	4.948	0.830 (0.011)	21.235	2.599 (0.014)	18.326
	Late	1.914 (0.033)		0.788 (0.011)		2.563 (0.015)	
New	Early	1.849 (0.048)	1.192	0.808 (0.012)	7.087	2.593 (0.017)	5.364
	Late	1.899 (0.030)		0.788 (0.010)		2.567 (0.015)	

Standard errors are reported in parentheses and we report the Wald test statistic (T) for a nominal test of the null hypothesis that the average causal effect is zero

[a]Length-adjusted AUC of HIV-1 RNA level, logarithm scale

[b]Proportion of time spent with HIV-1 RNA below limit of detection

[c]Length-adjusted AUC of CD4 T-cell counts, logarithm scale

augmented doubly robust semi-parametric estimators suggest there were statistically significant differences in cumulative HIV-1 RNA, proportion of time spent below a limit of detection, and cumulative CD4 cell counts between patients switching early versus late. At the same time, parameter estimates were not statistically different using simple and naive methods. When evaluating differences using the non-parametric causal ensemble, we found statistically significant differences in the limit of detection endpoint and cumulative CD4 endpoint but not the cumulative HIV-1 RNA endpoint.

4 Simulation Studies

We performed simulation studies to compare the non-parametric causal ensembles with semi-parametric estimators. To simplify the study design, we considered the missing data estimators in Sect. 2.1 rather than the two-stage problem in Sect. 2.2. Our simulation setup followed an outline similar to experiments in [5, 20]. For each $i = 1, \ldots, n$, let Z_{ij}, $j = 1, \ldots, 4$ be independent, standard normal random variables and then define $X_{i1} = \exp(Z_{i1}/2)$, $X_{i2} = Z_{i2}/(1 + \exp(Z_{i1})) + 10$, $X_{i3} = (Z_{i1}Z_{i3}/25 + 0.6)^3$ and $X_{i4} = (Z_{i1} + Z_{i2})^2$. The true propensity score model is $\pi_0(Z_i) = \text{expit}(-Z_{i1} + 0.5Z_{i2} - 0.25Z_{i3} - 0.1Z_{i4})$. The true outcome regression model is a linear model of (Z_1, Z_2, Z_3, Z_4) with a coefficient vector of $(210, 27.4, 13.7, 13.7, 13.7)$.

We evaluated the performances of the estimators in four scenarios, as described in Table 11.2. In the first scenario, both PS and OR models are correctly specified and we expected all estimators to perform well. In the second scenario, the working PS model is correct while the working OR model is not. True outcomes are generated from an exponential distribution; however, the working outcome regression model assumes a normal distribution on the outcome. Moreover, covariates in the working model include $X_1, \ldots, X_4$ instead of $Z_1, \ldots, Z_4$. In scenario 3, in addition, working PS model is built on $X_1, \ldots, X_4$. In Scenario 4 we considered 50 confounders in the true OR model. Because the semi-parametric methods do not perform variable selection automatically, we expected the finite sample performance of the semi-parametric estimators to be worse than the causal ensemble.

Table 11.2 Simulation scenarios

Scenario	True PS	True OR	$P(Y_a^* \leq y)$ Skewed	No. of confounders	Correct PS	Correct OR
1	$Z_1, \ldots, Z_4$	$Z_1, \ldots, Z_4$	No	4	Yes	Yes
2	$Z_1, \ldots, Z_4$	$Z_1, \ldots, Z_4$	Yes	4	Yes	No
3	$Z_1, \ldots, Z_4$	$Z_1, \ldots, Z_4$	Yes	4	No	No
4	$Z_1, \ldots, Z_{50}$	$Z_1, \ldots, Z_{50}$	No	50	Yes	Yes

Table 11.3 Simulation results based on 200 Monte Carlo replications

Scenario	IPW	AIPW	RRZ	Tan	New
1	1.7 (18.7)	0.1 (3.8)	2.9 (8.0)	0.2 (2.7)	1.4 (4.7)
2	2.3 (47.5)	1.3 (41.2)	8.0 (38.9)	1.8 (42.8)	8.9 (28.3)
3	0.4 (66.7)	2.9 (48.9)	8.6 (38.1)	0.4 (59.4)	10.5 (31.5)
4	–	–	–	–	0.02 (12.9)

Table entries are Monte Carlo bias and standard deviation and the true value is 210

Our simulation results are summarized in Table 11.3. Table entries are the Monte Carlo bias and standard deviation from 200 Monte Carlo datasets. For the first scenario where both PS and OR models are correct specified, the non-parametric estimator had similar bias to the semi-parametric estimators but was less precise than AIPW estimator and Tan estmator. For the second and third scenario where the OR models were incorrectly specified, IPW, AIPW, and Tan had small Monte Carlo bias while the non-parametric estimator had bias similar to the RRZ estimator. The variance of the non-parametric estimator was smaller than any semi-parametric estimator in scenarios 2–3. In the last scenario where the PS and OR models depend on 50 covariates, the semi-parametric estimators failed while the non-parametric causal ensemble had small bias and variance.

5 Discussion

We have proposed a non-parametric estimator for missing data in Sect. 2.1 and for a two-stage causal estimand in Sect. 2.2. The latter two-stage estimator is a non-parametric alternative to the semi-parametric estimators proposed in our earlier work [20] for comparing early versus late switch from a failing combined antiretroviral therapy. In general, doubly robust semi-parametric estimators of the causal estimand in [20] require that either one or both of the propensity score or outcome regression models are correctly modeled. The non-parametric estimator proposed here is based on blackbox boosted ensemble methods, does not model the propensity score, and places minimal assumptions on the outcome regression model. Our estimator uses readily available software via the `mboost` package in `R`.

Our simulation studies suggest the non-parametric estimator performs similar to the semi-parametric methods under linear regression with normal error models, but better than semi-parametric methods when the number of potential confounders is large. When the errors are heterogeneous or are otherwise non-normal, we found that doubly robust semi-parametric methods have the potential to offer smaller bias and variance when at least one of the treatment or outcome models is correctly specified. In the data analysis, non-parametric estimates of the mean potential outcome for the switch-early group were better than those for the delayed-switch group. This conclusion is similar to what we found using semi-parametric methods in Table 11.1 and reported in our earlier report [20].

Appendix

Boosting is machine learning algorithm from a theory that attempts to construct a strong learner from a series or collection of weak learners and the earliest substantial contribution is widely attributed to [9, 10]. Freund and Schapire developed an early version of an adaptive resampling and combination scheme that became adaptive boosting or *AdaBoost*. Breiman [3] showed that AdaBoost can be viewed as functional gradient descent in function space while Friedman et al. [12, 13] linked AdaBoost and other boosting algorithms to a statistical framework in function estimation. The work by Breiman [3] and Friedman et al. [12, 13] brought boosting to a wide array of statistical regression and prediction applications beyond classification and our proposed estimator builds on this idea of function estimation.

Bühlmann and Hothorn [4] recently reviewed the literature in boosting and aggregation and their review informed our outline here. Boosting algorithms can be written as functional gradient descent techniques [3, 12, 13] and we adopt this view here. Briefly, the goal of functional gradient descent is to estimate a function by minimizing an expected loss

$$E[\rho\{Y, f(X)\}],$$

where $\rho(\cdot,\cdot)$ is a (loss) function of data $O \equiv \{(X_1, Y_1), \ldots, (X_n, Y_n)\}$ and convex with respect to the second argument. Friedman [11] provided a generic outline of a descent algorithm through the following steps:

1. Initialize $\hat{f}^0(\cdot)$ with an offset value. A common choice is

$$\hat{f}^0(\cdot) = \text{argmin} \frac{1}{n} \sum_{i=1}^{n} \rho(Y_i, c);$$

for a constant c or let $\hat{f}^0(\cdot) = 0$. Set $m = 0$.

2. Increase m by 1. Compute the negative gradient $(\partial/\partial f)\rho(Y, f)$ and evaluate it at $\hat{f}^{m-1}(X_i)$:

$$U_i = -\left.\frac{\partial \rho(Y_i, f)}{\partial f}\right|_{f=\hat{f}^{m-1}(X_i)}, i = 1, \ldots, n.$$

3. Fit the negative gradient vector $U_1, \ldots, U_n$ to $X_1, \ldots, X_n$ by the real-valued base procedure

$$(X_i, U_i)_{i=1}^{n} \xrightarrow{\text{base procedure}} \hat{g}^m(\cdot).$$

4. Update $\hat{f}^m(\cdot) = \hat{f}^{m-1}(\cdot) + \upsilon \hat{g}^m(\cdot)$, where υ is a step-length factor.
5. Iterate steps 2–4 until $m = m_{\text{stop}}$ for some stopping iteration m_{stop}.

We need to determine two user-defined parameters in the algorithm above; namely, m_{stop} in step 5 and the step-length factor υ in step 4. The stopping iteration m_{stop} is determined via cross-validation or some information criterion, such as corrected AIC criterion. The choice of the step-length factor υ is chosen to be sufficiently small (e.g., $\upsilon = 0.1$). Popular loss functions $\rho(y,f)$ are $\exp\{-(2y-1)f\}$ or $\log_2[1+\exp\{-(2y-1)f\}]$ for binary outcomes and squared error loss for continuous outcomes.

BlackBoost was developed by Friedman [11] and uses regression trees as the base learner. Bühlmann and Hothorn [4] reviewed both theory and applications and have highlighted the advantage that estimates will be invariant under monotone transformations of variables. In addition, regression trees can handle continuous and categorical covariates in a unified way.

References

1. Binder, H., Tutz, G.: A comparison of methods for the fitting of generalized additive models. Stat. Comput. **18**, 87–99 (2008)
2. Borra, S., Ciaccio, A.: Improving nonparametric regression methods by bagging and boosting. Comput. Stat. Data Anal. **38**, 407–420 (2002). doi:10.1016/S0167-9473(01)00068-8
3. Breiman, L.: Prediction Games and Arcing Algorithms. Technical Report 504. Statistics Department, University of California, Berkeley (1997/1998), revised. http://stat-www.berkeley.edu/tech-reports/index.html
4. Bühlmann, P., Hothorn, T.: Boosting algorithms: regularization, prediction and model fitting. Stat. Sci. **22**, 477–505 (2007). doi:10.1214/07-STS242
5. Cao, W., Tsiatis, A.A., Davidian, M.: Improving efficiency and robustness of the doubly robust. Biometrika **96**, 723–734 (2009)
6. Cheng, P.E.: Nonparametric estimation of mean functionals with data missing at random. J. Am. Stat. Assoc. **89**, 81–87 (1994)
7. Efron, B., Tibshirani, R.: Bootstrap methods for standard errors, confidence intervals, and other measures of statistical accuracy. Stat. Sci. **1**, 54–75 (1986). doi:10.1214/ss/1177013815
8. Fan, J., Gijbels, I.: Local polynomial fitting. In: Smoothing and Regression. Approaches, Computation and Application (M.G. Schimek), pp. 228–275. Wiley, New York (2000)
9. Freund, Y., Schapire, R.E.: A decision-theoretic generalization of on-line learning and an application to boosting. J. Comput. Syst. Sci. **55**, 119–139 (1997)
10. Freund, Y., Schapire, R.E.: A short introduction to boosting. J. Jpn. Soc. Artif. Intell. **14**, 771–780 (1999)
11. Friedman, J.: Greedy function approximation: a gradient boosting machine. Ann. Stat. **29**, 1189–1232 (2001)
12. Friedman, J., Hastie, T., Tibshirani, T.: Additive logistic regression: a statistical view of boosting. Ann. Stat. **28**, 337–374 (2000)
13. Friedman, J., Hastie, T., Tibshirani, T.: Rejoiner for additive logistic regression: a statistical view of boosting. Ann. Stat. **28**, 400–407 (2000)
14. Gu, C.: Smoothing Spline ANOVA Models. Springer, New York (2002)
15. Gulick, R.M., Ribaudo, H.J., Lustgarten, S., Squires, K.E., Meyer, W.A., Acosta, E.P., Schackman, B.R., Pilcher, C.D., Murphy, R.L., Maher, W.L., Witt, M.D., Reichman, R.C., Snyder, S., Klingman, K.L., Kuritzkes, D.R.: Triple-nucleoside regimens versus efavirenz-containing regimens for the initial treatment of HIV-1 infection. N. Engl. J. Med. **350**, 1850–1861 (2004)

16. Gulick, R.M., Ribaudo, H.J., Shikuma, C.M., Lalama, C., Schackman, B.R., Meyer, W.A. 3rd., Acosta, E.P., Schouten, J., Squires, K.E., Pilcher, C.D., Murphy, R.L., Koletar, S.L., Carlson, M., Reichman, R.C., Bastow, B., Klingman, K.L., Kuritzkes, D.R., AIDS Clinical Trials Group (ACTG) A5095 Study Team: Three- vs four-drug antiretroviral regimens for the initial treatment of HIV-1 infection: a randomized controlled trial. J. Am. Med. Assoc. **296**(7), 768–781 (2006)
17. Hastie, T., Tibshirani, R.: Generalized Additive Models, 1st edn. Monographs on Statistics and Applied Probability. Chapman and Hall/CRC, Boca Raton (1990)
18. Hastie, T., Tibshirani, R., Friedman, J.: The Elements of Statistical Learning: Data Mining, Inference, and Prediction, 2nd edn. Springer, New York (2001)
19. Johnson, B.A., Ribaudo, H., Gulick, R.M., Eron, J.J.: Modeling clinical endpoints as a function of time of switch to second-line ART with incomplete data on switching times. Biometrics **69**, 732–740 (2013)
20. Li, L., Eron, J., Ribaudo, H., Gulick, R.M., Johnson, B.A.: Evaluating the effect of early versus late ARV regimen change after failure on the initial regimen: results from the AIDS clinical trials group study A5095. J. Am. Stat. Assoc. **107**, 542–554 (2012)
21. Lunceford, J., Davidian, M., Tsitatis, A.: Estimation of survival distributions of treatment policies in two-stage randomization designs in clinical trials. Biometrics **58**, 48–57 (2002)
22. McCullagh, P., Nelder, J.A. :Generalized Linear Models, 1st edn. Chapman and Hall, London (1983)
23. Nadaraya, E.A.: On estimating regression. Theory Probab. Appl. **9**(1), 141–142 (1964). doi:10.1137/1109020
24. Petersen, M.L., van der Laan, M.J., Napravnik, S., Eron, J., Moore, R., Deeks, S.: Long term consequences of the delay between virologic failure of highly active antiretroviral therapy and regimen modification: a prospective cohort study. AIDS **22**, 2097–106 (2008)
25. Riddler, S., Jiang, H., Tenorio, A., Huang, H., Kuritzkes, D., Acosta, E., Landay, A., Bastow, B., Haas, D., Tashima, K., Jain, M., Deeks, S., Bartlett, J.: A randomized study of antiviral medication switch at lower- versus higher-switch thresholds: AIDS clinical trials group study A5115. Antivir. Ther. **12**, 531–541 (2007)
26. Robins, J.M., Rotnitzky, A., Zhao, L.P.: Estimation of regression coefficients when some regressors are not always observed. J. Am. Stat. Assoc. **89**, 846–866 (1994)
27. Robins, J.M., Rotnitzky, A., Zhao, L.P.: Analysis of semiparametric regression models for repeated outcomes in the presence of missing data. J. Am. Stat. Assoc. **90**, 106–121 (1995)
28. Rosenbaum, P.R., Rubin, D.B.: The central role of the propensity score in observational studies for causal effects. Biometrika **70**, 41–55 (1983)
29. Shao, J., Sitter, R.R.: Bootstrap for imputed survey data. J. Am. Stat. Assoc. **91**, 1278–1288 (1996)
30. Simonoff, J.: Smoothing Methods in Statistics. Springer Science and Business Media, New York (1996)
31. Stone, R.M., Berg, D.T., George, S.L., Dodge, R.K., Paciucci, P.A., Schulman, P., Lee, E.J., Moore, J.O., Powell, B.L., Schiker, C.A.: Granulocyte- macrophage colony-stimulating factor after initial chemotherapy for elderly patients with primary acute myelogenous leukemia. N. Engl. J. Med. **322**, 1671–1677 (1995)
32. Tan, Z.: A distributional approach for causal inference using propensity scores. J. Am. Stat. Assoc. **101**, 1619–1637 (2006)
33. Tan, Z.: Understanding OR, PS and DR. Stat. Sci. **22**, 560–568 (2007)
34. Watson, G.S.: Smooth regression analysis. Sankhyā Indian J. Stat. Ser. A **26**(4), 359–372 (1964) [JSTOR 25049340]

Chapter 12
Structural Functional Response Models for Complex Intervention Trials

Pan Wu and Xin M. Tu

Abstract Estimating causal effect under different treatment exposures in empirical research is sometimes difficult because of lack of control for the distribution of such exposure in either randomly assigned or self-selected treatment groups. In clinical studies, when the treatment doesn't follow standard design under a consistent and single-layer intervention for each subject, the estimation and inference on causal treatment effect would become more complicated than the standard intervention for the most of the statistical models. In this book chapter, we are interested in introducing a new class of structural models in estimation of causal treatment effect, the structural functional response models (SFRM), which is an extended version of existing structural mean models (SMM), but more effectively used in addressing imperfect of treatment compliance in clinical trials. In contrast with SMM, the SFRM has a flexible model structure and is naturally adaptive to complex intervention design for both experimental and non-experimental studies. The computation of the SFRM is more straightforward than the G-estimation algorithm that is widely used by SMM. Moreover, the SFRM is ready to be generalized to binary and count outcomes through logit and log-linear functions. Simulation studies are conducted to illustrate its strength and superiority of model performance. Then, the SFRM is applied to a randomized clinical trial in comparison of a new intervention with standard therapy in improvement of teenage's mental health to estimate the causal treatment effect under the multi-layered intervention design.

P. Wu (✉)
Value Institute, Christiana Care Health System, 4755 Ogletown-Stanton Road, Newark, DE 19718, USA
e-mail: PWu@Christianacare.org

X.M. Tu
Department of Biostatistics and Computational Biology, University of Rochester, 265 Crittenden Boulevard, Rochester, NY 14642, USA
e-mail: Xin_Tu@urmc.rochester.edu

H. He et al. (eds.), *Statistical Causal Inferences and Their Applications in Public Health Research*, ICSA Book Series in Statistics, DOI 10.1007/978-3-319-41259-7_12

1 Introduction

The randomized controlled trials (RCTs) has been treated as the gold standard in causal inference since the effect of randomization ensures that no pre-treatment variables could potentially confound both treatment assignment and outcomes of interest. This effort is rewarded by a simple design with robust results that is easily understood and implementable in the general public. The RCTs, however, may not always guarantee the causality of treatment on the outcome of interest when the after-randomization treatment suffers imperfect or non-compliance issue in practices, such as the inconsistent exposure of intervention for each individual subject in active treatment arms or less control on other variables (mediators) related to both treatment and outcomes. The traditional intention to treat (ITT) approach is recommended to use in RCTs for its simplicity in study design, control, and implementation, but isn't capable of addressing the post-treatment confounding and may lead to biased inference and make analytic results without causal interpretation.

In the past decades, the problem of estimating the causal effect of compliance with active treatment in randomized trials has received much attention in statistical literature. Efron and Feldman [3] introduced a one-one monotone mapping between compliance and treatment and implemented with a full parametric model. Angrist et al. [1] used the Instrumental Variables approach to calculate the complier average treatment effect for the placebo-controlled trials and generated this further to binary compliance on binary outcome. Frangakis and Rubin [7] developed the Principal Stratification (PS) method to adjust post-treatment compliance within the stratified covariate groups and estimate causal effect for each strata using Bayesian approach. Robins [23] proposed the structural nested mean models (SMM) to find the causal parameters in a quite robust semiparametric framework with (repeated) continuous outcomes. Goetghebeur and Lapp [8] and Vansteelandt and Goetghebeur [29] applied this theory to address confounding issue of treatment compliance in placebo-controlled trials and then extended to generalized SMM in accommodating binary and count data. SMM give a precise but subtle meaning in efficient estimation of causal treatment effect. This model can be seen as a robust regression with unpaired data.

Although RCTs remain as a benchmark for clinical research and practice, observational studies with self-selected treatment and semi-RCTs (trials that initiate treatment dynamically when needed) have become more popular, especially in studies in the behavioral and social sciences, epidemiological studies, and healthcare research, because of the large amount of data generated by new web technologies and social media. Even within the standard RCTs, we have found that single intervention design is becoming less attractive in empirical research due to its simplicity of treatment structural. More and more studies would prefer using complex design, such as multi-level, multi-layered, or multi-modal, dynamic interventions to take advantage of both static (e.g., genetic traits) and dynamic (e.g., treatment response) information during the treatment.

In this chapter, we focus on community-based multi-layered RCTs and introduce a new class of structural functional response models (SFRM) to address complex design with treatment compliance issues when evaluating intervention effects. This SFRM can be treated as an extended version of SMM, but is more flexible to establish the treatment–effect relationships under complex intervention design when the treatment exposures are not completely controlled.

The proposed approach is motivated by a community-based multi-layered RCT—the child resilience project (CRP), where post-treatment noncompliance arises from both the primary (subject) and supportive (support group) layer. The CRP is designed to promote behaviorally and emotionally healthy trajectories in 1st–3rd grade urban children who are showing aggressive-disruptive and school socialization problems, a group at elevated risk for future mental health disorders, substance abuse problems, reduced educational outcomes, and costly services. The study involved 401 children randomized to the intervention and control groups. In addition, the study interventionists also worked with parents to teach children a set of skills to strengthen emotion self-regulation, adaptive social behavior, and classroom conduct. Parent visits focus first on identifying parent goals for the child, then on introducing and preparing parents to use activity sets that teach and reinforce children's use of emotion self-regulation skills and incorporating those skills into their everyday relationship.

The initial intention-to-treat (ITT) analyses failed to show any treatment effect for the primary behavior outcomes. Since ITT estimates are defined based on treatment assignment at randomization, rather than what actually goes on during the trial, such estimates completely ignore issues pertaining to violations of treatment protocols such as treatment noncompliance. For example, had only a small fraction of subjects in the intervention condition taken the treatment as prescribed, ITT would unduly underestimate the effect of receiving the intervention. However, child participation over 18 months was, as expected, high due to skill lessons being delivered in the school setting; 97 % of children in the intervention condition completed all 14 lessons in the first year, and 81 % completed all ten lessons in the second year. Of the 39 non-completers, 33 were children relocating to non-study schools. Non-participation was unrelated to any baseline outcome measure.

Parent participation, however, was significantly lower; as shown in Table 12.1, with only 63.4 % of parents (128 of 203 enrolled) participating in one or more intervention visits (Sessions > 0) and few completing the 15 scheduled sessions. Under this condition of lower participation, ITT analyses are less informative about

Table 12.1 Child resilience complete dataset

Total sessions attended by parents of children in intervention										
Sessions	0	1	2	3	4	5	6	7	8	9+	Total
Frequency	74	38	26	11	9	10	3	1	2	30	202
Percent	36.6	18.8	12.9	5.4	4.5	5.0	1.5	0.5	1	15	100
Cumulative percent	36.6	55.4	68.3	73.8	78.2	83.2	84.7	85.1	86.1	100.0	100

the true causal effects of parent involvement in the program, especially if the effect of treatment on child outcomes is achieved in part through parental participation.

As we introduced, a number of approaches for addressing treatment noncompliance in RCTs have been developed based on the counterfactual outcome framework. Unfortunately, none of the available methods is able to address treatment noncompliance in multi-layered intervention studies. The new approach we have developed is to extend the principles in these approaches to this new setting with treatment noncompliance from multiple layers of the intervention. In Sect. 2, we briefly review the counterfactual outcome based causal framework and introduce a class of SFRM to address both pre- and post-treatment confounding. In Sect. 3, the SFRM is extended to address treatment noncompliance in multi-layered interventions within a longitudinal study setting. Simulation studies are presented in Sect. 4 to evaluate the performance of the proposed SFRM. In Sect. 5, we apply the approach to address the variability in parent participation in the two-layered CRP study. We conclude with a discussion in Sect. 6.

2 Structural Functional Response Models for Causal Inference

2.1 Counterfactual Outcomes

The concept of *counterfactual outcome*, the underpinning of the modern causal inference paradigm, addresses the fundamental question of causal treatment effect [25]. Under this framework, associated with every patient is a *potential outcome* for each treatment condition, and the treatment effect is defined by the difference between the outcomes in response to the respective treatments from the same individual, thereby free of any confounding effect and providing a conceptual basis for causal effect without relying on the notation of randomization.

For example, if the two potential outcomes for the *i*th child in the CRP Study are y_{i1} and y_{i0} for the intervention and control condition, the difference $\Delta_i = y_{i1} - y_{i0}$ is the treatment effect for the child. Since this difference is based on the outcomes from the same child, it must be the result of the intervention. Unfortunately, since only the outcome from the treatment condition actually assigned is observed, this difference is unobservable. A large part of the causal inference literature centers on how to estimate the *average*, or *population-level*, causal treatment effect, $\Delta = E\left(y_{i1} - y_{i0}\right)$.

In RCTs, treatment assignment is independent of potential outcomes, i.e., $y_{ik} \perp z_i$, where z_i denotes a binary indicator for treatment assignment and $\perp$ denotes stochastic independence. In this case, the average causal effect $E\left(y_{i1} - y_{i0}\right)$ can be estimated by the difference between the two sample means from the intervention and control group:

$$\widehat{\Delta} = \overline{y}_{\cdot 1} - \overline{y}_{\cdot 0}, \quad \overline{y}_{\cdot 1} = \frac{1}{n_1} \sum_{i=1}^{n_1} y_{i_1 1}, \quad \overline{y}_{\cdot 0} = \frac{1}{n_0} \sum_{i=1}^{n_0} y_{i_0 0}, \tag{12.1}$$

where n_k denotes the number of subjects assigned to the kth treatment group such that $n = n_1 + n_0$ and i_k denotes the ith subject within the kth treatment group. Note that $y_{i_k k}$ refers to the observed outcome for the i_kth subject in the assigned kth treatment, while y_{ik} denotes the potential outcome corresponding to the kth treatment.

The above shows that standard statistical models such as linear regression and mixed-effects models can be applied to RCTs to infer causal treatment effects. Randomization is key to the transition from the unobserved individual level difference, $y_{i1} - y_{i0}$, to the estimable average treatment effect by the computable sample means in (12.1). For non-randomized trials such as most epidemiological studies, exposure to treatment or agent is non-random, in which case (12.1) generally does not estimate the average causal effect $\Delta = E(y_{i1} - y_{i0})$. Thus, associations found in observational studies generally do not imply causation.

2.2 *Structural Functional Response Models*

Since only one of the potential outcomes y_{ik} is observable, we cannot model the y_{ik}'s directly using conventional regression models. One way around this is to model the observed outcomes such as $y_{i_k k}$ as in the preceding section. Alternatively, we can circumvent this difficulty by constructing an observable response based on the unobserved y_{ik} and relate the response created to the mean of y_{ik} as follows:

$$E\left(\frac{z_i^k (1-z_i)^{1-k} y_{ik}}{\pi^k (1-\pi)^{1-k}}\right) = \mu_k, \quad E(z_i) = \pi, \quad z_i = 0, 1, \quad 1 \le i \le n, \quad k = 0, 1, \tag{12.2}$$

where $\mu_k = E(y_{ik})$ is the mean of potential outcome y_{ik}, since it is readily checked that

$$E\left(\frac{z_i^k (1-z_i)^{1-k} y_{ik}}{\pi^k (1-\pi)^{1-k}}\right) = \frac{1}{\pi^k (1-\pi)^{1-k}} E\left(z_i^k (1-z_i)^{1-k} y_{ik}\right) = \mu_k.$$

Although y_{ik} are not both observed, the *functional response*, $f(y_{i0}, y_{i1}, z_i) = \frac{z_i^k (1-z_i)^{1-k} y_{ik}}{\pi^k (1-\pi)^{1-k}}$, in (12.2) is still well defined. If π is known as in most RCTs, it is unnecessary to model z_i and (12.2) reduces to the first equation.

The model in (12.2) is not a conventional regression model such as the generalized linear or non-linear models, since $f(y_{i0}, y_{i1}, z_i)$ is not a single linear response such as y_{ik} or z_i. Rather, this model is a member of the following class of functional response models (FRM):

$$E\left[f\left(y_{i_1}, \ldots, y_{i_q}, \boldsymbol{\theta}\right) \mid \mathbf{x}_{i_1}, \ldots, \mathbf{x}_{i_q}\right] = h\left(\mathbf{x}_{i_1}, \ldots, \mathbf{x}_{i_q}; \boldsymbol{\theta}\right), \quad \left(i_1, \ldots, i_q\right) \in C_q^n, \tag{12.3}$$

where $f(\cdot)$ is some function, $h(\cdot)$ is some smooth function (e.g., continuous second-order derivatives), y_i and $\mathbf{x}_i$ denote some response and explanatory variables, C_q^n denotes the set of $\binom{n}{q}$ combinations of q distinct elements $(i_1, \ldots, i_q)$ from the integer set $\{1, \ldots, n\}$, and $\boldsymbol{\theta}$ a vector of parameters. The response $f(y_{i_1}, \ldots, y_{i_q}, \boldsymbol{\theta})$ in (12.3) for the general FRM can be quite a complex function of multiple outcomes [e.g., y_{ik}, z_i in (12.2)] from different subjects as well as unknown parameters $\boldsymbol{\theta}$ [e.g., π in (12.2)]. By generalizing the response variable in this fashion, (12.3) provides a general framework for modeling a broad set of problems involving higher-order moments and between-subject attributes. The FRM has been applied to a range of methodological issues involving multi-subject responses such as extensions of the Mann-Whitney-Wilcoxon rank sum test to longitudinal and causal inference settings [2, 31], social network analysis [4, 14, 32], gene expression analysis [11], reliability coefficients [10, 12, 15–18, 27], and complex response functions such as models for population mixtures [33] and structural equation models [9].

Because of its relationship to (12.3), the model in (12.2) will be referred to as the structural FRM (SFRM):

$$E\left(f_{ik}\left(y_{i0}, y_{i1}, z_i\right)\right) = h_{ik}\left(\boldsymbol{\theta}\right), \quad f_{i1} = \frac{z_i y_{i1}}{\pi}, \quad f_{i2} = \frac{\left(1 - z_i\right) y_{i0}}{1 - \pi}, \quad f_{i3} = z_i, \tag{12.4}$$
$$h_{i1}\left(\boldsymbol{\theta}\right) = \mu_1, \quad h_{i2}\left(\boldsymbol{\theta}\right) = \mu_0, \quad h_{i3}\left(\boldsymbol{\theta}\right) = \pi,$$

where $\boldsymbol{\theta} = (\mu_1, \mu_0, \pi)^\top$ denotes the collection of the parameters for this SFRM. Before adding more complexity to this SFRM to address treatment noncompliance within our context, let us first extend it to address selection bias in observational studies.

2.2.1 Selection Bias by Pre-treatment Confounders

If subjects are not randomized with respect to the treatment condition (or exposure) as in observational studies (e.g., survey, epidemiologic studies), $y_{ik} \perp z_i$ is generally not true. In the presence of such *selection bias*, if $\mathbf{w}_i$ is a vector of covariates containing all sources of confounding such that the *ignorability condition* [26], $y_{ik} \perp z_i \mid \mathbf{w}_i$, holds, then we have

$$E\left(\frac{z_i^k\left(1 - z_i\right)^{1-k} y_{ik}}{\pi\left(\mathbf{w}_i\right)^k \left(1 - \pi\left(\mathbf{w}_i\right)\right)^{1-k}}\right) = E\left[E\left(\frac{z_i^k\left(1 - z_i\right)^{1-k} y_{ik}}{\pi\left(\mathbf{w}_i\right)^k \left(1 - \pi\left(\mathbf{w}_i\right)\right)^{1-k}} \mid \mathbf{w}_i\right)\right] = \mu_k. \tag{12.5}$$

where $\pi(\mathbf{w}_i) = E(z_i \mid \mathbf{w}_i)$. We may model z_i using a generalized linear model such as logistic regression:

$$E\left(z_i \mid \mathbf{w}_i\right) = \pi\left(\mathbf{w}_i; \boldsymbol{\eta}\right), \quad \operatorname{logit}\left(\pi\left(\mathbf{w}_i; \boldsymbol{\eta}\right)\right) = \boldsymbol{\eta}^\top \mathbf{w}_i, \quad 1 \leq i \leq n. \tag{12.6}$$

By combining (12.5) and (12.6), we have the following SFRM to provide valid inference about $\boldsymbol{\theta} = \left(\mu_1, \mu_0, \boldsymbol{\eta}^\top\right)^\top$ under selection bias:

$$E\left(f_{ik}\left(y_{i0}, y_{i1}, z_i, \mathbf{w}_i\right) \mid F_k\right) = h_{ik}\left(\boldsymbol{\theta}\right)$$

$$f_{i1} = \frac{z_i y_{i1}}{\pi\left(\mathbf{w}_i; \boldsymbol{\eta}\right)}, \quad f_{i2} = \frac{\left(1 - z_i\right) y_{i0}}{1 - \pi\left(\mathbf{w}_i; \boldsymbol{\eta}\right)}, \quad f_{i3} = z_i, \quad 1 \le i \le n \qquad (12.7)$$

$$h_{i1}\left(\boldsymbol{\theta}\right) = \mu_1, \quad h_{i2}\left(\boldsymbol{\theta}\right) = \mu_0, \quad h_{i3}\left(\mathbf{w}_i; \boldsymbol{\theta}\right) = \pi\left(\mathbf{w}_i; \boldsymbol{\eta}\right),$$

$$\pi\left(\mathbf{w}_i; \boldsymbol{\eta}\right) = \text{logit}^{-1}(\boldsymbol{\eta}^\top \mathbf{w}_i), \quad F_1 = F_2 = \{0\}, \quad F_3 = \{\mathbf{w}_i\},$$

where $F_k = \{0\}$ $(k = 1, 2)$ denotes the sigma field generated by the constant 0 and $F_3 = \mathbf{w}_i$ denotes the sigma field generated by $\mathbf{w}_i$. Note that $E(f_{ik}(y_{i0}, y_{i1}, z_i, \mathbf{w}_i)|F_k) = E(f_{ik}(y_{i0}, y_{i1}, z_i, \mathbf{w}_i))$, since F_k is contained in F_3 for $k = 1, 2$ (e.g., see Kowalski and Tu [12]).

2.2.2 Treatment Noncompliance as Post-treatment Confounders

In many RCTs, even well-planned and executed ones, treatment effect may be significantly modified by levels of exposure of intervention (e.g., compliance or dosage) due to treatment noncompliance. One popular approach for addressing this primary post-treatment confounder is the structural mean model (SMM) [8, 23, 29]. Other competing approaches also address treatment noncompliance such as the instrumental variable [1] and principal stratification methods [7]. However, only SMM models treatment compliance on a continuous scale, which is more appropriate for session attendance within our context. We first frame this model within the FRM framework and then discuss its extensions to accommodate complex intervention design study, such as multi-layered treatments and missing data in Sect. 3.

Consider a randomized medication vs. placebo study and let d_{i1} denote a continuous potential outcome of medication use, if the ith subject is assigned to the medication condition. The SMM models the dose effect on treatment difference as follows:

$$E\left(y_{i1} - y_{i0} \mid d_{i1}, \mathbf{x}_i\right) = g\left(d_{i1}, \mathbf{x}_i\right), \qquad (12.8)$$

where $g\left(\cdot\right)$ is known up to a set of parameters (i.e., only the functional form of $g\left(d_{i1}, \mathbf{x}_i\right)$ is specified) and $\mathbf{x}_i$ is the baseline covariates. However, the above model cannot be fit directly using conventional statistical methods, since only one of the potential outcomes $\left(y_{i1}, y_{i0}\right)$ is observed. For RCTs, we have $y_{i1}, y_{i0} \perp z_i$ and the above Eq. (12.8) follows that

$$E\left(y_{i1} \mid d_{i1}, \mathbf{x}_i, z_i = 1\right) = g\left(d_{i1}, \mathbf{x}_i\right) + E\left(y_{i0} \mid d_{i1}, \mathbf{x}_i, z_i = 0\right). \qquad (12.9)$$

By conditioning on the assigned treatment $z_i = k$, y_{ik} in (12.9) represents the observed outcome from the kth treatment group ($k = 0, 1$). Thus, $E(y_{i0} \mid d_{i1}, \mathbf{x}_i, z_i = 0)$ cannot be modeled directly, since d_{i1} is not observed for the subjects assigned to the placebo condition.

If treatment compliance is tracked for the subjects in the placebo group, then d_{i0}, the potential outcome of placebo use if the subject is assigned to the placebo condition, is observed. Because of randomization and the fact that subjects cannot distinguish between medication and placebo, d_{i0} has the same distribution as d_{i1}. Thus, we may replace d_{i1} by d_{i0} in $E(y_{i0} \mid d_{i1}, z_i = 0)$ to re-express (12.9) as

$$E(y_{i1} \mid d_{i1}, \mathbf{x}_i, z_i = 1) = g(d_{i1}, \mathbf{x}_i) + E(y_{i0} \mid d_{i0}, \mathbf{x}_i, z_i = 0). \tag{12.10}$$

Under this *treatment compliance explainable* condition, we will be able to model the right side to obtain estimates of dose–response relationships $g(d_{i1}, \mathbf{x}_i)$ [5].

Although applicable to medication studies, the SMM in (12.10) in general does not apply to psychosocial research. Many psychosocial intervention studies do offer attention or information controls for both treatment arms such that subjects in the control groups may also be tracked for their treatment noncompliance. However, unlike medication studies, compliance observed in the control group d_{i0} generally does not have the same distribution as d_{i1}. For example, consider a HIV prevention intervention study for teenage girls at high risk for HIV infection, in which the intervention condition contains information on HIV infection, condom use, and safe sex, while the control condition consists of nutritional and dietary information. Subjects with high compliance in the intervention group are generally different from their counterparts in the control condition; sexually active girls may form a majority of those with high attendance in the intervention group, while such girls might have low attendance rates, had they been assigned to the control condition. Thus, when assessing the effect of prevention intervention using outcomes of HIV risk behavior such as the number of unprotected vaginal sex over the past month, it is not meaningful to compare compliant subgroups between the two treatment conditions.

Thus for psychosocial research studies, we cannot simply replace d_{i1} in $E(y_{i0} \mid d_{i1}, \mathbf{x}_i, z_i = 0)$ by a measure of treatment compliance such as session attendance in the control group d_{i0} as in medication trials. In many studies, it is reasonable to assume that there is sufficient information to predict d_{i1}, i.e., given a set of covariates $\mathbf{x}_i$, d_{i1} is independent of y_{i0}. For example, if $\mathbf{x}_i$ contains information on sexuality and other information on a subject's interest to attend sessions in the intervention condition of the HIV study example above, y_{i0} may no longer depend on d_{i1} given $\mathbf{x}_i$. In this case, $E(y_{i0} \mid d_{i1}, \mathbf{x}_i, z_i = 0) = E(y_{i0} \mid \mathbf{x}_i, z_i = 0)$. Thus, under this *ignorability condition*, $y_{i0} \perp d_{i1} \mid \mathbf{x}_i$ (12.9) becomes

$$E(y_{i1} \mid \mathbf{x}_i, d_{i1}, z_i = 1) = g(d_{i1}, \mathbf{x}_i) + E(y_{i0} \mid \mathbf{x}_i, z_i = 0). \tag{12.11}$$

Note that the SMM in this case is essentially the same as the Principal Stratification Model, except that it requires neither discretization of d_{i1} nor parametric distribution models for y_{ik}, since (12.11) only specifies the conditional mean of y_{ik} given d_{i1}, $\mathbf{x}_i$, and z_i.

By modeling $E(y_{i0} \mid \mathbf{x}_i, z_i = 0)$ and casting (12.11) in the form of FRM, we obtain the following SFRM for modeling treatment compliance measured by a continuous dose variable d_{i1} (for the intervention condition only):

$$f_{i1} = \frac{z_i y_{i1}}{\pi}, \quad f_{i2} = \frac{(1 - z_i)\, y_{i0}}{1 - \pi}, \quad f_{i3} = z_i, \quad 1 \leq i \leq n, \tag{12.12}$$

$$h_{i1}(\mathbf{x}, \boldsymbol{\beta}) = h(\mathbf{x}_i, \boldsymbol{\beta}), \quad h_{i2}(\mathbf{x}_i, d_{i1}, \boldsymbol{\theta}) = g(d_{i1}, \mathbf{x}_i, \boldsymbol{\gamma}) + h(\mathbf{x}_i, \boldsymbol{\beta}), \quad \boldsymbol{\theta} = \left(\boldsymbol{\beta}^\top, \boldsymbol{\gamma}^\top, \pi\right)^\top$$

where $h(\mathbf{x}, \boldsymbol{\beta})$ $(g(d, \mathbf{x}, \boldsymbol{\gamma}))$ is some function of $\mathbf{x}$ (d) parameterized by $\boldsymbol{\beta}$ $(\boldsymbol{\gamma})$. As before, n is the sample size of the study, i.e., the sample size of the intervention plus the control group. Although for RCTs it is not necessary to include π as a parameter, the general SFRM in (12.12) allows us to extend this model to observational studies. For example, for non-randomized studies, $y_{ik} \perp z_i$ in general is not true. If $y_{ik} \perp z_i$ holds conditional on a set of covariates $\mathbf{w}_i$ (possibly overlapping with $\mathbf{x}_i$), then by modeling π as a function of $\mathbf{w}_i$ as in (12.6), the following SFRM still provides consistent estimates in the face of selection bias:

$$f_{i1} = \frac{z_i y_{i1}}{\pi(\mathbf{w}_i; \boldsymbol{\eta})}, \quad f_{i2} = \frac{(1 - z_i)\, y_{i0}}{1 - \pi(\mathbf{w}_i; \boldsymbol{\eta})}, \quad f_{i3} = z_i, \quad 1 \leq i \leq n, \tag{12.13}$$

$$h_{i1} = h_{i1}(\mathbf{x}_i, \boldsymbol{\beta}), \quad h_{i2}(\mathbf{x}_i, d_{i1}, \boldsymbol{\beta}, \boldsymbol{\gamma}) = g(d_{i1}, \mathbf{x}_i, \boldsymbol{\gamma}) + h_{i1}(\mathbf{x}_i, \boldsymbol{\beta}),$$

$$h_{i3} = \pi(\mathbf{w}_i; \boldsymbol{\eta}), \quad \operatorname{logit}(\pi(\mathbf{w}_i; \boldsymbol{\eta})) = \boldsymbol{\eta}^\top \mathbf{w}_i, \quad \boldsymbol{\theta} = \left(\boldsymbol{\beta}^\top, \boldsymbol{\gamma}^\top, \boldsymbol{\eta}^\top\right)^\top.$$

We can model $h(\mathbf{x}_i, \boldsymbol{\beta})$ and $g(d_{i1}, \boldsymbol{\gamma})$ in various ways. For example, we may simply model both as a linear function: $h_1(\mathbf{x}_i, \boldsymbol{\beta}) = \mathbf{x}_i^\top \boldsymbol{\beta}$ and $g(d_{i1}, \mathbf{x}_i, \boldsymbol{\gamma}) = d_{i1}\gamma$. By specifying an appropriate form for $g(d_{i1}, \mathbf{x}_i, \boldsymbol{\gamma})$, we may also extend (12.12) to non-continuous dose variables such as categorical variables. Further, by appropriately specifying $h_1(\mathbf{x}_i, \boldsymbol{\beta})$ and $h_2(\mathbf{x}_i, d_{i1}, \boldsymbol{\beta})$, we can also generalize (12.12) to non-continuous responses. For example, for a binary y_i, we may specify $h_1(\mathbf{x}_i, \boldsymbol{\beta})$ and $h_2(\mathbf{x}_i, d_{i1}, \boldsymbol{\beta})$ as follows:

$$h_1(\mathbf{x}_i, \boldsymbol{\beta}) = \operatorname{logit}^{-1}\left(\mathbf{x}_i^\top \boldsymbol{\beta}\right), \quad h_2(\mathbf{x}_i, d_{i1}, \boldsymbol{\beta}) = \operatorname{logit}^{-1}\left(g(d_{i1}, \mathbf{x}_i, \boldsymbol{\gamma}) + h_1(\mathbf{x}_i, \boldsymbol{\beta})\right).$$

2.2.3 Inference for Structural Functional Response Models

We focus on inference about $\boldsymbol{\theta} = \left(\boldsymbol{\beta}^\top, \boldsymbol{\gamma}^\top, \boldsymbol{\eta}^\top\right)^\top$ for the SFRM in (12.13), from which (12.7) and (12.12) follow as a special case. Let

$$\mathbf{f}(\mathbf{y}_i; z_i) = (f_{i1}, f_{i2}, f_{i3})^\top, \quad \mathbf{h}_i(\boldsymbol{\theta}) = (h_{i1}, h_{i2}, h_{i3})^\top, \quad 1 \leq i \leq n,$$

where f_{ik} and h_{ik} are defined in (12.13). Then, consistent estimates of $\boldsymbol{\theta}$ are readily obtained by using the generalized estimating equations (GEE) for FRM [9, 12, 33]:

$$\mathbf{U}(\boldsymbol{\theta}) = \sum_{i=1}^{n} D_i V_i^{-1} S_i = \mathbf{0}, \quad S_i = \mathbf{f}_i - \mathbf{h}_i, \quad D_i = \frac{\partial}{\partial \boldsymbol{\theta}} \mathbf{h}_i, \tag{12.14}$$

$$V_i = A_i^{\frac{1}{2}} R(\boldsymbol{\alpha}) A_i^{\frac{1}{2}}, \quad A_i = \operatorname{diag}_t \left(\operatorname{Var}\left(f_{it} \mid F_{it}\right)\right),$$

where $R(\boldsymbol{\alpha})$ denotes a choice of working correlation matrix.

The choice of $R(\boldsymbol{\alpha})$ and associated properties for the GEE estimate of $\boldsymbol{\theta}$ have been extensively discussed in the literature, which are stated for ease of reference without justifications [6, 21]. In particular, the GEE estimate may not be consistent in the presence of time-varying covariates under working correlation structures other than the working independence model [21]. Thus, the working independence model may be used in general to ensure valid inference. Although this simple working correlation structure may incur some loss of efficiency for time-dependent covariates [6] and thus other models such as the uniform compound symmetry matrix may be used in some specific applications to improve power, it suffices for the purpose of illustrating the proposed approach. We focus on the working independence model in what follows unless otherwise stated.

3 Extension to Complex Studies

We first extend the SFRM in Sect. 2 to longitudinal data and then to multi-layered intervention studies.

3.1 Longitudinal Data with Missing Values

Let $\mathbf{y}_{it} = (y_{it1}, y_{it0})^{\top}$ $(\mathbf{x}_{it})$ denote the potential outcomes of y_{it} (a vector of explanatory variables) of interest with $i\,(t)$ indexing the subject (assessment time) for $1 \leq i \leq n$ and $1 \leq t \leq T$. By applying (12.13) to each time point, we obtain a longitudinal version of the SFRM:

$$\mathbf{f}_i = \left(\mathbf{f}_{i1}^{\top}, \ldots, \mathbf{f}_{iT}^{\top}, z_i\right)^{\top}, \quad \mathbf{h}_i = \left(\mathbf{h}_{i1}^{\top}, \ldots, \mathbf{h}_{iT}^{\top}, \pi_i\right)^{\top}, E\left(\mathbf{f}_i \mid \mathbf{x}_i\right) = \mathbf{h}\left(\mathbf{x}_i, \boldsymbol{\theta}\right), \tag{12.15}$$

$$\mathbf{f}_{it} = \left(f_{it1}, f_{it2}\right)^{\top}, \quad f_{it1} = \frac{z_i}{\pi_i} y_{it1}, \quad f_{it2} = \frac{1 - z_i}{1 - \pi_i} y_{it0}, \quad 1 \leq i \leq n,$$

$$\mathbf{h}_{it} = \left(h_{it1}, h_{it2}\right)^{\top} \quad h_{it1} = h_1\left(\mathbf{x}_{it}, \boldsymbol{\beta}\right), \quad h_{it2} = g_t\left(d_{i1}, \boldsymbol{\gamma}\right) + h_{it1},$$

$$\pi_i = \operatorname{logit}^{-1}\left(\boldsymbol{\eta}^{\top} \mathbf{w}_i\right), \quad \boldsymbol{\theta} = \left(\boldsymbol{\beta}^{\top}, \boldsymbol{\gamma}^{\top}, \boldsymbol{\eta}^{\top}\right)^{\top}.$$

Inference for the FRM above is based on the following GEE for FRM [9, 12, 33]:

$$\mathbf{U}(\boldsymbol{\theta}) = \sum_{i=1}^{n} D_i V_i^{-1} S_i = \mathbf{0}, \quad S_i = \mathbf{f}_i - \mathbf{h}_i, \quad D_i = \frac{\partial}{\partial \boldsymbol{\theta}} \mathbf{h}_i, \tag{12.16}$$

$$V_i = A_i^{\frac{1}{2}} R(\boldsymbol{\alpha}) A_i^{\frac{1}{2}}, \quad A_i = \operatorname{diag}_t \left(\operatorname{Var}\left(\mathbf{f}_{it} \mid \mathbf{x}_{it}\right)\right),$$

where D_i and V_i are readily computed given (12.15) and $R(\boldsymbol{\alpha})$ denotes a choice of working correlation matrix.

Missing data is a common issue in longitudinal studies. The GEE in (12.16) generally yields biased estimates under the missing at random (MAR) mechanism [13, 24, 30]. The *weighted generalized estimating equations* (WGEE), a common approach for addressing this issue, has been extended to the FRM [9, 33]. We adapt this approach to the current context, with an alternative implementation to simplify the inference procedure. As in the literature, we assume monotone missing data patterns (MMDP) to facilitate inference [9, 13, 24, 30, 33].

Let y_{it} denote the observed potential outcome, i.e., $y_{it} = y_{itk}$ if the subject is assigned the kth treatment. Let

$$\mathbf{y}_{it^-} = \left(y_{i1}, \ldots, y_{i(t-1)}\right)^\top, \quad \mathbf{x}_{it^-} = \left(\mathbf{x}_{i1}^\top, \ldots, \mathbf{x}_{i(t-1)}^\top\right)^\top, \quad 1 \leq t \leq m,$$

denoting the all individual responses ($\mathbf{y}_{it^-}$) and explanatory variables ($\mathbf{x}_{it^-}$) prior to time t. Let

$$r_{it} = \begin{cases} 1 \text{ if } i \text{ th subject is observed at time } t \\ 0 \text{ otherwise} \end{cases}, \tag{12.17}$$

$$p_{it} = \begin{cases} 1 & \text{if } t = 1 \\ E\left(r_{it} = 1 \mid r_{i(t-1)} = 1, \mathbf{x}_{it^-}, \mathbf{y}_{it^-}\right) & \text{if } t > 1 \end{cases},$$

$$p_{it} = \operatorname{logit}^{-1}(\xi_{0t} + \boldsymbol{\xi}_{xt}^\top \mathbf{x}_{it^-} + \boldsymbol{\xi}_{yt}^\top \mathbf{y}_{it^-}),$$

$$\Psi_{it} = \left(\prod_{s=1}^{t} p_{it}\right)^{-1} r_{it} \mathbf{I}_2, \quad \Psi_i(\boldsymbol{\xi}) = \operatorname{diag}_t(\Psi_{it}),$$

$$\boldsymbol{\xi}_t = \left(\xi_{0t}, \boldsymbol{\xi}_{xt}^\top, \boldsymbol{\xi}_{yt}^\top\right)^\top, \quad \boldsymbol{\xi} = \left(\boldsymbol{\xi}_2^\top, \ldots, \boldsymbol{\xi}_T^\top\right)^\top.$$

We assume no missing data at baseline such that $r_{i1} \equiv 1$ $(1 \leq i \leq n)$. Under MAR and MMDP assumptions, p_{it} in (12.17) is well defined for $1 \leq t \leq T$. By integrating the weights Ψ_i into the GEE in (12.16), we obtain the following WGEE for inference about $\boldsymbol{\beta}$:

$$\mathbf{U}(\boldsymbol{\theta}, \boldsymbol{\xi}) = \sum_{i=1}^{n} D_i V_i^{-1} \Psi_i S_i = \mathbf{0}. \tag{12.18}$$

In the extant literature, an estimate $\widehat{\boldsymbol{\xi}}$ of $\boldsymbol{\xi}$, obtained from a separate set of estimating equations, is substituted into the WGEE and (12.18) is then solved for $\boldsymbol{\theta}$ to obtain the WGEE estimate $\widehat{\boldsymbol{\theta}}$ of $\boldsymbol{\theta}$. Since $\widehat{\boldsymbol{\theta}}$ is conditional upon $\widehat{\boldsymbol{\xi}}$, its asymptotic variance is then adjusted to account for the sampling variability of $\widehat{\boldsymbol{\xi}}$. If $\boldsymbol{\alpha}$ is $\sqrt{n}$-consistent and $\widehat{\boldsymbol{\xi}}$ is asymptotically normal, the WGEE estimate $\widehat{\boldsymbol{\theta}}$ obtained from (12.17) is consistent and asymptotically normal [9, 30, 33]. The procedure for adjusting the sampling variability of $\widehat{\boldsymbol{\xi}}$ in the asymptotic variance is quite complex and thus we discuss an alternative approach to estimate $\boldsymbol{\xi}$ and $\boldsymbol{\theta}$ simultaneously.

Let

$$\mathbf{f}_i = \left(\mathbf{f}_{i1}^\top, \ldots, \mathbf{f}_{iT}^\top, z_i, r_{i2}, \ldots, r_{iT}\right)^\top, \quad \mathbf{h}_i = \left(\mathbf{h}_{i1}^\top, \ldots, \mathbf{h}_{iT}^\top, \pi_i, p_{i2}, \ldots, p_{iT}\right)^\top, \quad (12.19)$$

$$\boldsymbol{\theta} = \left(\boldsymbol{\beta}^\top, \boldsymbol{\eta}^\top, \boldsymbol{\gamma}^\top\right)^\top, \quad 1 \le i \le n, \quad 1 \le t \le T,$$

where $\mathbf{f}_{it}$, $\mathbf{h}_{it}$, and π_i are defined in (12.15), and r_{it} and p_{it} are defined in (12.17). Consider the WGEE in (12.18), but with D_i and Ψ_i redefined as follows to provide estimates for both $\boldsymbol{\theta}$ and $\boldsymbol{\xi}$:

$$D_i = \frac{\partial}{\partial \boldsymbol{\theta}} \mathbf{h}_i, \quad V_i = \begin{pmatrix} V_{i11} & 0 & 0 \\ 0 & V_{i22} & 0 \\ 0 & 0 & V_{i33} \end{pmatrix}, \quad V_{i11} = A_i^{\frac{1}{2}} R(\boldsymbol{\alpha}) A_i^{\frac{1}{2}}, \quad V_{i22} = \pi_i (1 - \pi_i),$$

$$V_{i33} = \begin{pmatrix} p_{i2}(1-p_{i2}) & \cdots & 0 \\ & \ddots & \vdots \\ & \cdots & p_{iT}(1-p_{iT}) \end{pmatrix}, \quad \Psi_i = \begin{pmatrix} \Psi_{i11} & 0 & 0 \\ 0 & \Psi_{i22} & 0 \\ 0 & 0 & \Psi_{i33} \end{pmatrix}, \quad (12.20)$$

$$\Psi_{i11} = \operatorname{diag}(\Psi_{it}), \quad \Psi_{it} = r_{it} \left(\prod_{s=1}^{t} p_{is}\right)^{-1} \mathbf{I}_2, \quad \Psi_{i22} = 1, \quad \Psi_{i33} = \begin{pmatrix} r_{i1} & \cdots & 0 \\ & \ddots & \vdots \\ & & r_{i(T-1)} \end{pmatrix},$$

where A_i is defined in (12.17). Unlike (12.18), the WGEE in (12.19) makes joint inference about $\boldsymbol{\theta}$ and $\boldsymbol{\xi}$. Thus, no adjustment is necessary for the asymptotic variance of the WGEE estimate of $\boldsymbol{\theta}$ to account for the sampling variability of $\widehat{\boldsymbol{\xi}}$ as in the standard approach above.

3.2 Multi-layered Intervention Study

We now extend the SFRM above to multi-layered interventions to address treatment noncompliance from different intervention layers, such as the child and parent layers of the CRP. For notational brevity, we focus on two-layered interventions, since extensions to multi-layered interventions with more than two layers are straightforward.

Consider a two-layered intervention study and let u_{i1} denote some (continuous) treatment compliance measure for the second layer. By taking into account both compliance measures d_{i1} and u_{i1}, we obtain from (12.11) the following dose–response relationship :

$$E\left(y_{i1} \mid \mathbf{x}_i, d_{i1}, u_{i1}, z_i = 1\right) = g\left(d_{i1}, u_{i1}\right) + E\left(y_{i0} \mid \mathbf{x}_i, z_i = 0\right). \tag{12.21}$$

where $g(d_{i1}, u_{i1})$ does not depend on $\mathbf{x}_i$ for model simplicity. We assume that the covariates $\mathbf{x}_i$ sufficiently explain treatment compliance patterns for both the primary and secondary layers of the multi-layered intervention, i.e., $d_{i1}, y_{i0} \perp \mathbf{x}_i$ and $u_{i1}, y_{i0} \perp \mathbf{x}_i$. In some studies, treatment noncompliance may be limited to some intervention layers, in which case $\mathbf{x}_i$ is only required to explain the affected layers. For example, in the CRP, noncompliance is a major issue only for the second parent support layer and the ignorability condition only needs to be assumed for parent participation.

By formulating (12.21) as an FRM as in the case of single-layered intervention study, we obtain the following SFRM for modeling the effect of treatment noncompliance on the outcome in a two-layered intervention study:

$$\begin{aligned}
f_{i1} &= \frac{z_i y_{i1}}{\pi\left(\mathbf{w}_i; \boldsymbol{\eta}\right)}, \quad f_{i2} = \frac{\left(1 - z_i\right) y_{i0}}{1 - \pi\left(\mathbf{w}_i; \boldsymbol{\eta}\right)}, \quad f_{i3} = z_i, \quad 1 \leq i \leq n, \\
h_{i1} &= h_i\left(\mathbf{x}_i, \boldsymbol{\beta}\right), \quad h_{i2}\left(\mathbf{x}_i, d_{i1}, u_{i1}\right) = g\left(d_{i1}, u_{i1}, \boldsymbol{\gamma}\right) + h_1\left(\mathbf{x}_i, \boldsymbol{\beta}\right), \\
\pi_i &= \pi\left(\mathbf{w}_i; \boldsymbol{\eta}\right), \quad E\left(z_i \mid \mathbf{x}_i, d_{i1}, u_{i1}, \boldsymbol{\theta}\right) = \pi_i, \quad \boldsymbol{\theta} = \left(\boldsymbol{\beta}^\top, \boldsymbol{\gamma}^\top, \boldsymbol{\eta}^\top\right)^\top.
\end{aligned} \tag{12.22}$$

where $1 \leq i \leq n$. The above has essentially the same form as the single-layered SFRM, except that the treatment effect $g\left(d_{i1}, u_{i1}, \boldsymbol{\gamma}\right)$ is a function of compliance from both the primary and secondary intervention layers. Note that (12.22) applies to observational studies well, in which case $\mathbf{w}_i$ is assumed to account for all sources of selection bias.

We can model treatment effect $g\left(d_{i1}, u_{i1}, \boldsymbol{\gamma}\right)$ to reflect treatment compliance in both layers. For example, we may specify an additive effect function, $g\left(d_{i1}, u_{i1}, \boldsymbol{\gamma}\right) = \gamma_1 d_{i1} + \gamma_2 u_{i1}$ or we may also include a between-layer treatment compliance interaction $d_{i1} u_{i1}$. If the treatment effect is moderated by some covariate x_i, we may also include treatment moderating effect by setting $g\left(d_{i1}, u_{i1}, x_i, \boldsymbol{\gamma}\right) = x_i\left(\gamma_1 d_{i1} + \gamma_2 u_{i1}\right)$. If the moderating effect only occurs to one of the intervention layers, we may model $g\left(d_{i1}, u_{i1}, x_i, \boldsymbol{\gamma}\right)$ as $\gamma_1 x_i d_{i1} + \gamma_2 u_{i1}$ or $\gamma_1 d_{i1} + \gamma_2 x_i u_{i1}$, depending on whether the moderating effect operates at the primary or secondary layer of the intervention.

As in the case of single-layered intervention study, the cross-sectional SFRM in (12.22) is readily extended to longitudinal studies. For example, by replacing the treatment effect function $g_t\left(d_{i1}, \boldsymbol{\gamma}\right)$ in (12.15) by $g_t\left(d_{i1}, u_{i1}, \boldsymbol{\gamma}\right)$ in (12.22), the SFRM in (12.15) can be applied to model the effect of treatment compliance for two-layered observational studies. As well, by modeling the missing data under MAR using (12.17), we can make joint inference about $\boldsymbol{\theta}$ in (12.22) and $\boldsymbol{\xi}$ for the missing data model using a WGEE akin to (12.18), but with D_i, V_i, Ψ_i, and S_i in (12.20) redefined based on (12.22).

In the above, we have assumed that both d_{i1} and u_{i1} are continuous. The models are easily extended to non-continuous compliance variables, if either d_{i1} or u_{i1} or both are non-continuous.

4 Simulation Studies

We carried out a series of simulation studies to assess the performance of the proposed SFRM for multi-layered intervention studies for the most general case under both pre-treatment and post-treatment confounders. Since our CRP is a two-layered intervention study, we only considered this special case for the simulation study. We assessed the performance of the models under both cross-sectional and longitudinal data.

We considered continuous and binary outcomes y_i for both cross-sectional and longitudinal data settings, with a continuous treatment noncompliance variable for both the primary and secondary layer. For space consideration, we only report results for two sample sizes $n = 50$ and 200 for a continuous response in cross-sectional data (Model I) and $n = 100$ and 400 for a binary response in longitudinal data (Model II). The increase in sample size for the binary outcome is to achieve more reliable estimates because of data sparseness in this binary response case, especially in the presence of missing data in the longitudinal data setting. All simulations were performed with a Monte Carlo (MC) sample of 1000. All analyses were carried out using codes developed by the authors for implementing the models considered using the R software platform [22].

For the cross-sectional data scenario, let y_{ik} $(k = 0, 1)$ be a continuous outcome in Model I and let d_i (u_i) denote a continuous treatment noncompliance variable for the primary (secondary) intervention layer. Model I for the continuous y_{ik} is defined as follows:

Model I—Continuous y_{ik} for Cross-sectional Data (12.23)

$$y_{i0} \mid x_i, b_i = \mu(x_i; \boldsymbol{\beta}) + b_i + e_{i0}, \quad \mu(x_i; \boldsymbol{\beta}) = \beta_0 + \beta_1 x_i,$$

$$y_{i1} \mid \begin{cases} d_i, u_i, x_i, b_i = g_1(d_i, u_i, x_i; \boldsymbol{\gamma}) + \mu(x_i; \boldsymbol{\beta}) + b_i + e_{i1} \\ d_i, u_i, x_i, c_i, b_i = g_2(d_i, u_i, x_i, c_i; \boldsymbol{\gamma}) + \mu(x_i; \boldsymbol{\beta}) + b_i + e_{i1} \end{cases},$$

$$g_1(d_i, u_i, x_i; \boldsymbol{\gamma}) = \gamma_0 d_i + \gamma_1 u_i + \gamma_2 u_i d_i, \quad g_2(d_i, u_i, x_i, c_i; \boldsymbol{\gamma}) = c_i g_1(d_i, u_i, x_i; \boldsymbol{\gamma}),$$

$$\pi_i = \text{logit}^{-1}(\eta_0 + \eta_1 x_i), \quad d_i, u_i \sim U(0, 5), \quad x_i, c_i \sim N(0, 1),$$

$$b_i \sim (\chi_1^2 - 1)\sqrt{\sigma_b^2/2}, \quad e_{i1}, e_{i0} \sim (\chi_1^2 - 1)\sqrt{\sigma^2/2},,$$

$$\boldsymbol{\beta} = (\beta_0, \beta_1)^\top = (5, 2), \quad \boldsymbol{\gamma} = (\gamma_0, \gamma_1, \gamma_2)^\top = (0.5, 0.5, 0.4),$$

$$\boldsymbol{\eta} = (\eta_0, \eta_1)^\top = (0, -1), \quad \sigma_b^2 = \sigma^2 = 1, \quad \boldsymbol{\theta} = \left(\boldsymbol{\beta}^\top, \boldsymbol{\gamma}^\top, \boldsymbol{\eta}^\top\right)^\top,$$

where z_i is the indicator of treatment assignment, x_i is a confounding variable (for both pre- and post-treatment), c_i is a treatment moderator, g_1 (g_2) is a function modeling the effect of treatment noncompliance without (with) the treatment moderator, $U(a,b)$ denotes a uniform over the interval between a and b, and χ_p^2 denotes a χ^2 distribution with p degrees of freedom. Since (y_{i0}, y_{i1}) share the same random effect b_i, they are not independent. Note that to demonstrate robustness of the SFRM, both the random effect b_i and model error e_{ik} followed non-normal distributions. In (12.23), we considered two treatment effect functions, $g_1(d_i, u_i, x_i; \boldsymbol{\gamma})$ and $g_2(d_i, u_i, x_i, c_i; \boldsymbol{\gamma})$, with the latter including a moderating effect of the former by a treatment moderator c_i. This moderator c_i can be associated with either the primary or secondary layer of the multi-layered intervention.

Shown in Table 12.2 are the estimates of $\boldsymbol{\theta}$, along with their model-based (Mod. S.E.) and empirical (Emp. S.E.) standard errors for Model I. The model-based standard errors were computed based on the estimated asymptotic variance, while their empirical counterparts were calculated from the MC replicates. At the larger sample size $n = 200$, all parameter estimates were quite close to the true values of the respective parameters. The model-based standard errors also matched their empirical counterparts quite well. Although the difference all increased between the parameter estimates and their true values and between the model-based and empirical standard errors for the smaller sample size $n = 50$, the SFRM still performed quite well.

For the longitudinal data, as noted earlier, we only report results for a binary response. We extended both the mean for the control group, $\mu_t(x_i; \boldsymbol{\beta})$, and the treatment effect function, $g_t(d_i, u_i, x_i, c_i; \boldsymbol{\gamma})$, in the cross-sectional case to include a temporal trend. In addition, to reflect the treatment noncompliance patterns in the CRP study, where treatment noncompliance only occurred in the supportive parent layer, we only considered treatment noncompliance in the second layer. As in the cross-sectional data setting, we also included a treatment moderator $c_i$in $g_t(d_i, u_i, x_i, c_i; \boldsymbol{\gamma})$. For notational brevity, we only considered one treatment effect function and two assessments, with $t = 1$ (2) denoting the baseline (follow-up). We created about 22 % missing data at the follow-up.

We discussed two approaches for longitudinal data analysis. The first employs the conventional WGEE that conditions on the estimates of the missing data model and adjusts the variance estimates of parameter estimates to account for the sampling variability in the estimates of the missing data model. Since the adjustment part is quite complex, we also discussed an alternative that utilized the flexibility of FRM to model both missing data and treatment effect simultaneously. We used this latter approach in the simulation study.

Table 12.2 Parameter estimates and standard errors for Model I with a cross-sectional continuous response

Parameter estimates and standard errors for Model I with treatment effect functions g_1/g_2

	$n = 50$			$n = 200$		
Parameter	Est.	Mod. S.E.	Emp. S.E.	Est.	Mod. S.E.	Emp. S.E.
$\gamma_0 = 0.5$	0.475/0.514	0.629/0.566	0.744/0.758	0.496/0.495	0.318/0.274	0.326/0.289
$\gamma_1 = 0.5$	0.459/0.535	0.655/0.553	0.808/0.760	0.505/0.515	0.315/0.271	0.323/0.300
$\gamma_2 = 0.4$	0.428/0.377	0.369/0.313	0.453/0.438	0.402/0.395	0.179/0.163	0.196/0.175
$\beta_0 = 5$	5.107/4.981	0.289/0.304	0.338/0.394	4.998/5.012	0.158/0.157	0.175/0.172
$\beta_1 = 2$	1.995/2.013	0.341/0.348	0.402/0.509	2.002/1.976	0.189/0.189	0.201/0.195
$\eta_0 = 0$	0.033/-0.003	0.325/0.326	0.329/0.343	0.001/-0.003	0.150/0.157	0.158/0.150
$\eta_1 = -1$	$-1.086/-1.107$	0.394/0.395	0.422/0.461	$-1.017/-1.016$	0.189/0.189	0.188/0.199

For the binary response y_{ik}, the SFRM is given by

$$\text{Model II—Binary } y_{ik} \text{ for Longitudinal Data Setting} \tag{12.24}$$

$$\begin{aligned}
&y_{it0} \mid x_i = \text{logit}^{-1}\left(\mu_t(x_i;\boldsymbol{\beta})\right), \mu_t(x_i;\boldsymbol{\beta}) = \beta_0 + \beta_1 t + \beta_2 x_i + \beta_3 x_i t,\\
&y_{it1} \mid d_i, u_i, x_i, c_i = \text{logit}^{-1}\left(g_t(d_i, u_i, x_i, c_i;\boldsymbol{\gamma})\right) + \mu_t(x_i;\boldsymbol{\beta})\},\\
&\pi_i = \text{logit}^{-1}\left(\eta_0 + \eta_1 x_i\right), \quad g_t(d_i, u_i, x_i, c_i;\boldsymbol{\gamma}) = \gamma_0 u_i t + \gamma_2 c_i u_i t,\\
&p_i = \text{logit}^{-1}\left(\xi_0 + \xi_1 y_{i0}^o\right), \quad d_i, u_i \sim U(0,4), \quad x_i, c_i \sim N(0.1)\\
&\boldsymbol{\beta} = (\beta_0, \beta_1, \beta_2, \beta_3)^\top = (-1, 1, 1, -1), \quad \boldsymbol{\gamma} = (\gamma_0, \gamma_1)^\top = (1, 1),\\
&\boldsymbol{\eta} = (\eta_0, \eta_1)^\top = (0, -1), \quad \boldsymbol{\xi} = (\xi_0, \xi_1)^\top = (1, 1), \quad \boldsymbol{\theta} = \left(\boldsymbol{\beta}^\top, \boldsymbol{\gamma}^\top, \boldsymbol{\eta}^\top, \boldsymbol{\xi}^\top\right)^\top.
\end{aligned}$$

where $p_i = E(r_{i1} = 1 \mid y_{i1})$ is the probability of missing data at the follow-up $t = 2$ for both the treatment and control groups. For the control group, we included a time as well as a time by covariate interaction. As indicated earlier, the treatment effect function $g_t(d_i, u_i, x_i, c_i;\boldsymbol{\gamma})$ also included a treatment moderator c_i. Since the probability of missing response at post-treatment p_i depends on the baseline y_{i1}, the missing data mechanism follows the MAR. Under the specified $\boldsymbol{\xi}$, there was about 22 % missing data. The correlated y_{itk} were created by the copula methods [20, 34]. The correlation between the two potential outcomes with each assessment time as well as between two assessments within the same potential outcome in our simulation study was set at about 0.5, uncontrolled for any of the explanatory variables.

Shown in Table 12.3 are the estimates of $\boldsymbol{\theta}$, along with their model-based (Mod. S.E.) and empirical standard (Emp. S.E.) errors for Model II. In comparison with the cross-sectional data case, Table 12.3 contains estimates for the additional parameters

Table 12.3 Parameter estimates and standard errors for Model II with a longitudinal binary response

Parameter estimates and standard errors for Model II						
	$n = 100$			$n = 400$		
Parameter	Est.	Mod. S.E.	Emp. S.E.	Est.	Mod. S.E.	Emp. S.E.
$\gamma_0 = 1$	0.873	0.497	0.508	1.047	0.290	0.321
$\gamma_1 = 1$	0.964	0.512	0.586	1.070	0.284	0.317
$\beta_0 = -1$	−1.067	0.468	0.491	−1.022	0.216	0.213
$\beta_1 = 1$	1.128	0.461	0.525	1.025	0.205	0.217
$\beta_2 = 1$	1.089	0.589	0.605	1.016	0.264	0.279
$\beta_3 = -1$	−1.182	0.583	0.690	−1.044	0.261	0.275
$\eta_0 = 0$	0.021	0.209	0.224	−0.001	0.108	0.109
$\eta_1 = -1$	−1.058	0.291	0.303	−1.008	0.144	0.148
$\xi_0 = 1$	1.022	0.273	0.292	1.010	0.131	0.135
$\xi_1 = 1$	1.088	0.689	0.702	1.033	0.325	0.346

$\boldsymbol{\xi} = (\xi_0, \xi_1)^\top$ for the missing data model. As in the case of cross-sectional data, both the parameter estimates and model-based standard errors were quite good when compared to their true values or empirical counterparts.

5 Child Resilience Project

The Child Resilience Project (CRP) is a randomized two-layered intervention study with significant treatment noncompliance by the parent, whose participation forms the second supportive layer of the intervention. The study's enrollment began in Fall 2006, with data collection for the final cohort completed by June 2011. There were 401 students from first up to third grade from Rochester City School District elementary schools. The study examines how children with a higher risk of developing behavioral problems in the intervention condition improve as compared to the control condition over a 30-month period. Each child was assessed at baseline, and 6, 18, and 30 months post baseline.

Since treatment compliance was quite good for the children in the study, we only considered variability in the parent participation. In order to apply the proposed SFRM to analyze the data in this study, we first examined the baseline covariates to see if any of these variables effectively predicted the patterns of treatment noncompliance. We treated the second-layer noncompliance measure, u_i, the number of session attendance by the parent, as a continuous variable and applied linear regression.

Shown in Table 12.4 are the estimated coefficients, standard errors, and p-values of the variables that significantly predicted the number of session attendance u_i from the linear regression model. The variable School Number represents the different schools which the children attended. The variable PNC stands for the Perceived Need for Care scale, assessing frequency over the past six months that parent viewed her child as needing help for behavior or emotional problems, including from communication with others about child [19]. The DomEX

Table 12.4 Estimates, standard errors, and p-values for significant predictors of parent participation for the Rochester Resilience Project from generalized linear models

Significant predictors for parent participation			
Explanatory variable	Estimate	Standard error	p-value
PNC	0.9191	0.2698	0.0008
Parent age	0.0882	0.0293	0.0030
DomEX baseline	0.9127	0.0373	0.0154
School number			<.0001
School 19	−4.1065	0.8446	<.0001
School 22	−3.3860	0.9122	0.0003
School 30	−3.1342	0.9873	0.0018
School 45	−4.3626	0.8440	<.0001
School 50	0.0000		

Baseline denotes the baseline value of the subscale of the Dominic Interactive self-report, assessing symptoms of three externalizing (oppositional defiant, conduct problems, and ADHD) problems [28]. The results from the regression show that session participation was significantly different across the different schools and children with different PNC and DomEX baseline values. In addition, parent age also significantly predicted the session attendance.

For our illustrations of the model, we focused on two primary behavior outcomes of the study, the Teacher ratings of aggressive behavior (AthAcc) and Parent rating of internalizing behavior problem (PIntD). For both outcomes, higher values indicate fewer problems. For each of these behavior outcomes y_{it}, let y_{it1} and y_{it0} denote the potential outcomes of y_{it} at baseline ($t = 1$) and each of the three follow-ups ($2 \leq t \leq 4$). We modeled the causal treatment effect as a function of treatment compliance from the parent layer using an SFRM as follows:

$$E\left(\frac{1-z_i}{1-\pi} y_{it0} \mid u_i\right) = \mu_{it}, \quad E\left(\frac{z_i}{\pi} y_{it1} \mid u_i\right) = g_{it} + \mu_{it}, \quad E(z_i) = \pi, \qquad (12.25)$$

$$\begin{aligned} \mu_{it} &= \beta_0 + \beta_1 t + x_{i1}\beta_2 + \beta_3 x_{i1} t + \beta_4 x_{i2} + \beta_5 x_{i3} + \beta_6 x_{i4} + \\ &\quad + \beta_7 x_{i5} + \beta_8 x_{i6} + \beta_9 x_{i7} + \beta_{10} x_{i8}, \\ g_{it} &= \gamma u_i t, \quad 1 \leq t \leq 4, \end{aligned}$$

where z_i is the indicator variable of treatment assignment with $z_i = 1$ (0) for intervention (control), x_{i1} denotes the age of the child at baseline, $x_{i2} - x_{i5}$ denote the four binary indicators of School 19, 22, 30, 45, and x_{i6}, x_{i7}, and x_{i8} denote the PNC, DomEXT Baseline, and Parent Age variables, respectively. In addition, we included Age and Age by time interaction, since our theory and preliminary analyses show that these behavioral outcomes have different trajectories for children of different ages.

Prior to fitting the SFRM, we examined the missing data mechanism using logistic regression to determine whether missing data at each of the follow-up times, 6, 18, and 30 months post-baseline, depended on the observed outcomes at prior assessment times. Results indicated that missing data was not associated with the observed data for any of the two behavior outcomes considered. Thus, we assumed the dropouts for these two behavior outcomes in this CRP study followed the Missing Complete at Random (MCAR) mechanism. The MCAR mechanism was also consistent with the excellent treatment compliance observed for the study subjects (children), since unlike parent participation both the intervention and assessment were performed during the regular school time.

Shown in Table 12.5 are the estimates (Est.), standard errors (S.E.), and p-values (p-value) for the parameter γ in the treatment effect function g_{it} in (12.25) for the two behavior outcomes analyzed. Within the context of the study, this parameter γ measures the rate of change of the behavior outcome per month for each additional session attended by the parent. The results show that for both behavior outcomes γ was quite significant, with the positive estimate indicating that the intervention

Table 12.5 Child resilience complete dataset

Estimation results of treatment time effect (γ)						
	Causal effect			ITT effect		
	Est.	S.E.	p-value	Est.	S.E.	p-value
AthAcc	0.0167	0.0014	< 0.0001	0.0053	0.0069	0.2235
PIntD	0.1640	0.0163	< 0.0001	0.0476	0.0663	0.2365

improved the child's behaviors and reduced the risk for future mental disorder and substance abuse. With the SFRM in (12.25), causal treatment effect is given by γu_i. For example, if the parent of the child attended all the planned 15 sessions, then $u_i = 15$ and the causal effect is $\beta_4 u_i = 0.25$ per month time in the scale of the AthAcc outcome. Thus, in 18 months post-baseline, for instance, the intervention will on average increase the child AthAcc outcome by 4.32 points.

For comparison purposes, we also performed the intent-to-treat (ITT) analysis for the two behavior outcomes by setting $u_i = 1$ in g_{it} of the SFRM in (12.25). The estimated γ, standard errors (S.E.), and p-values (p-value) are shown in Table 12.5 under the column "ITT Effect." As seen, γ was not significant for either outcome. Thus, parent support played a significant role in improving the two child behavior outcomes in this two-layered intervention study.

6 Discussion

We developed an approach to address treatment noncompliance in multi-layered intervention studies. This approach extends the structural mean model (SMM) to multi-layered intervention and longitudinal data settings. We selected the SMM to develop our approach because of the need to model treatment noncompliance on a continuous scale. Other competing approaches such as the Principal Stratification method characterize variability in treatment noncompliance using categorical outcomes. However, within the context of multi-layered intervention study, such methods yield a large number of noncompliance categories, limiting their applications. For example, if a four-level categorical outcome is used to characterize treatment noncompliance for each layer of a two-layered intervention, we will need a 16-level categorical outcome to understand treatment noncompliance when considering interactions of noncompliance patterns between the two intervention layers. The larger number of levels of a categorical outcome may cause problems for fitting models, if there are a limited number of subjects in one or more strata (defined by the levels of the categorical outcome). With the freedom to choose a continuous or categorical noncompliance measure as in the SMM and proposed SFRM, we can consider between-layer interactions in a more parsimonious and reliable fashion.

We also adopted the distribution-free framework of SMM for inference for our proposed model. Using the framework of FRM, we are able to provide robust inference about model parameters like the SMM and accommodate noncompliance from multiple intervention layers as well as missing data under MAR. Our simulation studies show that the proposed approach perform quite well even for a sample size as small as 50 (for combined intervention and control groups). As well, applications of the proposed model to the Rochester Resilience Project demonstrate the importance to consider treatment noncompliance from the supportive parent layer in this two-layered intervention study.

Acknowledgements This research was partially supported by NIMH R01 MH091354.

References

1. Angrist, J., Imbens, G.W., Rubin, D.B.: Identification of causal effects using instrumental variables (with discussion). J. Am. Stat. Assoc. **91**, 444–472 (1996)
2. Chen, R., Chen, T., Lu, N., Zhang, H., Wu, P., Feng, C., Tu, X.M.: Extending the Mann-Whitney-Wilcoxon rank sum test to longitudinal data analysis with covariates. J. Appl. Stat. **41**(12), 2659–2675 (2014)
3. Efron, B., Feldman, D.: Compliance as an explanatory variable in clinical trials. J. Am. Stat. Assoc. **91**, 444–472 (1991)
4. El-Sayed, A.M., Scarborough, P., Seemann, L., Galea, S.: Social network analysis and agent based modeling in social epidemiology. Epidemiol. Perspect. Innov. **9**, 1–9 (2012)
5. Fischer, K., Goetghebeur, E.: Structural mean effects of noncompliance. J. Am. Stat. Assoc. **99**(468), 918–928 (2004)
6. Fitzmaurice, G.M.: A caveat concerning independence estimating equations with multiple multivariate binary data. Biometrics **51**, 309–317 (1995)
7. Frangakis, C.E., Rubin, D.B.: Principal stratification in causal inference. Biometrics **58**, 21–29 (2002)
8. Goetghebeur, E., Lapp, K.: The effect of treatment compliance in a placebo-controlled trials: regression with unpaired data. J. R. Stat. Soc. Ser. C Appl. Stat. **46**, 351–364 (1997)
9. Gunzler, D., Tang, W., Lu, N., Wu, P., Tu, X.M.: A class of distribution-free models for longitudinal mediation analysis. Psychometrika **79**(4), 543–568 (2013)
10. King, T.S., Chinchilli, V.M.: A generalized concordance correlation coefficient for continuous and categorical data. Stat. Med. **20**, 2131–47 (2001)
11. Kowalski, J., Powell, J.: Nonparametric inference for stochastic linear hypotheses: application to high-dimensional data. Biometrika **91**(2), 393–408 (2004)
12. Kowalski, J., Tu, X.M.:Modern Applied U Statistics. Wiley, New York (2007)
13. Lu, N., Tang, W., He, H., Yu, Q., Crits-Christoph, P., Zhang, H., Tu, X.M.: On the impact of parametric assumptions and robust alternatives for longitudinal data analysis. Biom. J. **51**, 627–643 (2009)
14. Lu, N., White, A.M., Wu, P., He, H., Hu, J., Feng, C., Tu, X.M.: Social network endogeneity and its implications for statistical and causal inferences. In: Lu, N., White, A.M., Tu, X.M. (eds.) Social Networking: Recent Trends, Emerging Issues and Future Outlook. Nova Science, New York (2013)
15. Lu, N., Chen, T., Wu, P., Gunzler, D., Zhang, H., He, H., Tu, X.M.: Functional response models for intraclass correlation coefficients. J. Appl. Stat. **41**(11), 2539–2556 (2014)
16. Ma, Y., Tang, W., Feng, C., Tu, X.M.: Inference for Kappas for longitudinal study data: applications to sexual health research. Biometrics **64**, 781–789 (2008)
17. Ma, Y., Tang, W., Yu, Q., Tu, X.M.: Modeling concordance correlation coefficient for longitudinal study data. Psychometrika **75**, 99–119 (2010)

18. Ma, Y., Alejandro, G.D., Hui, Z., Tu, X.M.: A U-statistics based approach for modeling Cronbach Coefficient Alpha within a longitudinal data setting. Stat. Med. **29**(6), 659–670 (2011)
19. Meadows, G., Burgess, P., Fossey, E., Harvey, C.: Perceived need for mental health care, findings from the Australian National Survey of Mental Health and Wellbeing. Psychol. Med. **30**, 645–656 (2000)
20. Nelsen, R.B.: An Introduction to Copulas. Springer, New York (2006)
21. Pepe, M.S., Anderson, G.L.: A cautionary note on inference for marginal regression models with longitudinal data and general correlated response data. Commun. Stat. Simul. **23**, 939–951 (1994)
22. R Development Core Team: R: A Language and Environment for Statistical Computing. R Foundation for Statistical Computing, Vienna (2010). ISBN 3-900051-07-0. http://www.R-project.org
23. Robins, J.M.: Correcting for noncompliance in randomized trials using structural nested mean models. Commun. Stat. **23**, 2379–2412 (1994)
24. Robins, J.M., Rotnitzky, A., Zhao, L.P.: Analysis of semiparametric regression models for repeated outcomes in the presence of missing data. J. Am. Stat. Assoc. **90**, 106–121 (1995)
25. Rubin, D.B.: Estimating causal effects of treatments in randomized and nonrandomized studies. J. Educ. Psychol. **66**, 688–701 (1974)
26. Rubin, D.B.: Bayesian inference for causal effects: the role of randomization. Ann. Stat. **6**, 34–58 (1978)
27. Tu, X.M., Feng, C., Kowalski, J., Tang, W., Wang, H., Wan, C., Ma, Y.: Correlation analysis for longitudinal data: applications to HIV and psychosocial research. Stat. Med. **26**, 4116–4138 (2007)
28. Valla, J.P., Bergeron, L., Smolla, N.: The Dominic-R: a pictorial interview for 6- to11-year old children. J. Am. Acad. Child Adolesc. Psychiatry **39**, 85–93 (2000)
29. Vansteelandt, S., Goetghebeur, E.: Causal inference with generalized structural mean models. J. R. Stat. Soc. Ser. B **65**, 817–835 (2003)
30. Wu, P., Tu, X.M., Kowalski, J.: On assessing model fit for distribution-free longitudinal models under missing data. Stat. Med. **33**(1), 143–157 (2014)
31. Wu, P., Han, Y., Chen, T., Tu, X.M.: Causal inference for Mann-Whitney-Wilcoxon rank sum and other nonparametric statistics. Stat. Med. **33**(8), 1261–1271 (2014)
32. Yu, Q., Tang, W., Kowalski, J., Tu, X.M.: Multivariate U-Statistics: a tutorial with applications. Wiley Interdiscip. Rev. Comput. Stat. **3**, 457–471 (2011)
33. Yu, Q., Chen, R., Tang, W., He, H., Gallop, R., Crits-Christoph, P., Hu, J., Tu, X.M.: Distribution-free models for longitudinal count responses with over-dispersion and structural zeros. Stat. Med. **32**, 2390–2405 (2013)
34. Zhang, H., Lu, N., Feng, C., Thurston, S.W., Xia, Y., Tu, X.M.: On fitting generalized linear mixed-effects models for binary responses using different statistical packages. Stat. Med. **30**, 2562–2572 (2011)

Part IV
Structural Equation Models for Mediation Analysis

Chapter 13
Identification of Causal Mediation Models with an Unobserved Pre-treatment Confounder

Ping He, Zhenguo Wu, Xiaohua Douglas Zhang, and Zhi Geng

Abstract In this paper, we discuss identifiability of mediation, direct and indirect effects of treatment on outcome. The mediation effects are represented by a causal mediation model which includes an unobserved confounder (i.e., a common cause of the mediator and the outcome variable), and the direct and indirect effects are represented by the mediation effects. Without requiring the sequential ignorability assumption or the exclusion restriction assumption (i.e., the absence of direct effect of treatment on outcome), we require that only treatment is randomized and that the degree of equation nonlinearity for the treatment effect on the mediator is higher than that for the outcome. If the requirement of nonlinearity degree is not satisfied, we may use a covariate as an instrumental variable to improve the identifiability. In this paper, we focus on the identifiability of parameters, although, to illustrate our identifiability results, we describe estimation approaches. The simulations show good estimation performance by our approach compared to the standard mediation approach.

1 Introduction

A main goal of mediation analysis is to investigate how an independent variable (or a treatment variable) changes an intermediate variable (or a mediator) and how this change in the mediator in turn affects a dependent variable (or an outcome variable). In causal mediation models, the indirect and direct effects are the effects of treatment on the outcome through and not through the mediator, respectively [2, 11, 12]. Baron and Kenny [1] discussed the concept of mediator and its distinction to moderator and proposed methods for examining mediator effects. MacKinnon et al. [9] reviewed three main approaches to statistical mediation analysis: (a) causal steps, (b) difference in coefficients, and (c) product of coefficients, and Li et al. [8]

P. He • Z. Wu • Z. Geng
School of Mathematical Sciences, Peking University, Beijing 100871, China
e-mail: sunhp@pku.edu.cn; wuzhenguo@gmail.com; zhigeng@pku.edu.cn

X.D. Zhang (✉)
Faculty of Health Sciences, University of Macau, Macau, China
e-mail: douglaszhang@umac.mo

H. He et al. (eds.), *Statistical Causal Inferences and Their Applications in Public Health Research*, ICSA Book Series in Statistics, DOI 10.1007/978-3-319-41259-7_13

presented an approach for a binary mediator. All of these approaches use three linear models which include three variables: a treatment, a mediator, and an outcome variable, but do not include an unobserved confounder which affects both the mediator and the outcome variables (i.e., a strong form of sequential ignorability). Jo [6] compared two different mediation analysis approaches: the structural equation modeling approach and the principal stratification model. The former assumes that there is no confounder which affects both the mediator and the dependent variable, that is, the ignorability assumption of the mediator status. The latter assumes that the effect of treatment on the outcome is completely mediated through the mediator, that is, no direct effect of treatment on the outcome, also called the exclusion restriction assumption. VanderWeele [15] discussed the estimation of direct and indirect effects under the assumptions of no unobserved variable which confounds the treatment–outcome relationship or the mediator–outcome relationship. Imai et al. [5] discussed the identification of causal mediation effects under the sequential ignorability assumption, which is different from the no observed confounder assumption. Sobel [13] discussed the identification and estimation of causal effects using an instrumental variable (IV) which satisfies the exclusion restriction assumption. However the exclusion restriction assumption means no direct treatment effect on the outcome variable which requires that all treatment effects on the outcome variable are blocked by the mediator, and it may be too strong in many real applications. For these approaches, the required assumptions are untestable from observed data and may be very restrictive or impractical in observational studies and even in experimental studies where only the treatment assignments can be manipulated. Herting [4] and Kaufman et al. [7] pointed out, respectively, that the models with and without direct effect of treatment on the outcome are statistically indistinguishable and that the parameters are not identifiable when there exists an unobserved confounder between the mediator and the outcome. For models with unobserved confounders, Ten Have et al. [14] presented an approach for estimating direct and indirect effects via G-estimation equations which requires an additional covariate satisfying some conditions.

In this paper, we describe models of the outcome and the mediator which include an unobserved pre-treatment confounder (i.e., a common cause of the mediator and the outcome). For an experimental study of randomized treatment assignment or an observational study where the assignment of treatment is ignorable conditionally on observed covariates, we propose an approach for identifying parameters in the models. Without requiring the sequential ignorability assumption or the exclusion restriction assumption, we require that the degree of equation nonlinearity for the treatment on the mediator is higher than that for the treatment on the outcome. For example, the mediator model is nonlinear with respect to treatment, and the outcome model is linear with respect to treatment. Especially when the mediator is a binary variable and it has a logistic regression model, then the nonlinearity condition may be generally satisfied. As an example, let an binary variable indicate whether an irregular heartbeat is corrected as the intermediator between a treatment variable and the outcome of survival time. The nonlinear requirement can be considered as a parametric and functional assumption on the model of treatment effect on mediator. Unlike the untestability of sequential ignorability assumption, the nonlinearity assumption on the model of treatment effect on mediator is testable by using the observed data. This testability is one advantage of our approach. If the nonlinearity is not satisfied, we may try to use a covariate to improve the identifiability. The covariate Z requires a model assumption like that for an instrumental variable which can be used to remove the confounding bias generated by an unobserved confounder. In our models, the outcome variable is continuous, the treatment may be continuous or ordinal or discrete, and the mediator may be discrete or continuous. In this paper, we first discuss the identifiability of parameters in mediation models, and then to illustrate our identifiability results, we describe an instrumental variable estimation approach via the efficient instrument variable proposed by Newey and McFadden [10] and the generalized method of moments (GMM) estimators developed by Hansen [3].

Section 2 gives the notation and definitions of mediation models and direct and indirect effects. The conditions for identifiability of parameters in the models are presented in Sect. 3. An estimation approach of parameters is presented in Sect. 4. In Sect. 5, we compare our approach with the ordinary least squares (OLS) regression via simulations. In Sect. 6, we extend the results to more general models, such as the moderated-mediation model which has an interaction of treatment and mediator on the outcome. Finally we give discussions in Sect. 7. Most proofs of theoretical results are presented in the supplementary material.

2 Notation and Definitions

Let Y be an outcome, X be a treatment, and M be a mediator. Two main equations of a standard mediation model are

$$\begin{aligned} Y &= b_0 + b_1 M + b_2 X + \varepsilon_Y, \\ M &= \psi(X, \varepsilon_M), \end{aligned}$$

where $\psi(\cdot)$ is an arbitrary function (usually a linear function), and ε_Y and ε_M are two mutually independent random errors with means 0 and variances σ_Y^2 and σ_M^2, respectively [8, 9]. In the linear structural model, the model for M is

$$M = a_0 + a_1 X + \varepsilon_M.$$

The mediation effect of X on Y is interpreted as $a_1 b_1$ [9]. Since the standard mediation model requires that there is not any unobserved confounders which affect both M and Y, the parameters a_1, b_1, b_2 can be identified and consistently estimated by two linear regressions. Unfortunately, the no unobserved confounder assumption generally does not hold in practical studies. To identify the parameters of the standard mediation model, we can design a two-step randomized experiment to obtain two data sets:

- Step I, only treatment X is randomized and we obtain one data set of X and M. Then the parameters a_0 and a_1 of the model for M can be identified and estimated by this data set;
- Step II, both treatment X and mediator M are randomized and we obtain the other data set of X, M, and Y. Then the parameters b_0, b_1, and b_2 in the model for Y can be identified and estimated by this data set.

Thus the mediation effect $a_1 b_1$ can be identified by the two data sets. Notice that the randomization of both the treatment and the mediator is not sufficient for the sequential ignorability assumption. However, such a two-step randomized experiment may not be practical for many applications. Especially the mediator (e.g., the blood pressure as a mediator in a clinical trial) is almost impossible to be randomized since it cannot be manipulated or controlled directly.

Below we consider how to identify and estimate the mediation effect without the requirement of no unobserved confounder assumption. Allowing the existence of an unobserved confounder between mediator and outcome, we introduce the following causal model. Let U be an unobserved pre-treatment confounder which is a common cause of Y and M, and U may be a continuous or discrete variable or a variable vector. Hereafter we assume that treatment X is randomized or that the assignment of X is ignorable conditionally on observed covariates and we omit these covariates for simplicity. Then X is independent of the confounder U, and the relationships among X, U, M, and Y can be depicted as a causal diagram in Fig. 13.1.

Below we use the potential outcome framework to describe the causal model of X, M, U, Y. Let M_x and Y_{xm} be the values of M and Y that would have been observed if X and M were set to x and m, respectively [15]. Consider the following linear models for M_x and Y_{xm} with the unobserved pre-treatment confounder U

$$Y_{xm} = b_0 + b_1 m + b_2 x + \phi(U, \varepsilon_Y), \tag{13.1}$$

$$M_x = \psi(x, U, \varepsilon_M), \tag{13.2}$$

Fig. 13.1 A causal diagram with a randomized X and an unobserved confounder U

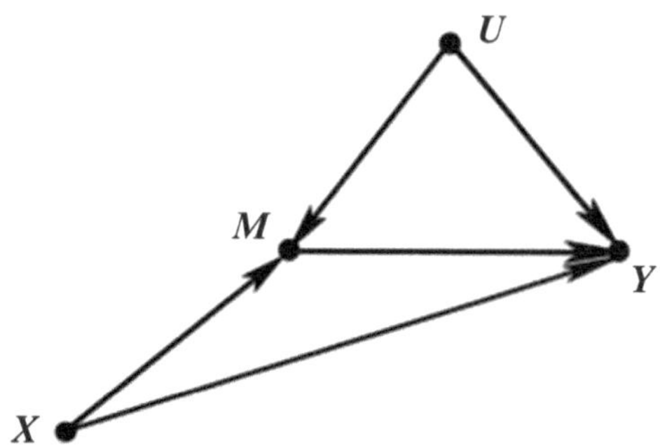

where $\psi(\cdot)$ and $\phi(\cdot)$ are arbitrary functions, and ε_Y and ε_M with means 0 are independent of (X, M, U) and (X, U), respectively.

With the definitions and notation of Pearl [11], the average total effect $\tau(x, x')$ of treatment X on outcome Y, the average controlled direct effect $\eta_C(x, x'; m)$ of X on Y when controlling M, and the average natural direct effect $\eta_N(x, x')$ of X on Y are defined for treatment levels x versus x', respectively, as follows:

$$\tau(x, x') = E(Y_x - Y_{x'}) = b_1[E(M_x) - E(M_{x'})] + b_2(x - x'), \quad (13.3)$$

$$\eta_C(x, x'; m) = E(Y_{xm} - Y_{x'm}) = b_2(x - x'), \quad (13.4)$$

$$\eta_N(x, x') = E(Y_{x,M_{x'}} - Y_{x'}) = b_2(x - x'), \quad (13.5)$$

where $M_{x'}$ in (13.3) denotes the value of M if X were set to x'. The average controlled direct effect $\eta_C(x, x'; m)$ means the average effect of x versus x' on Y when the mediator M is fixed at a value m; The average natural direct effect $\eta_N(x, x')$ means the average effect of x versus x' on Y when the mediator M is fixed at the value of $M_{x'}$ which would have been set naturally under $X = x'$ [11, 16]. For model (13.1), we have that the controlled direct effect η_C does not depend on m and that the controlled direct effect equals the natural direct effect, that is, $\eta_C(x, x'; m) = \eta_N(x, x')$, hereafter denoted as $\eta(x, x')$.

Pearl [11] defines the average natural indirect effect as

$$\nu(x, x') = E(Y_{xM_x} - Y_{xM_{x'}}).$$

It represents the average difference between the potential outcome $Y_x = Y_{xM_x}$ that would result under treatment status x, and the potential outcome $Y_{xM_{x'}}$ that would occur if the treatment status is the same and yet the mediator takes a value $M_{x'}$ that would result under the other treatment status x', called the average causal mediation effect [5].

Since model (13.1) is a linear model, the average natural indirect effect $\nu(t, t')$ of T on Y is equal to the difference of the average total effect and the average direct effect

$$\nu(x, x') = \tau(x, x') - \eta(x, x') = b_1[E(M_x) - E(M'_x)], \quad (13.6)$$

which is b_1 times the average causal effect of X on M. Thus the identifiability of average total, direct, and indirect effects is equivalent to the identifiability of parameters b_1 and b_2, while $E(M_x) = E(M|X = x)$ is identifiable for a randomized treatment experiment.

Note that if both treatment X and mediator M are randomized, then b_1, b_2 and the average direct effect of X on Y can be identified, but the effect of treatment on mediator is not identifiable. Similar to the two-step randomized experiment, to identify the average indirect effect, we need identify the effect of treatment X on mediator M using an additional data set from an experiment where we only randomize the treatment X and observe the mediator M.

For observed variables X, M, and Y, we have $M = M_X$ and $Y = Y_{XM}$. Thus the models (13.1) and (13.2) imply that the models for the observed variables X, M, and Y should be

$$Y = b_0 + b_1 M + b_2 X + \phi(U, \varepsilon_Y), \tag{13.7}$$

$$M = \psi(X, U, \varepsilon_M). \tag{13.8}$$

Parameters b_1 and b_2 are called mediation effects in Jo [6]. For (13.7), the OLS estimates of parameters b_0, b_1, and b_2 are inconsistent because U is correlated to M. It will be shown in Sect. 3 that these parameters in (13.7) are not identifiable if the function $\psi(\cdot)$ in (13.8) is linear with respect to X, as assumed in the traditional IV method.

3 Identifiability of Parameters in Mediation Models

In this paper, we assume that treatment X is randomly assigned or that the assignment of treatment X is ignorable conditionally on an observed covariate and we omit the observed covariate for simplicity. But we do not assume that the assignment of mediator M is ignorable or sequentially ignorable. According to the equations (13.3), (13.4), (13.5), and (13.6), the average direct and indirect effects are identifiable if the parameters in model (13.7) are identifiable.

In this section, we first present a general condition for identifiability of these parameters, then we discuss two special cases where the mediator M is discrete or continuous, and finally we discuss an approach to improve identifiability via a covariate when the identifiability condition is not satisfied.

3.1 General Conditions for Identifiability

To avoid the mathematical complexity, we first consider model (13.7) and then extend the result to more general models, such as moderated-mediation and nonlinear direct effect models in Sect. 6, without any essential difficulty. Suppose that treatment X is a continuous variable or an ordinal discrete variable. Without loss of generality, assume that $E[\phi(U, \varepsilon_Y)] = 0$. By randomization of X, we have that X is independent of (U, ε_Y) and then $E[\phi(U, \varepsilon_Y)|X = x] = 0$. Hereafter let $E(\cdot|x)$ denote $E(\cdot|X = x)$ for simplicity. Thus from model (13.7), we have the following equation

$$E(Y|x) = b_0 + b_1 E(M|x) + b_2 x + E[\phi(U, \varepsilon_Y)|x] = b_0 + b_1 E(M|x) + b_2 x. \tag{13.9}$$

Comparing different treatment levels $x_1, \ldots, x_K$, we obtain

$$\begin{bmatrix} E(M|x_1) - E(M|x_2) & x_1 - x_2 \\ \vdots & \vdots \\ E(M|x_{K-1}) - E(M|x_K) & x_{K-1} - x_K \end{bmatrix} \begin{bmatrix} b_1 \\ b_2 \end{bmatrix} = \begin{bmatrix} E(Y|x_1) - E(Y|x_2) \\ \vdots \\ E(Y|x_{K-1}) - E(Y|x_K) \end{bmatrix}. \tag{13.10}$$

Let A denote the $(K-1)\times 2$ matrix on the left-hand side. According to this equation, parameters b_1 and b_2 are identifiable if there exist $K(K \geq 3)$ different levels of treatment X such that the matrix A has full column rank.

If M has a linear model $M = a_0 + a_1X + a_2U + \varepsilon_M$ as usually assumed in simultaneous equation models or X is a binary treatment, then the matrix A is not full rank, and thus parameters b_1 and b_2 are not identifiable, although a_0 and a_1 can be identified via an ordinary method by treating $a_2U + \varepsilon_M$ as an error term since X and (U, ε_M) are independent. Alternatively applying the IV method to model (13.7), we obtain

$$\text{Cov}(X, Y) = b_1\text{Cov}(X, M) + b_2\text{Var}(X) = (b_1a_1 + b_2)\text{Var}(X).$$

Parameters b_1 and b_2 are also not identifiable since there are two parameters but only one equation.

Below we discuss the necessary and sufficient condition for identifiability of parameters in the model (13.7), and we show that parameters are identifiable if and only if the conditional expectation $E(M|X)$ of M given X is not linear with respect to X. Trivially, $E(M|X)$ is linear with respect to X when X is binary.

Theorem 1. *Assume that treatment X is randomly assigned. Parameters b_1 and b_2 in model (13.7) are identifiable if and only if the conditional expectation of M given X is not linear with respect to X, that is, $|\rho(E(M|X), X)| < 1$, where $\rho(E(M|X), X)$ is the correlation coefficient of $E(M|X)$ and X.*

The proof of Theorem 1 is given in Appendix 1. From Theorem 1, we immediately have the following corollary for a continuous or discrete X.

Corollary 1. *Assume that treatment X is randomly assigned. Parameters b_1 and b_2 in model (13.7) are identifiable if*

1. *for a continuous X, $\partial^2 E(M|x)/\partial x^2|_{x=x_0} \neq 0$ for some x_0 in the support of the distribution of X, or*
2. *for a discrete X,*

$$\frac{E(M|x_i) - E(M|x_j)}{x_i - x_j} \neq \frac{E(M|x_j) - E(M|x_k)}{x_j - x_k}$$

for some $x_i < x_j < x_k$.

In the following two subsections, we discuss the identifiability via models of the mediator M for the cases of a discrete or continuous M separately.

3.2 The Case of a Discrete Mediator M

For a discrete variable M with a logistic model, we discuss the identifiability of parameters b_1 and b_2 in model (13.7). When M is a nominal variable with L (> 2) categories, it can be represented by a vector of $L-1$ dummy variables, and a logistic model is used for each dummy variable. When M is an ordinal variable, a cumulative logistic model is often used. For the identifiability problem, we can treat the discrete mediator M as a binary variable without loss of generality. Consider the logistic model

$$\log \frac{P(M=1|x,u)}{1-P(M=1|x,u)} = \alpha_0 + \alpha_1 x + \alpha_2 u.$$

For a continuous X, we have

$$\frac{\partial^2 P(M=1|x)}{\partial x^2} = \alpha_1^2 \int P(M=1|x,u)P(M=0|x,u)[1-2P(M=1|x,u)]dP(u).$$

According to Corollary 1, a sufficient condition for identifiability of parameters b_1 and b_2 is (1) $\alpha_1 \neq 0$, and (2) there is some x_0 such that $P(M=1|x_0,u) \in (0, 0.5)$ for all u or $P(M=1|x_0,u) \in (0.5, 1)$ for all u. For the condition (2), we generally have the probability $P(M=1|x_0,u) < 0.5$ for all u when $M=1$ denotes a rare event (say a kind of rare disease). When U is a normal variable, we can show that the nonlinear condition in Theorem 1 generally holds, that is, $\partial^2 P(M=1|x)/\partial x^2 \neq 0$ for all x except a special value $x = -[\alpha_0 + E(U)]/\alpha_1$. When X is an ordinal discrete variable, the differential with respect to X is replaced by the difference between two adjacent levels of X and we can obtain a similar result.

3.3 The Case of a Continuous Mediator M

As an example of a nonlinear model of M with respect to X, we consider the following quadratic model:

$$M = a_1 X + a_2 X^2 + \psi(U, \varepsilon_M). \tag{13.11}$$

Then parameters a_1 and a_2 can be estimated without bias via an ordinary method by treating $\psi(U, \varepsilon_M)$ as an error term since X is independent of (U, ε_M). Since

$\partial^2 E(M|x)/\partial x^2 = a_2$, according to Corollary 1, parameters b_1 and b_2 in model (13.7) are identifiable if and only if $a_2 \neq 0$.

3.4 The Case of a Linear Model of M with Respect to X

When the nonlinearity of the expectation of M conditional on X required in Theorem 1 does not hold, that is, $E(M|X) = a_1X + a_0$, the parameters b_1 and b_2 in model (13.7) are unidentifiable, as shown in Sect. 3.1. In this case, we can try to find a covariate Z to improve the identifiability.

Introducing a pre-treatment covariate Z, model (13.7) can be rewritten as

$$Y = b_0 + b_1M + b_2X + \phi(U, Z, \varepsilon_Y). \tag{13.12}$$

First we use a simple example to show how covariate Z can be used to identify the parameters in (13.12). Suppose that the model of M has an interaction of X and Z:

$$M = a_0 + a_1X + a_2ZX + \phi(U, Z, \varepsilon_M). \tag{13.13}$$

Marginalizing the above model by ignoring Z, we have from the randomization assumption of X

$$E(M|X) = a_0 + [a_1 + a_2E(Z)]X,$$

and then the nonlinearity condition required for identifiability in Theorem 1 does not hold. Thus the parameters in (13.12) cannot be identified. Below we show how to use the covariate Z for identifying the parameters. If treatment X is randomly assigned conditionally on Z or not conditionally on Z, then we have $X \perp\!\!\!\perp (U, \varepsilon_Y)|Z$, and we obtain from (13.12)

$$\begin{bmatrix} E(M|x_1, z_1) - E(M|x_1', z_1) & x_1 - x_1' \\ E(M|x_2, z_2) - E(M|x_2', z_2) & x_2 - x_2' \\ \vdots & \vdots \\ E(M|x_K, z_K) - E(S|x_K', z_K) & x_K - x_K' \end{bmatrix} \begin{bmatrix} b_1 \\ b_2 \end{bmatrix}$$

$$= \begin{bmatrix} E(Y|x_1, z_1) - E(Y|x_1', z_1) \\ E(Y|x_2, z_2) - E(Y|x_2', z_2) \\ \vdots \\ E(Y|x_K, z_K) - E(Y|x_K', z_K) \end{bmatrix}, \tag{13.14}$$

where $K \geq 2$, $x_i \neq x_i'$ for different values of treatment X and $z_i \neq z_j$ for $i \neq j$. The matrix on the left-hand side of the equation has full column rank since $z_i \neq z_j$ for $i \neq j$ and $E(M|x_i, z_j) - E(M|x_i', z_j) = a_2 z_j (x_i - x_i')$ from model (13.13). Thus the parameters b_1 and b_2 in (13.12) is identifiable if $a_2 \neq 0$ in (13.13).

Next we present the following general result to improve identifiability via a covariate Z when M has a linear model of treatment X.

Theorem 2. *Assume that $X \perp\!\!\!\perp (U, \varepsilon_Y)|Z$. Then parameters b_1 and b_2 in model (13.12) are identifiable if and only if $E(M|X,Z) \neq cX + \mu(Z)$, where c is a constant and $\mu(Z)$ is an arbitrary function.*

The proof of Theorem 2 is given in Appendix 2. The assumption in Theorem 2 that $X \perp\!\!\!\perp (U, \varepsilon_Y)|Z$ is looser than the completely randomization assumption of X in Theorem 1. The necessary and sufficient condition for identifiability given in Theorem 2 implies that the conditional expectation of M given X and Z is nonlinear with respect to X. For a binary treatment X, we obtain the following special result from Theorem 2.

Corollary 2. *Assume that $X \perp\!\!\!\perp (U, \varepsilon_Y)|Z$. For a binary treatment X, the parameters in model (13.12) are identifiable if and only if $E(M|X = 1, z) - E(M|X = 0, z)$ depends on z.*

The necessary and sufficient conditions in Theorem 2 and Corollary 2 mean that there is an interaction between X and Z on M, which implies but is not equivalent to $E(M|X = 1, z) \neq E(M|X = 0, z)$. The condition is similar to the requirement of a valid estimation in [14], although their estimation equation requires the randomization assumption of X while our approach can relax the randomization assumption to $X \perp\!\!\!\perp (U, \varepsilon_Y)|Z$.

4 Estimation of Parameters

The identifiability discussed in the previous section requires that the distribution of observed variables has sufficient information on parameters. After confirming the identifiability, we can use various estimation approaches to estimate these parameters, such as the moment estimation, and the maximum likelihood estimation if we can assume the parametric models of $\phi(U, \varepsilon_Y)$ and $\psi(X, U, \varepsilon_M)$ and the distributions of random errors ε_Y and ε_M. In this section, we try to find an efficient estimation of parameters in the semi-parametric model (13.7).

In our estimation approach, the pivotal condition is independency between randomized treatment X and $(U, \varepsilon_M, \varepsilon_Y)$, which implies the following equation:

$$E[(Y - b_0 - b_1 M - b_2 X)\mathbf{f}(X)] = E[\phi(U, \varepsilon_Y)]E[\mathbf{f}(X)] = \mathbf{0}, \qquad (13.15)$$

where $\mathbf{f}(\cdot) = (f_1(\cdot), \cdots, f_K(\cdot))'$ is an arbitrary vector function and $\mathbf{0}$ is a $K \times 1$ zero vector.

In the following two subsections, we first present a simple but efficient estimator for the case of a three-value treatment X, and then we describe a GMM estimator with the efficient instrument (a function of X) for the case of a general treatment X proposed by Newey and McFadden [10], which has the minimum variance among all estimators satisfying the Eq. (13.15).

4.1 The Case of a Three-Value Treatment

For this case of a three-value treatment X, we choose the function $\mathbf{f}^*(X) = (\delta(X=1), \delta(X=2), \delta(X=3))'$. From Eq. (13.15), we have

$$E[(Y - b_0 - b_1 M - b_2 X)\mathbf{f}^*(X)] = 0. \tag{13.16}$$

Define $\beta = (b_0, b_1, b_2)'$,

$$G^* = \begin{bmatrix} E[\delta(X=1)] & E[M\delta(X=1)] & E[X\delta(X=1)] \\ E[\delta(X=2)] & E[M\delta(X=2)] & E[X\delta(X=2)] \\ E[\delta(X=3)] & E[M\delta(X=3)] & E[X\delta(X=3)] \end{bmatrix},$$

$$H^* = \begin{bmatrix} E[Y\delta(X=1)] \\ E[Y\delta(X=2)] \\ E[Y\delta(X=3)] \end{bmatrix}.$$

The Eq. (13.16) can be rewritten as $H^* - G^*\beta = 0$. Then we can estimate β by

$$\hat{\beta}^* = \widehat{G^*}^{-1}\widehat{H^*},$$

where the elements of $\widehat{G^*}$ and $\widehat{H^*}$ are sample means of the corresponding elements of G^* and H^*. Thus $\hat{\beta}^*$ is a valid estimator only when G^* has full rank. Now we show that the nonlinearity of $E(M|X)$ with respect to X can ensure G^* has full rank. The determinant of the matrix G^* is

$$\begin{aligned} \det(G^*) &= P(X=1)P(X=2)P(X=3) \\ &\quad \cdot [2E(M|X=2) - E(M|X=1) - E(M|X=3)]. \end{aligned}$$

Thus G^* has full rank if and only if $2E(M|X=2) - E(M|X=1) - E(M|X=3) \neq 0$. This inequality is equivalent to that $E(M|X)$ is nonlinear with respect to X.

In Appendix 3, we show for a three-value treatment that any $\mathbf{f}(\cdot)$ which makes the Eq. (13.15) have the unique solution leads to the same estimator of parameters as that obtained by $\mathbf{f}^*(X)$. Thus for a three-value treatment, our estimator is efficient and it is not necessary to choose a complicated $\mathbf{f}(\cdot)$ to improve the efficiency.

4.2 The Case of a General Treatment

Different from the case of a three-value treatment, for a general treatment X with more values, different $\mathbf{f}(\cdot)$'s for X in (13.15) lead to different estimators. In this section, we derive a GMM estimator with the efficient instrument proposed in [10]. Equation (13.15) can be rewritten as

$$\begin{bmatrix} E[f_1(X)] & E[Mf_1(X)] & E[Xf_1(X)] \\ \vdots & \vdots & \vdots \\ E[f_K(X)] & E[Mf_K(X)] & E[Xf_K(X)] \end{bmatrix} \beta = \begin{bmatrix} E[Yf_1(X)] \\ \vdots \\ E[Yf_K(X)] \end{bmatrix}. \tag{13.17}$$

Let G denote the $K \times 3$ matrix on the left-hand side of Eq. (13.17) and H denote the vector on the right-hand side. The Eq. (13.17) can be denoted as $G\beta = H$.

Define $\mathbf{m}(\beta) = E[(Y - b_0 - b_1 M - b_2 X)\mathbf{f}(X)]$. Note that β can be identified only when $r(G) = 3$, where $r(\cdot)$ denotes the rank of a matrix. Then for any $\mathbf{f}(\cdot)$ that makes $r(G) = 3$, a GMM estimate of β is

$$\hat{\beta} = \arg\min_{\beta}\{\widehat{\mathbf{m}}(\beta)'\widehat{W}\widehat{\mathbf{m}}(\beta)\} = (\widehat{G}'\widehat{W}\widehat{G})^{-1}\widehat{G}'\widehat{W}\widehat{H}, \tag{13.18}$$

where $\widehat{W}$ is required to be a positive semi-definite weighting matrix for $K > 3$ and to be positive definite for $K = 3$) whose elements are functions of observed data, and the elements of $\widehat{\mathbf{m}}(\beta)$, $\widehat{G}$, and $\widehat{H}$ are sample means of the corresponding elements of $\mathbf{m}(\beta)$, G, and H, respectively [3].

Let N denote the sample size. According to [10], if $\widehat{W} \to W$ in probability where W is positive semi-definite, then the GMM estimator $\hat{\beta}$ has the following properties:

1. $\hat{\beta} \to \beta_0$ in probability as $N \to \infty$, where β_0 denotes the true parameter, and
2. $\sqrt{N}(\hat{\beta} - \beta_0)$ converges in distribution to a normal variable with mean zero and variance $(G'WG)^{-1}G'WE[\mathbf{f}(X)\mathbf{f}(X)']WG(G'WG)^{-1}\sigma^2_{\text{res}}$, where σ^2_{res} is the variance of $\phi(U, \varepsilon_Y)$.

Newey and McFadden [10] also showed that the estimator $\hat{\beta}$ has a minimum variance when the instrument $\mathbf{f}(X)$ is defined as

$$\mathbf{f}^{\text{eff}}(X) = \left\{E\left[\frac{\partial\varphi(X, M, Y, \beta)}{\partial\beta}|_{\beta_0}|X\right]\right\}' = (1, E(M|X), X)'.$$

From (13.18), the GMM estimator of β with the efficient instrument is

$$\hat{\beta}^{\text{eff}} = [(\widehat{G}^{\text{eff}})'\widehat{W}\widehat{G}^{\text{eff}}]^{-1}(\widehat{G}^{\text{eff}})'\widehat{W}\widehat{H}^{\text{eff}},$$

where $\widehat{G}^{\text{eff}}$ and $\widehat{H}^{\text{eff}}$ are the sample means of

$$G^{\text{eff}} = \begin{bmatrix} 1 & E(M) & E(X) \\ E(M) & E[E(M|X)^2] & E(XM) \\ E(X) & E(XM) & E(X^2) \end{bmatrix}, H^{\text{eff}} = \begin{bmatrix} E(Y) \\ E[YE(M|X)] \\ E(XY) \end{bmatrix}.$$

When X is discrete, $E(M|x)$ can be estimated by sample means of M given $X = x$. When X is continuous, $E(M|x)$ can be estimated by parametric (e.g., a polynomial fitting) or nonparametric (e.g., a kernel smoothing) approaches.

It is proved in Appendix 4 that the matrix G^{eff} equals $E[\mathbf{f}^{\text{eff}}(X)\mathbf{f}^{\text{eff}}(X)']$ and has full rank if the nonlinearity condition of identifiability in Theorem 1 holds. Thus $\hat{\beta}^{\text{eff}}$ is a valid estimator. Since G^{eff} and the positive definite $\widehat{W}$ for $K = 3$ are full rank, $\hat{\beta}^{\text{eff}}$ can be simplified as

$$\hat{\beta}^{\text{eff}} = (\widehat{G}^{\text{eff}})^{-1}\widehat{W}^{-1}[(\widehat{G}^{\text{eff}})']^{-1}(\widehat{G}^{\text{eff}})'\widehat{W}\widehat{H}^{\text{eff}} = (\widehat{G}^{\text{eff}})^{-1}\widehat{H}^{\text{eff}}, \qquad (13.19)$$

which does not rely on the choice of $\widehat{W}$. From the above property 2 of the GMM estimator, we can obtain the asymptotic variance of $\hat{\beta}^{\text{eff}}$

$$V(\hat{\beta}^{\text{eff}}) = (G^{\text{eff}})^{-1}E[\mathbf{f}^{\text{eff}}(X)\mathbf{f}^{\text{eff}}(X)'](G^{\text{eff}})^{-1}\sigma_Y^2 = (G^{\text{eff}})^{-1}\sigma_{\text{res}}^2.$$

The estimate of $V(\hat{\beta}^{\text{eff}})$ can be obtained by $\widehat{G}^{\text{eff}}$ and $\widehat{\sigma}_{\text{res}}^2$ which is the sample variance of residuals of linear model (13.7).

For the case of a three-value treatment, as shown in the previous subsection, we have $\hat{\beta}^* = \hat{\beta}^{\text{eff}}$. From property 2 of the GMM estimator, the variance of $\hat{\beta}^*$ can be estimated by

$$\widehat{V}(\hat{\beta}^*) = \widehat{G^*}^{-1}\widehat{E}[\mathbf{f}^*(X)\mathbf{f}^*(X)'](\widehat{G^{*\prime}})^{-1}\widehat{\sigma}_{\text{res}}^2,$$

where

$$\widehat{E}[\mathbf{f}^*(X)\mathbf{f}^*(X)'] = \begin{bmatrix} \widehat{P}(X=1) & 0 & 0 \\ 0 & \widehat{P}(X=2) & 0 \\ 0 & 0 & \widehat{P}(X=3) \end{bmatrix}$$

and $\widehat{P}(X = i)$ is estimated by the observed frequency of $X = i$.

5 Simulation Study

In this section, we compare our estimates with the OLS estimates via simulations. In our simulations, data are generated from the causal diagram depicted in Fig. 13.1, and the underlying model is

$$Y = b_1 M + b_2 X + dU + \varepsilon_Y,$$

$$\text{Logit}\, P(M = 1|X, U) = -1 + 3X + cU,$$

where the true parameters $b_1 = 1.0$ and $b_2 = 0.6$.

We conduct 3×5 scenarios for all combinations of parameter $c = 0$, 2, and 4 and the parameter $d = 0, \pm 1$ and ± 2. Nonzero parameters c and d mean that there is the unobserved confounder U which affects both M and Y. For each scenario, we replicate 1000 samples with sizes $n = 300$ and 600. The data are generated for each individual in the following process:

1. Draw X from 1, 2, and 3 with equal probabilities (i.e., $P(X = 1) = P(X = 2) = P(X = 3) = 1/3$).
2. Draw U from a normal distribution $N(0, 0.3^2)$.
3. Draw M from a Bernoulli one with $P(M = 1|X, U)$ which has a logistic model

$$\text{Logit}\, P(M = 1|X, U) = -1 + 3X + cU.$$

4. Draw ε_Y from $N(0, 0.3^2)$, and then compute

$$Y = M + 0.6X + dU + \varepsilon_Y.$$

Since X is a three-value treatment, we used the estimators given in Sect. 4.1. The means of estimates obtained in 1000 simulations for each scenario are given in Table 13.1. $(\hat{b}_1, \hat{b}_2)$ denotes our estimates and $(\tilde{b}_1, \tilde{b}_2)$ denotes the OLS estimates. The average lengths of the 95 % confidence intervals and the rates of the confidence intervals covering the true value of a parameter are given in Table 13.2. For the scenarios 1 to 5, $c = 0$ means that the latent variable U does not affect the mediator M, and thus U is not a confounder and can be treated as an error term in the models of Y and M. From Tables 13.1 and 13.2, we can see that both our approach and the OLS approach performed well in these scenarios. For the scenarios 6–10 with $c = 2$ and $d \neq 0$, the confounder U is correlated to M in the model of Y, and thus U is a confounder and cannot be treated as an error term in the models. It can be seen from Table 13.1 that the biases of OLS estimates $(\tilde{b}_1, \tilde{b}_2)$ increase greatly as d departs from 0. From Table 13.2, it can also be seen that for OLS estimates, the coverage rates of 95 % confidence intervals decrease as c and d departs from 0, and few of the 1000 confidence intervals of b_2 covered the true parameter b_2 for the scenarios with $d \neq 0$. Our estimates $(\hat{b}_1, \hat{b}_2)$ are close to the true parameters (b_1, b_2), and the coverage rates of 95 % confidence intervals always are around 95 % in all scenarios.

Table 13.1 Means of estimates for 1000 simulations (The true values $b_1 = 1.0$ and $b_2 = 0.6$) $c = 0$ or $d = 0$ means that U is not a confounder

		$N = 300$				$N = 600$			
		OLS estimates		Our estimates		OLS estimates		Our estimates	
		$\tilde{b}_1$	$\tilde{b}_2$	$\hat{b}_1$	$\hat{b}_2$	$\tilde{b}_1$	$\tilde{b}_2$	$\hat{b}_1$	$\hat{b}_2$
$c = 0$	$d = -2$	0.999	0.601	0.994	0.602	0.995	0.602	0.996	0.602
	$d = -1$	0.999	0.601	0.996	0.602	0.996	0.601	0.996	0.602
	$d = 0$	1.000	0.600	0.998	0.601	0.997	0.601	0.994	0.602
	$d = 1$	1.001	0.600	1.001	0.600	0.998	0.600	0.993	0.602
	$d = 2$	1.001	0.599	1.003	0.600	0.998	0.599	0.992	0.601
$c = 2$	$d = -2$	0.700	0.693	1.008	0.598	0.698	0.694	1.005	0.599
	$d = -1$	0.849	0.647	1.003	0.600	0.848	0.647	1.000	0.600
	$d = 0$	0.999	0.600	0.997	0.601	0.998	0.600	0.994	0.602
	$d = 1$	1.148	0.554	0.991	0.602	1.148	0.554	0.989	0.603
	$d = 2$	1.297	0.508	0.985	0.604	1.299	0.507	0.983	0.604
$c = 4$	$d = -2$	0.443	0.767	1.072	0.582	0.443	0.768	1.017	0.596
	$d = -1$	0.721	0.684	1.035	0.591	0.720	0.684	1.006	0.599
	$d = 0$	0.998	0.601	0.997	0.601	0.997	0.601	0.993	0.602
	$d = 1$	1.276	0.517	0.959	0.610	1.274	0.517	0.979	0.605
	$d = 2$	1.553	0.434	0.921	0.620	1.551	0.433	0.967	0.608

Comparing the results for two different sample sizes, we can see for the larger size $N = 600$ that our estimates are closer to the true values and have smaller standard errors but that the OLS estimates become even worse, have lower coverage rates, and do not reduce the biases. We also did simulations for other sample sizes and got the similar results.

6 Extension

In the previous sections, we discussed the model (13.7) of Y which is linear with respect to M and X. These results can be extended to more general models, such as the presence of an interaction term or a nonlinear direct effect of treatment on outcome in (13.1). First we consider the model of moderated-mediation analysis which has an interaction of X and M on Y as an example to illustrate the extension. Consider the following moderated-mediation model:

$$\begin{aligned} Y &= b_0 + b_1 M + b_2 X + b_3 XM + \phi(U, \varepsilon_Y), \\ M &= \psi(X, U, \varepsilon_M). \end{aligned} \tag{13.20}$$

From the model (13.20), we get the following equation:

Table 13.2 Coverage rates of 95 % confidence intervals and estimated standard deviations in brackets for 1000 simulations

		$N = 300$				$N = 600$			
		OLS estimates		Our estimates		OLS estimates		Our estimates	
		$\tilde{b}_1$	$\tilde{b}_2$	$\hat{b}_1$	$\hat{b}_2$	$\tilde{b}_1$	$\tilde{b}_2$	$\hat{b}_1$	$\hat{b}_2$
$c = 0$	$d = -2$	0.955	0.943	0.985	0.966	0.957	0.950	0.966	0.967
		(0.135)	(0.067)	(0.378)	(0.128)	(0.095)	(0.047)	(0.255)	(0.088)
	$d = -1$	0.945	0.938	0.982	0.969	0.956	0.959	0.968	0.959
		(0.085)	(0.042)	(0.240)	(0.081)	(0.060)	(0.030)	(0.161)	(0.055)
	$d = 0$	0.957	0.953	0.984	0.974	0.957	0.950	0.967	0.956
		(0.060)	(0.030)	(0.170)	(0.058)	(0.043)	(0.021)	(0.114)	(0.039)
	$d = 1$	0.952	0.943	0.979	0.963	0.953	0.952	0.976	0.962
		(0.085)	(0.042)	(0.240)	(0.082)	(0.060)	(0.030)	(0.161)	(0.056)
	$d = 2$	0.950	0.950	0.979	0.971	0.957	0.954	0.976	0.962
		(0.134)	(0.067)	(0.379)	(0.129)	(0.095)	(0.047)	(0.255)	(0.088)
$c = 2$	$d = -2$	0.368	0.701	0.984	0.970	0.088	0.459	0.970	0.970
		(0.131)	(0.065)	(0.426)	(0.141)	(0.092)	(0.046)	(0.283)	(0.095)
	$d = -1$	0.548	0.776	0.983	0.974	0.265	0.638	0.972	0.961
		(0.083)	(0.042)	(0.270)	(0.089)	(0.059)	(0.029)	(0.179)	(0.060)
	$d = 0$	0.952	0.945	0.988	0.979	0.949	0.950	0.971	0.962
		(0.059)	(0.030)	(0.192)	(0.063)	(0.042)	(0.021)	(0.126)	(0.042)
	$d = 1$	0.565	0.802	0.982	0.974	0.269	0.626	0.977	0.967
		(0.083)	(0.042)	(0.271)	(0.089)	(0.059)	(0.029)	(0.179)	(0.060)
	$d = 2$	0.388	0.708	0.978	0.972	0.092	0.485	0.982	0.967
		(0.131)	(0.065)	(0.428)	(0.141)	(0.092)	(0.046)	(0.284)	(0.095)
$c = 4$	$d = -2$	0.005	0.227	0.981	0.968	0	0.031	0.971	0.973
		(0.121)	(0.062)	(0.758)	(0.224)	(0.086)	(0.044)	(0.373)	(0.118)
	$d = -1$	0.054	0.437	0.987	0.973	0.002	0.149	0.975	0.970
		(0.078)	(0.040)	(0.469)	(0.139)	(0.055)	(0.028)	(0.235)	(0.074)
	$d = 0$	0.952	0.941	0.995	0.988	0.941	0.947	0.982	0.970
		(0.056)	(0.029)	(0.305)	(0.092)	(0.040)	(0.020)	(0.165)	(0.052)
	$d = 1$	0.060	0.451	0.989	0.982	0	0.169	0.985	0.970
		(0.078)	(0.040)	(0.434)	(0.132)	(0.055)	(0.028)	(0.236)	(0.074)
	$d = 2$	0.003	0.239	0.982	0.980	0	0.033	0.984	0.973
		(0.121)	(0.062)	(0.724)	(0.217)	(0.086)	(0.044)	(0.374)	(0.118)

$$E(Y|x) = b_0 + b_1 E(M|x) + b_2 x + b_3 x E(M|x).$$

For different levels $x_1, \ldots, x_K$ of treatment, we obtain the following equations:

$$\begin{bmatrix} 1 & E(M|x_1) & x_1 & x_1 E(M|x_1) \\ \vdots & \vdots & \vdots & \vdots \\ 1 & E(M|x_K) & x_K & x_K E(M|x_K) \end{bmatrix} \begin{bmatrix} b_0 \\ b_1 \\ b_2 \\ b_3 \end{bmatrix} = \begin{bmatrix} E(Y|x_1) \\ \vdots \\ E(Y|x_K) \end{bmatrix}.$$

Both $E(M|x_i)$ and $E(Y|x_i)$ can be estimated from data by a parametric or nonparametric approach. Parameters b_i's are identifiable if $K \geq 4$ and the $K \times 4$ matrix on the left-hand side has full column rank. It can be shown that the matrix has full column rank if and only if $E(M|x)$ is not a linear function of x, which is the same as the condition of Theorem 1. Notice that this identifiability condition can also be checked by observed data. It is obvious that under the commonly used assumption of a linear regression of M on X in the simultaneous equation model, we cannot identify these parameters in mediation models.

Next we consider the case of a nonlinear direct effect of treatment on outcome. For example, consider a quadratic equation of X

$$Y = b_0 + b_1 M + b_2 X + b_3 X^2 + \phi(U, \varepsilon_Y).$$

To identify the parameters, we need a higher degree of equation nonlinearity for M, such as

$$M = a_0 + a_1 X + a_2 X^2 + a_3 X^3 + \psi(X, U, \varepsilon_M),$$

and we need to manipulate the treatment X for $K(\geq 4)$ levels. The higher degree nonlinearity of treatment effect on mediator M can be used to distinguish the indirect effect of treatment on outcome from the lower-degree direct effect of treatment on outcome. This essentially means that the change rates of outcome through the direct arrow $X \to Y$ and through the path $X \to M \to Y$ are different, and thus we can separate the direct effect from the indirect effect.

Similarly for a more general model of Y, to identify the parameters in the model, we require that the treatment X has the number K of levels larger than or equal to the number of the parameters in the model of Y and that the nonlinearity degree of treatment effect on mediator is higher than the nonlinearity degree of the direct effect of treatment on outcome such that the equations of the expectations of Y conditional on these levels have a unique solution for the parameters.

7 Discussions

To identify and estimate parameters in mediation models, different approaches require different assumptions or manipulation experiments. The structural equation modeling approach requires the sequential ignorability assumption of the mediator status, the principal stratification approach and the instrumental variable approach require the exclusion restriction assumption, and the experimental approach requires that the mediator is manipulatable. When a mediation model has a single mediator, it is difficult to satisfy the exclusion restriction assumption. The sequential ignorability assumption is hardly satisfied even if the mediator could be manipulated, and the manipulation experiment of the mediator may not be practical in many applications.

Removing these untestable assumptions, the approach proposed in this paper requires that the regression equation of the mediator on the treatment variable is nonlinear, otherwise a covariate is necessary for the identifiability. For the case of a binary mediator, a logistic regression equation is commonly used and the nonlinearity may be generally satisfied. The important difference between our nonlinear requirement and the assumptions of other approaches is that our nonlinearity requirement of the mediator M with respect to the treatment X is testable by the observed data of M and X, while the assumptions required by other approaches are untestable by observed data. This testability is an advantage of our approach.

When the nonlinearity required in Theorem 1 is not satisfied, we may try to find a covariate Z such that the slope of the regression of M on X depends on Z, see Theorem 2. This covariate is essentially used to try a possible nonlinearity between the treatment and the mediator such that the effects of treatment on the mediator are different conditionally on different levels of the covariate. In a sense, the covariate Z requires a model assumption like an instrumental variable so that we can remove the confounding bias generated by an unobserved confounder. Our approach may be more realistic for observational studies and experimental studies in which we cannot manipulate the mediator. For a pure observational study in which the treatment X cannot be manipulated, we need the commonly used assumptions for causal inference, such as the ignorability assumption of the treatment assignment.

Acknowledgements This research was supported by NSFC (11171365, 11021463, 10931002), 863 Program of China (2015AA020507) and a project founded by Merck (China).

Appendix 1: Proof of Theorem 1

We separately show the necessity and sufficiency for the identifiability of parameters in model (13.7). For necessity, suppose that the non-linearity condition does not hold, that is, $|\rho(E(M|X), X)| = 1$. This implies that there exist some a_0 and a_1 satisfying $E(M|X) = a_1X + a_0$ almost everywhere. Then from the model (13.7) we have

$$\begin{aligned} E(Y|X) &= b_0 + b_1E(M|X) + b_2X \\ &= (b_0 + a_0b_1) + (a_1b_1 + b_2)X. \end{aligned}$$

The above equation implies that Y is marginally linear with respect to X. For this linear model, only the intercept $(b_0+a_0b_1)$ and the slope $(a_1b_1+b_2)$ are identifiable as a whole, while parameters b_0, b_1, and b_2 cannot be distinguished each other.

For sufficiency, if M is not marginally linearly related with respect to X, then we can find 3 levels: x_1, x_2, and x_3, which satisfy $[E(M|x_1) - E(M|x_2)]/(x_1 - x_2) \neq [E(M|x_2) - E(M|x_3)]/(x_2 - x_3)$. Hence the matrix in (13.10) has full rank. Thus parameters b_1 and b_2 can be identified, and then parameter b_0 can be identified from $b_0 = E(Y|x_i) - b_1E(M|x_i) - b_2x_i$.

Appendix 2: Proof for Theorem 2

For sufficiency, when $E(M|X,Z) \neq cX + \psi(Z)$, there are two situations: (i) $E(M|X,Z) = \Psi(X) + \psi(Z)$, where $\Psi(\cdot)$ is a nonlinear function of X; (ii) $E(M|X,Z)$ is not additive with respect to X and Z.

For situation (i), since $\Psi(\cdot)$ is not a linear function, we can choose three levels of X (say x_1, x_2, x_3) and some z satisfying $[E(M|x_1,z) - E(M|x_2,z)]/(x_1 - x_2) \neq [E(M|x_2,z) - E(M|x_3,z)]/(x_2 - x_3)$. Then the following equation from model (13.12) has a unique solution because the coefficient matrix has full rank:

$$\begin{bmatrix} E(M|x_1,z) - E(M|x_2,z) & x_1 - x_2 \\ E(M|x_2,z) - E(M|x_3,z) & x_2 - x_3 \end{bmatrix} \begin{bmatrix} b_1 \\ b_2 \end{bmatrix} = \begin{bmatrix} E(Y|x_1,z) - E(Y|x_2,z) \\ E(Y|x_2,z) - E(Y|x_3,z) \end{bmatrix}.$$

Thus the parameters can be identified.

For situation (ii), since $E(M|X,Z)$ is not additive with respect to X and Z, we can find two levels of X (say x_1, x_2) and two levels of Z (say z_1, z_2) satisfying $E(M|x_1,z_1) - E(M|x_2,z_1) \neq E(M|x_1,z_2) - E(M|x_2,z_2)$. The following equation derived from model (3.4) has a unique solution because the coefficient matrix has full rank:

$$\begin{bmatrix} E(M|x_1,z_1) - E(M|x_2,z_1) & x_1 - x_2 \\ E(M|x_1,z_2) - E(S|x_2,z_2) & x_1 - x_2 \end{bmatrix} \begin{bmatrix} b_1 \\ b_2 \end{bmatrix} = \begin{bmatrix} E(Y|x_1,z_1) - E(Y|x_2,z_1) \\ E(Y|x_1,z_2) - E(Y|x_2,z_2) \end{bmatrix}.$$

Thus the parameters can be identified.

For necessity, if $E(M|X,Z) = cX + \psi(Z)$ for some constant c and $\psi(\cdot)$, then from model (13.12), we have

$$\begin{aligned} E(Y|X,Z) &= b_0 + b_1E(M|X,Z) + b_2X + E[\phi(U,Z,\varepsilon_Y)|Z] \\ &= b_0 + (b_1c + b_2)X + \Phi(Z), \end{aligned}$$

where $\Phi(Z) = b_1\psi(Z) + E[\phi(U,Z,\varepsilon_Y)|Z]$. We can easily see that only c, $b_1c + b_2$ and $\Phi(Z)$ can be identified given observed data of (Z,X,M,Y). b_1 and b_2 cannot be identified because (1) $E(M|X,Z)$ is linear with respect to X, and (2) $E[\phi(U,Z,\varepsilon_Y)|Z]$ cannot be identified since U and ε_Y are never observed. Thus the parameters in model (13.12) are identifiable only if $E(M|X,Z) \neq cX + \psi(Z)$.

Appendix 3: Proof for the Equivalence of Different Choices of f(·) in Eq. (13.15) for the Estimation When the Identifiability Condition in Theorem 1 Holds

We want to show that an arbitrary vector function $\mathbf{f}(\cdot)$ that identifies β via Eq. (13.15) leads to the same estimator as that based on the function $\mathbf{f}^*(\cdot)$. For an arbitrary vector function $\mathbf{f}(\cdot) = (f_1(\cdot), f_2(\cdot), \cdots, f_K(\cdot))'$ $(K > 2)$, we can denote it as

$$\mathbf{f}(X) = \begin{bmatrix} f_1(1) & f_1(2) & f_1(3) \\ \vdots & \vdots & \vdots \\ f_K(1) & f_K(2) & f_K(3) \end{bmatrix} \begin{bmatrix} \delta(X=1) \\ \delta(X=2) \\ \delta(X=3) \end{bmatrix}. \tag{13.21}$$

Let Q denote the $K \times 3$ matrix on the right-hand side. Equation (13.15) can be rewritten as $G\beta = H$, where $G = E[\mathbf{f}(X), M\mathbf{f}(X), X\mathbf{f}(X)]$ and $H = E[Y\mathbf{f}(X)]$. Then the estimation equation for β is $\widehat{G}\hat{\beta} = \widehat{H}$. From (13.21), we have

$$\begin{aligned} \widehat{G} &= \widehat{E}[\mathbf{f}(X), M\mathbf{f}(X), X\mathbf{f}(X)] \\ &= \widehat{E}[Q\mathbf{f}^*(X), MQ\mathbf{f}^*(X), XQ\mathbf{f}^*(X)] \\ &= Q\widehat{E}[\mathbf{f}^*(X), M\mathbf{f}^*(X), X\mathbf{f}^*(X)] \\ &= Q\widehat{G^*}, \end{aligned}$$

where $\widehat{E}(\cdot)$ denotes the sample mean of the corresponding variable. Similarly, we have

$$\widehat{H} = Q\widehat{H^*}.$$

Then by the function $\mathbf{f}(\cdot)$, the estimation equation for β is equivalent to

$$\widehat{G}\hat{\beta} - \widehat{H} = Q(\widehat{G^*}\hat{\beta} - \widehat{H^*}) = 0.$$

Since $\hat{\beta}^*$ satisfies the equation $\widehat{G^*}\hat{\beta}^* - \widehat{H^*} = 0$, we have that $\hat{\beta}^*$ also satisfies $\widehat{G}\hat{\beta}^* - \widehat{H} = 0$. Thus we proved $\hat{\beta} = \hat{\beta}^*$ when Q has full rank, which means that the above equation of $\hat{\beta}$ has a unique solution.

Appendix 4: Proof for Matrix G^{eff} in Sect. 4.2 Equals $E[\mathbf{f}^{\text{eff}}(X)\mathbf{f}^{\text{eff}}(X)']$ and Has Full Rank When Non-linearity Condition in Theorem 1 Holds

(i) We show that $G^{\text{eff}} = E[\mathbf{f}^{\text{eff}}(X)\mathbf{f}^{\text{eff}}(X)']$. It is obvious that

$$E[\mathbf{f}^{\text{eff}}(X)\mathbf{f}^{\text{eff}}(X)'] = E\left\{\begin{bmatrix} 1 & E(M|X) & X \\ E(M|X) & E(M|X)^2 & E(M|X)X \\ X & XE(M|X) & X^2 \end{bmatrix}\right\}$$

$$= \begin{bmatrix} 1 & E(M) & E(X) \\ E(M) & E[E(M|X)^2] & E(XM) \\ E(X) & E(XM) & E(X^2) \end{bmatrix} = G^{\text{eff}}.$$

(ii) We prove that G^{eff} has full rank when non-linearity condition in Theorem 1 holds. To prove that G^{eff} has full rank, we only need show that $\det(G^{\text{eff}}) \neq 0$ when $|\rho(X, E(M|X))| < 1$. We have

$$\det(G^{\text{eff}}) = \begin{vmatrix} 1 & E(M) & E(X) \\ E(M) & E[E(M|X)^2] & E(XM) \\ E(X) & E(XM) & E(X^2) \end{vmatrix}$$

$$= \begin{vmatrix} 1 & E(M) & E(X) \\ 0 & E[E(M|X)^2] - [E(M)]^2 & E(XM) - E(X)E(M) \\ 0 & E(XM) - E(X)E(M) & E(X^2) - [E(X)]^2 \end{vmatrix}.$$

Since

$$\text{var}[E(M|X)] = E[E(M|X)^2] - [E(M)]^2, \qquad \text{var}(X) = E(X^2) - [E(X)]^2$$

and

$$\text{cov}(X, E(M|X)) = E[XE(M|X)] - E(X)E[E(M|X)] = E(XM) - E(X)E(M),$$

we have

$$\det(G^{\text{eff}}) = \begin{vmatrix} 1 & E(M) & E(X) \\ 0 & \text{var}[E(M|X)] & \text{cov}(X, E(M|X)) \\ 0 & \text{cov}(X, E(M|X)) & \text{var}(X) \end{vmatrix}$$

$$= \text{var}[E(M|X)]\text{var}(X)(1 - [\rho(X, E(M|X)]^2) > 0,$$

since $|\rho(X, E(M|X))| < 1$.

References

1. Baron, R.M., Kenny, D.A.: The moderator-mediator variable distinction in social psychological research: conceptual, strategic, and statistical considerations. J. Pers. Soc. Psychol. **51**, 1173–1182 (1986)
2. Frangakis, C.E., Rubin, D.B.: Principle stratification in causal inference. Biometrics **58**, 21–29 (2002)
3. Hansen, L.S.: Large sample properties of generalized method of moments estimators. Econometrica **50**, 1029–1054 (1982)
4. Herting, J.R.: Evaluating and rejecting true mediation models: a cautionary note. Prev. Sci. **3**, 285–289 (2002)
5. Imai, K., Keele, L., Yamamoto, T.: Identification, inference and sensitivity analysis for causal mediation effects. Stat. Sci. **25**(1), 51C71 (2010)
6. Jo, B.: Causal inference in randomized experiments with mediational processes. Psychol. Methods **13**(4), 314–336 (2008)
7. Kaufman, S., Kaufman, J.S., MacLehose, R., Greenland, S., Poole, C.: Improved estimation of controlled direct effects in the presence of unmeasured confounding of intermediate variables. Stat. Med. **24**, 1683–1702 (2005)
8. Li, Y., Schneider, J.A., Bennet, D.A.: Estimation of the mediation effect with a binary mediator. Stat. Med. **26**, 3398–3414 (2007)
9. MacKinnon, D.P., Fairchild, A.J., Fritz, M.S.: Mediation analysis. Annu. Rev. Psychol. **58**, 593–614 (2007)
10. Newey, W.K., McFadden, D.: Large sample estimation and hypothesis testing. In: R. F. Engle, R.F., McFadden, D. (eds.) Handbook of Econometrics, vol. IV, pp. 2111–2245. Elsevier, Amsterdam (1994)
11. Pearl, J.: Direct and indirect effects. In: Proc. 17th Conf. Uncertainty in Artificial Intelligence, pp. 411–420 (2000)
12. Rubin, D.B.: Direct and indirect causal effects via potential outcomes. Scand. J. Stat. **31**, 161–170 (2004)
13. Sobel, M.E.: Identification of causal parameters in randomized studies with mediating variables. J. Educ. Behav. Stat. **33**, 230–251 (2008)
14. Ten Have, T.R., Joffe, M.M., Lynch, K.G., Brown, G.K., Maisto, S.A., Beck, A.T.: Causal mediation analyses with rank preserving models. Biometrics **63**, 926–934(2007)
15. VanderWeele, T.J.: Marginal structural models for the estimation of direct and indirect effects. Epidemiology **20**(1), 18–26 (2009)
16. VanderWeele, T.J.: Controlled direct and mediated effects: definition, identification and bounds. Scand. J. Stat. **38**, 551–563 (2011)

Chapter 14
A Comparison of Potential Outcome Approaches for Assessing Causal Mediation

Donna L. Coffman, David P. MacKinnon, Yeying Zhu, and Debashis Ghosh

Abstract Mediation occurs as part of a hypothesized causal chain of events: An intervention or treatment, T, has an effect on the mediator, M, which then affects an outcome variable, Y. Within the potential outcomes framework for causal inference, three different definitions of the mediation effects have been proposed: principal strata effects (e.g., Rubin, Scand. J. Stat. 31:161–170, 2004; Jo, Psychol. Methods 13:314–336, 2008), natural effects (e.g., Pearl, Proceedings of the Seventeenth Conference on Uncertainty in Artificial Intelligence, 2001; Imai et al., Psychol. Methods 15:309–334, 2010), and controlled effects (e.g., Robins and Greenland, Epidemiology 3:143–155, 1992; VanderWeele, Epidemiology 20:18–26, 2009). We illustrate that each of these definitions answers a different scientific question. We examine five different estimators of the various definitions and discuss identifying assumptions about unmeasured confounding, the existence of direct effects (i.e., the effect of T on Y that is not due to M), iatrogenic effects of T on M, the existence of post-treatment confounders, and the existence of interactions. We assess the robustness of each of the estimators to violations of the assumptions

---

Electronic supplementary material The online version of this chapter (doi: 10.1007/978-3-319-41259-7_14) contains supplementary material, which is available to authorized users.

Authors' note: Preparation of this article was supported by NIDA Center Grant P50 DA100075-15 and NIDA R01 DA09757. The content is solely the responsibility of the authors and does not necessarily represent the official views of the National Institute on Drug Abuse (NIDA) or the National Institutes of Health (NIH).

D.L. Coffman (✉)
The Methodology Center, Pennsylvania State University, 404 Health and Human Development Building, University Park, PA 16802, USA
e-mail: dlc30@psu.edu

D.P. MacKinnon
Department of Psychology, Arizona State University, Tempe, AZ 85281, USA

Y. Zhu
Department of Statistics and Actuarial Science, University of Waterloo, Waterloo, ON, Canada N2L 3G1

D. Ghosh
Department of Biostatistics and Informatics, University of Colorado, Aurora, CO 80045, USA

H. He et al. (eds.), *Statistical Causal Inferences and Their Applications in Public Health Research*, ICSA Book Series in Statistics, DOI 10.1007/978-3-319-41259-7_14

using a simulation study that systematically challenges different aspects of these assumptions. We found that when no assumptions were violated, as may be expected, each approach was unbiased for its respective population value and 95 % confidence interval (CI) coverage was maintained. However, when assumptions are violated, the effects may be severely biased and 95 % CI coverage is not maintained. We suggest that researchers choose the appropriate definition based on the scientific question to be addressed and the identifying assumptions that are plausible given their data.

Mediation is fundamental to many areas of research because many interventions attempt to change one variable in order to cause another variable to change [1]. Mediation analysis helps identify the intermediary processes by which an intervention achieves its effects; understanding the causal mediation pathway can help design interventions that are more effective and less expensive. Given a hypothesized theory regarding the effect of an intervention on a mediator and outcome, mediation analysis can evaluate whether the intervention status affects the mediator and whether the mediator affects the outcome as predicted by theory. Because of its practical and theoretical importance, mediation analysis is now commonly applied in many research disciplines [1]. Recently, more attention has been devoted to the causal aspects of mediation (e.g., [2–6]) and this work has identified several serious shortcomings of traditional mediation analysis (see also [7]). Fortunately, this work has also generated new methods to deal with the shortcomings of traditional mediation analysis. A primary goal of this paper is to introduce and compare these new methods to estimate causal mediation effects so that researchers can make informed decisions about which method to use.

Most of the new approaches to mediation analysis focus on the potential outcomes framework [8, 9]. Within this framework, three definitions of mediation effects have been proposed: natural, controlled, and principal strata effects. Within each of these definitions, different assumptions have been proposed for identifying and estimating the causal effects. Given the variety of choices, it is difficult for researchers to determine the ideal method for a research question. We compare the various definitions in terms of the assumptions typically used to identify and estimate causal effects and to examine how robust each approach is to violations of assumptions in a simulation study. The data generation for the simulation study is designed to be very general to avoid favoring one approach over another.

This article is organized as follows. First, we review the potential outcomes framework and notation. Second, we describe each definition of causal mediation effects under the potential outcomes framework. Next, for each of these definitions, we introduce the assumptions typically used to identify causal effects and the methods for estimating them. Finally, we turn to the simulation study, including data generation and results, followed by a general discussion.

1 Potential Outcomes Framework for Causal Inference

In the potential outcomes framework (see [8–10]), each individual has a potential outcome for each possible treatment condition, namely the value of the outcome that would have occurred had the individual received the given treatment condition. For simplicity, consider a binary treatment indicator, T_i, where $T_i = 1$ denotes the intervention condition and $T_i = 0$ denotes the control condition for participant i, $i = 1, \ldots, n$. The potential outcome if the individual receives the intervention is denoted $Y_i(1)$, and the potential outcome if the individual is in the control condition is denoted $Y_i(0)$. The individual causal effect is the difference between these two potential outcomes. Because each participant is observed in only one condition, only one of these potential outcomes is observed; the other is missing and, therefore, the individual causal effect cannot be computed. However, strategies have been implemented to estimate the causal effect averaged over participants in the study. This average causal effect (ACE) is defined as $E[Y_i(1) - Y_i(0)]$; that is, the expected (or average) difference between the two potential outcomes. Information on the potential outcomes framework outside of the context of mediation is provided by Little and Rubin [11], Schafer and Kang [12], and Winship and Morgan [13].

Extending the potential outcomes framework to mediation is more complicated because a mediator is an outcome of the intervention and, therefore, there are also potential values for the mediator under each treatment condition for each individual. The potential mediator under the intervention condition is denoted $M_i(1)$, and the potential mediator under the control condition is denoted $M_i(0)$. The notation for the potential outcomes is then expanded to include the potential mediators; this notation is referred to as nested potential outcomes. Thus, $Y_i(1,M_i(1))$ is the potential outcome if individual i receives the intervention and the potential mediator takes on the value that would have been obtained had they received the intervention; and $Y_i(0,M_i(0))$ is the potential outcome if individual i is in the control condition and the potential mediator takes on the value that would have been obtained had they been in the control condition. There are two other potential outcomes that can never be realized in practice and illustrate the challenge of identifying causal mediation effects. These two potential outcomes are needed to define the natural effects and correspond to $Y_i(1,M_i(0))$, the potential outcome if individual i receives the intervention and has the potential value of the mediator that would have been obtained had they been in the control condition, and $Y_i(0,M_i(1))$, the potential outcome if individual i is in the control condition and has the potential value of the mediator that would have been obtained had they received the intervention. The impossibility of ever observing these two potential outcomes is one of the reasons that causal mediation analysis is controversial.

Throughout the article, we use Y_i to denote the observed value of the outcome, M_i to denote the observed value for the mediator, and $Y_i(t,M_i(t))$ to denote the potential outcomes where t is one of the levels of treatment. We use X_0 to denote measured baseline (i.e., pre-treatment) confounders. We assume throughout that if an individual receives the intervention, then $Y_i = Y_i(1) = Y_i(1,M_i(1))$ and $M_i = M_i(1)$.

Likewise, if an individual is in the control condition, then $Y_i = Y_i(0) = Y_i(0,M_i(0))$ and $M_i = M_i(0)$. This assumption is usually referred to as the consistency assumption. In addition, the treatment variation irrelevant assumption [14] states that the potential mediator, $M_i(t)$, for individual i when exposed to treatment $T_i = t$ will be the same no matter what mechanism is used to assign treatment t to individual i. Similarly, the potential outcome, $Y_i(t,M_i(t))$, for individual i when exposed to treatment $T_i = t$ and mediator level $M_i(t) = m$ will be the same no matter what mechanism is used to assign t and m to individual i. The notation defined above is sufficient for describing the potential outcomes under each treatment level. Additionally we assume throughout that there is no interference among individuals, meaning that an individual's potential outcomes do not depend on another individual's treatment assignment. Thus, the potential outcome notation is a function of only T_i and not T_j, where i and j denote two different individuals. Finally, we assume common support, meaning that the probability of receiving the treatment, $P[T_i = 1]$, is between 0 and 1. If $P[T_i = 1] = 0$ or $P[T_i = 1] = 1$, then a causal effect is not meaningfully defined for that individual. This assumption is often referred to as positivity (see, e.g., [15]). Similarly for the mediator, we assume that all individuals have non-zero probability for all levels of mediator.

2 Using the Potential Outcomes Framework to Define Mediation Effects

There are several different definitions of mediation within the potential outcomes framework: natural effects, controlled effects, and principal strata effects. Before defining the effects using the potential outcomes framework, we define the effects as they have been traditionally defined in the social science literature. Briefly, in the social science literature, mediation has traditionally been assessed by fitting two linear regression models: one for the mediator,

$$E\left[M\middle|T = t\right] = \beta_{0M} + \beta_1 t \tag{14.1}$$

and one for the outcome,

$$E\left[Y\middle|T = t, M = m\right] = \beta_{0Y} + \beta_2 t + \beta_3 m. \tag{14.2}$$

The direct effect is defined as β_2, and the indirect effect is defined as the product of β_1 and β_3. Note that these definitions do not involve counterfactuals, as the models presented above are models for the observed mediator and outcome. These effects may be interpreted as causal effects only under certain assumptions to be discussed in the *Identification* section below.

Principal Strata Effects Principal stratification [16–18] was initially developed to handle non-compliance in intervention studies; recognizing that actual receipt of an intervention is a mediating variable between intervention assignment and the outcome, these methods have recently been applied to mediation analysis more broadly. Generally, the population is divided into subgroups, called principal strata, based on a cross-classification of the potential values for the mediator. A local ACE can then be defined within each principal stratum. Suppose that the mediator can take on values of 1 or 0. The four possible principal strata effects are defined as

1. $E[Y_i(1) - Y_i(0)|M_i(1) = M_i(0) = 1]$,
2. $E[Y_i(1) - Y_i(0)|M_i(1) = M_i(0) = 0]$,
3. $E[Y_i(1) - Y_i(0)|M_i(1) = 1, M_i(0) = 0]$, and
4. $E[Y_i(1) - Y_i(0)|M_i(1) = 0, M_i(0) = 1]$.

The effect in the first stratum is the causal effect of the intervention on the outcome, among those who would have a value 1 on the mediator regardless of intervention condition. In other words, in this stratum, the intervention had no causal effect on the mediator because the mediator would be 1 regardless of intervention condition. In the compliance literature, this principal stratum is referred to as the always-takers, since they would take the treatment whether they were randomized to it or not. The effect in the second stratum is the causal effect of the intervention on the outcome among those who would have a value 0 on the mediator regardless of intervention condition. In the second stratum, the mediator would be 0 regardless of intervention condition so the intervention had no causal effect on the mediator in this stratum either. In the compliance literature, this principal stratum is referred to as the never-takers. The effect in the third stratum is the causal effect of the intervention on the outcome among those who would have a value of 1 on the mediator if they received the intervention and 0 if they did not. In the compliance literature, this principal stratum is referred to as the compliers. The effect in the fourth stratum is the causal effect of the intervention on the outcome among those who would have a value of 0 on the mediator if they received the intervention and a 1 if not. In the compliance literature, this principal stratum is referred to as the defiers, since their treatment status reflects the opposite of their randomization. For the latter two strata, the intervention does have a causal effect on the mediator. Thus, the principal strata effects in these two strata represent the causal effects of the intervention on the outcome among those for whom the intervention had an effect on the mediator. The distinction between these two strata is that the effect of the intervention on the mediator is in opposite directions. All of these are causal effects of the intervention on the outcome among a latent subgroup or stratum of individuals: stratum membership is latent because only one of the potential mediators is observed. Finally, the ACE, $E[Y(1) - Y(0)]$, or total effect (TE), is defined as the sum of the four principal strata effects, $E[Y(1) - Y(0)|M(0) = M(1) = 1] * P[M(0) = M(1) = 1] + E[Y(1) - Y(0) | M(0) = M(1) = 0] * P[M(0) = M(1) = 0] + E[Y(1) - Y(0)|M(0) = 0, M(1) = 1] * P[M(0) = 0, M(1) = 1] + E[Y(1) - Y(0)|M(0) = 1, M(1) = 0] * P[M(0) = 1, M(1) = 0] = E[Y_i(1, M_i(1)) - Y_i(0, M_i(0))]$. It is referred to as the intent-to-treat effect in the compliance literature.

Note that the principal strata effects do not rely on nested potential outcomes of the form, $Y_i(t,M_i(t))$. Principal strata effects rely only on the potential outcomes, $Y_i(0)$, $Y_i(1)$, $M_i(1)$, and $M_i(0)$. Thus, principal strata effects do not rely on $Y_i(1,M_i(0))$ or $Y_i(0,M_i(1))$, which cannot be realized for any individual. This focus on only possible potential outcomes is both a strength and a limitation of this approach; we return to this point later.

Natural Effects Natural *direct* effects (NDEs) are defined by setting the mediator to one of its potential values and changing the intervention status. One NDE of interest, $E[Y_i(1,M_i(0)) - Y_i(0,M_i(0))]$, often called the *pure* NDE (e.g., [19]), defines a causal effect of the intervention on the outcome when the mediator is held to the value that would have been obtained had the individual not received the intervention (i.e., the effect of the intervention on the outcome if the intervention did not cause a change in the mediator or if the effect of the intervention on the mediator was in some way blocked). Additionally $E[Y_i(1,M_i(1)) - Y_i(0,M_i(1))]$, sometimes called the *total* NDE, defines a causal effect of the intervention on the outcome when the mediator is held to the value that would have been obtained had the individual received the intervention (i.e., the effect of the intervention on the outcome if absence of the intervention did not prevent a change in the mediator). Note that since each individual's set of potential mediators may be unique, setting the mediator to one of the potential mediators (i.e., $M_i(0)$ or $M_i(1)$) is not equivalent to setting the mediator to a given value of the mediator m. In other words, the value at which the mediator is set can be different for every individual. We will denote *pure* NDE and the *total* NDE as $\text{NDE}_{M(0)}$ and $\text{NDE}_{M(1)}$, respectively, where the subscript indicates the potential value the mediator is set to.

Natural *indirect* effects (NIEs) are defined by setting the intervention condition and changing the values of the potential mediator, $E[Y_i(1,M_i(1)) - Y_i(1,M_i(0))]$ or $E[Y_i(0,M_i(1)) - Y_i(0,M_i(0))]$. The former, sometimes referred to as the *total* NIE, defines the causal effect of receiving the intervention and having the value on the mediator that would be obtained under the intervention versus having the value on the mediator that would be obtained under the control condition; in other words, the effect of the intervention due to intervention-induced changes in the mediator. The latter, sometimes referred to as the *pure* NIE, defines the causal effect of receiving the control condition and having the value on the mediator that would be obtained under the intervention condition versus having the value on the mediator that would be obtained under the control condition. Note that again, these effects are defined with respect to potential mediators rather than a specific observed value of the mediator. Therefore, the value of the potential mediators may differ across individuals. We will denote the two NIEs as NIE_1 and NIE_0, where the subscript denotes the value to which the intervention status is set.

Note that for NDEs and NIEs, there is an effect for each level of the intervention. For example, in the case of a binary treatment, there are two NDEs and two NIEs. The TE, defined as $E[Y(1) - Y(0)] = E[Y_i(1,M_i(1)) - Y_i(0,M_i(0))]$, can be decomposed into $E[Y_i(1,M_i(1)) - Y_i(1,M_i(0))] + E[Y_i(1,M_i(0)) - Y_i(0,M_i(0))]$ or $E[Y_i(0,M_i(1)) - Y_i(0,M_i(0))] + E[Y_i(1,M_i(1)) - Y_i(0,M_i(1))]$. That is, the TE is the sum of the total NIE, NIE_1, and the pure NDE, $\text{NDE}_{M(0)}$; or of the pure NIE, NIE_0, and the total NDE, $\text{NDE}_{M(1)}$. The terms pure and total refer to whether interaction effects are included with the direct or indirect effect. Specifically, pure means that the interaction effects are not included and total means that they are. Therefore, the TE must include a total and a pure effect.

Controlled Effects The controlled direct effect (CDE; [20]) is the causal effect of the intervention on the outcome when setting the mediator to a specific value, m, for the entire population. That is, $E[Y_i(1,m) - Y_i(0,m)]$ where $Y_i(t,m)$ is the potential outcome when $T = t$ and $M = m$. We will denote the controlled direct effect as CDE_m, where the subscript m denotes the particular value to which m is held or set. Note the difference between the CDE and the NDE. For the CDE, the value at which the mediator is set (i.e., held constant) is the same for every individual. Also, for a binary treatment, there are two NDEs, but there are as many CDEs as there are possible values of the mediator. We have continued to use the i subscript through this section to emphasize that the CDE sets the value of the mediator to be the same for all individuals, whereas the NDE allows the value at which the mediator is set to vary across individuals.

There is not a controlled indirect effect that is comparable to the NIE without further assumptions, which will be discussed below. To illustrate, consider defining the effect $E[Y_i(1,m) - Y_i(1,m')]$ for two different values, for example $m = 0$ and $m' = 1$. We will denote this effect as $\theta_{M|t=1}$ and the corresponding $E[Y_i(0,m) - Y_i(0,m')]$ as $\theta_{M|t=0}$. The former is the effect of, for example, a one-unit change in the mediator on the outcome when $T_i = 1$. This effect does not tell us how the one-unit difference between m and m' has come about: it could have happened through the treatment intervention or through some other mechanism. On the other hand, consider the NIE, $E[Y_i(1,M_i(1)) - Y_i(1,M_i(0))]$, the effect of the intervention due to intervention-induced changes in the mediator. This effect, unlike $E[Y_i(1,m) - Y_i(1,m')]$, does indicate that the intervention caused the difference in $M_i(1)$ and $M_i(0)$ because these are potential outcomes under two different levels of the intervention. This distinction may seem subtle but it is extremely important. The NIE is what behavioral scientists typically think of as the mediation effect, commonly denoted ab in the behavioral science literature, whereas $E[Y_i(1,m) - Y_i(1,m')]$ is the causal effect of the mediator on the outcome, holding constant the intervention status, and is commonly denoted as b in the behavioral science literature. The effects $\theta_{M|t=1}$ and $\theta_{M|t=0}$ also imply that it is possible to set the mediator to the same value for all individuals as mentioned above. For elaboration of these conceptual issues, see VanderWeele and Vansteelandt [21].

It has been shown that under certain assumptions, the various definitions given above for the direct and indirect effects are equivalent (e.g., [5, 22, 23]). We will return to this point after discussing identification assumptions. These assumptions are summarized in Table 14.1.

Table 14.1 Summary of assumptions

	Effects				
	Natural effects	Controlled effects		Principal strata effects	
Assumptions	Imai et al. [4]	IPW	RPM	TSLS	Bayesian
No unmeasured confounders of					
(a) T & M	✓	✓	✓	✓	✓
(b) T & Y	✓	✓	✓	✓	✓
(c) M & Y	✓	✓			
No interactions between					
T & M on Y		✓	✓		
T & X on Y			✓		✓
M & X on Y			✓		
Interactions between T & X on M			✓		
(d) No post-T confounders	✓		✓	✓	✓
Monotonicity (no defiers)				✓	
Exclusion restriction (full mediation)				✓	

3 Identification

The causal effects defined above are written in terms of potential outcomes, not all of which can be observed. If all the potential outcomes were observed, then all of the above effects could be easily estimated. In order to estimate causal effects based on the observed data, assumptions must be made in order to identify the causal effects.

Principal Strata Effects Generally, principal strata effects are identified by assuming that there is no one for whom the intervention has an iatrogenic (i.e., undesirable) effect (e.g., $P[M(0)=1,\ M(1)=0]=0$), which is typically referred to as the monotonicity assumption. Note that the CACE is the causal effect of interest under the hypothesis that the intervention will increase the value of the mediator (i.e., increasing values of the mediator are desirable). If the hypothesis happens to be that the intervention decreases the value of the mediator (i.e., decreasing values of the mediator are desirable), the monotonicity assumption is that $P[M(0)=0,$ $M(1)=1]=0$, and thus, scientific interest lies in DACE. That is, the DACE would be the causal effect of interest.

Additionally, it is assumed that the only way in which the intervention can affect the outcome is through the mediator. This is known as the exclusion restriction and implies that $E[Y_i(1)-Y_i(0)|M_i(1)=M_i(0)]=0$. That is, among those for whom there is no causal effect of the intervention on the mediator, there is no causal effect of the intervention on the outcome. However, the exclusion restriction also means that there is no direct effect of the intervention on the outcome, among those for whom there is a causal effect of the intervention on the mediator (i.e., those in either stratum 3 or 4; the compliers or defiers). In fact, the only way that the principal strata effects for stratum 3 or 4 can be interpreted as an indirect effect is if the exclusion restriction holds. Otherwise, the causal effect estimated is the total effect of the intervention on the outcome among those for whom the intervention had a causal effect on the mediator. The exclusion restriction is particularly difficult to rationalize given that most interventions are designed to affect multiple mediators that are hypothesized to affect the outcome. In addition, an interaction between T and M is a violation of the exclusion restriction [5, 24].

Finally, it is assumed that there are no unmeasured confounders of T and Y (e.g., there is random assignment to T), which can be stated formally as $T \perp Y(0), Y(1)|X_0$. This assumption allows T to be used as an instrumental variable (IV) in the two-stage least-squares (TSLS) estimation to be described below. Note that unlike other causal mediation methods, the principal strata approach does not require a no-unmeasured-confounding assumption for M and Y (given the other assumptions stated above).

Note that the assumptions stated above are not the only set that could be used for identification. Gallop et al. [25] proposed alternative identification assumptions. They do not require the exclusion restriction or monotonicity assumption. Instead, baseline covariates, which predict the principal strata, are used to identify the stratum-specific ACE. In addition, they assume that there are no interactions between these baseline covariates and T within each principal stratum and that there are no unmeasured confounders of T and Y.

Natural Effects To identify the natural effects, it is usually assumed (e.g., [22, 26]) that (a) there are no unmeasured confounders of the intervention and the mediator, $T \perp M(0), M(1)|X_0$; (b) there are no unmeasured confounders of the intervention and the outcome; (c) there are no unmeasured confounders of the mediator and the outcome; and that (d) there are no measured or unmeasured confounders of the mediator and outcome that have themselves been influenced by the intervention (i.e., no post-treatment confounders, denoted $\boldsymbol{X_1}$). Note that the set of variables in $\boldsymbol{X_0}$ do not need to be the same for (a) and (b) and that if $\boldsymbol{X_1}$ is not affected by the intervention, then it does not violate (d) [27]. If individuals are randomized to the intervention, then (a) and (b) will typically hold as long as the randomization does not fail (e.g., individuals comply with the assigned intervention and there is no selective attrition). However, unless individuals are also randomized to levels of the mediator, which is typically impossible in practice, (c) is not guaranteed to hold. These are obviously very strong assumptions that cannot be tested in any empirical application. Nevertheless, if the researcher has given careful thought to all potential confounders, measured them, and properly adjusted for them, assumptions (a)–(c) are plausible. Furthermore, sensitivity analyses have been developed and conducted to assess the impact of violations of these assumptions (e.g., [4, 28, 29]).

Assumption (d) of no post-treatment confounders of the mediator and outcome is more difficult to rationalize. Note that confounders of the mediator and outcome that have been influenced by the intervention are essentially mediators themselves, although they may not be of scientific interest (i.e., the investigator is not interested in their effects and simply wishes to control for them). Assumption (d) is problematic given that most interventions target multiple mediators and because the assumption is that there are no measured or unmeasured variables such as these. Even if they are known to exist and have been measured, they must be assumed not to exist. The mathematical proof of this identification assumption is given in Avin et al. [30].

As with principal strata effects, other assumptions may be used to identify the natural effects ([31–33], but these assumptions do not relax assumption (d). Parametric assumptions, such as linearity, can be used to relax assumption (d).

Controlled Effects Identification of this approach for obtaining the indirect effect requires assuming that there are no unmeasured confounders of the intervention and the mediator (i.e., assumption (a) from above), the intervention and the outcome (i.e., assumption (b) from above), and the mediator and the outcome (i.e., assumption (c) from above); and (e) that there are no interactions between the intervention and the mediator. As discussed by VanderWeele [26], if there is no interaction between the intervention and the mediator, then the CDE is the same for every level of the mediator. In this case, the CDE is equal to the NDEs ($\text{NDE}_{M(0)} = \text{NDE}_{M(1)} = \text{CDE}_m$) and the CDE can be subtracted from the TE, via the decomposition for natural effects (e.g., $\text{TE-CDE}_m = \text{TE-NDE}_{M(0)} = \text{NIE}_1$), to obtain the indirect effect. If there is no interaction, the $\text{NIE}_1 = \text{NIE}_0$ and, therefore, the decomposition may also be written as $\text{TE-NDE}_{M(1)} = \text{NIE}_0$. Note that this approach does not, however, require assumption (d) but replaces it with a parametric assumption. As before, if individuals are randomized to levels of the intervention, then assumptions (a) and (b) will hold, and if individuals could be randomly assigned to levels of the mediator, then assumption (c) would also hold.

Assumptions (a) and (e) are not required for identification of the CDE or for θ_M, the causal effect of M on Y. These two assumptions are only needed to identify the indirect effect. Note that assumption (e) is not as innocuous as it may seem at first. For linear models, it requires the absence of a T by M interaction (i.e., a non-significant coefficient estimate for the product term, $T \times M$). In non-linear models, this assumption is more restrictive; the controlled direct effects at every level, m, of the mediator must be equal.

As with natural effects and principal strata effects, other assumptions may be used to identify the causal effects instead of (a)–(c) and (e). Specifically, assumption (c) can be replaced by assuming that (f) there are no interaction effects between baseline covariates and the mediator, and between baseline covariates and intervention assignment on the potential outcomes; and that (g) there are strong interaction effects between the baseline covariates and intervention assignment on the mediator. The latter two assumptions are key for using the G-estimator proposed by Ten Have et al. [34], described below. All assumptions are summarized in Table 14.1.

When certain conditions or assumptions are met, some of the estimands discussed may be equivalent. For example, as discussed above, if there are no interactions between the intervention and the mediator, the NDE will equal the CDE. Jo [5] and Sobel [24] showed that the traditional behavioral science definitions correspond to the principal strata definitions of effects if there are no unmeasured confounders of M and Y, of T and Y, and of T and M; no interactions between T and M; and the exclusion restriction, monotonicity assumption, and linearity hold. VanderWeele [23] discusses the relations between definitions of principal strata effects and natural effects, and between principal strata effects and controlled effects. Lynch et al. [35] compared and contrasted direct effect definitions in the Ten Have et al. [34] approach with those of the traditional [29] approach and the principal stratification approach. Ten Have and Joffe [33] reviewed identifying assumptions for direct effects under each of the three approaches. However, to our knowledge, the comparisons presented here are the first to focus on definitions, identification assumptions, and estimation methods for all of the effects defined under each approach.

4 Estimation

For each of the definitions, different estimators have been proposed using different sets of identifying assumptions described above. We will consider only a few estimators for each definition. For principal strata effects, we will consider a TSLS IV estimator [36] and a Bayesian estimator [25]. For natural effects, we will consider the estimator proposed by Imai et al. [4]. For controlled effects, we will consider the G-estimator proposed by Ten Have et al. [34] and an inverse propensity weighted (IPW) estimator [3, 26].

Principal Strata Effects Given the monotonicity and exclusion restriction identifying assumptions, the TSLS IV estimator [36], in which intervention assignment is the instrument, is typically used to estimate the principal strata effects. In order for the intervention assignment to be considered an instrumental variable, individuals should be randomly assigned to intervention conditions such that assumptions (a) and (b) hold. Further, for all practical purposes, the principal stratification framework requires a binary mediator.[1] Even for a mediator that takes on, say, 5 values, the number of latent principal strata grows tremendously. Specifically, for a mediator that takes on 5 possible values, there would be 25 latent strata or subgroups of individuals and thus it would be difficult to identify and estimate principal strata effects. Given a binary mediator, monotonicity, the exclusion restriction, and random assignment to the intervention (i.e., no unmeasured confounders of T and M or T and Y), the latent subgroups of individuals are no longer latent because all but one stratum is eliminated.

In the recent statistical literature, there have been attempts to use different identifying assumptions and Bayesian estimation procedures (e.g., [25, 37]) in order to relax the exclusion restriction. The Elliott et al. estimator is limited to both binary mediators and outcomes. We use the Bayesian estimator proposed by Gallop et al.

[1] Gallop [59] proposed Bayesian estimation of direct effects when the mediator is continuous.

to estimate the principal strata effect. This approach was developed to estimate the direct effect, although it estimates all four principal strata effects. Because the authors were not interested in an unbiased causal estimate of the indirect effect, they did not need an assumption of no unmeasured confounders of M and Y. However, if interest lies in a causal estimate of the indirect effect, then this assumption is required. In addition, both the TSLS IV and Bayesian estimators require assumption (d). Although not explicitly stated in the previous literature, a post-T confounder violates the exclusion restriction because there is pathway from T to Y that does not go through M.

Natural Effects Several estimators have now been proposed for estimating natural effects (e.g., [38, 39]) but we will focus on the estimator proposed by Imai and colleagues [4, 22] and implemented in the R package *mediation* [40], which uses identifying assumptions (a)–(d). This estimator involves generating bootstrapped samples and fitting models, which may be parametric or non-parametric, for the observed outcome and observed mediator. From these models, potential values of the mediator are simulated and then potential values of the outcome are simulated given the simulated values of the mediator. Once all of the potential values for the mediator and outcome have been simulated, the natural effects can be computed as defined previously.

Controlled Effects VanderWeele [26] proposed using a marginal structural model (MSM; [41]) with an IPW estimator for defining and estimating the controlled direct effect in the mediation context. MSMs are models for the potential outcomes and are used to define causal effects. For example, for a continuous outcome, the MSMs may be given as $E[M(t)] = \beta_{0M} + \beta_1 t$ and $E[Y(t,m)] = \beta_{0Y} + \beta_2 t + \beta_3 m$, where $\beta_2 = E[Y(1,m) - Y(0,m)] = (\beta_{0Y} + \beta_2 + \beta_3 m) - (\beta_{0Y} + \beta_3 m)$ is the CDE defined above, $\beta_1 = E[M(1) - M(0)] = (\beta_{0M} + \beta_1) - \beta_{0M}$ is the effect of the intervention on the mediator, and $\beta_3 = E[Y(t,m) - Y(t,m')]$ is the effect of the mediator on the outcome for $T = t$. A $T \times M$ interaction term can also be included in the MSM. MSMs are fit by choosing an appropriate model for the observed outcome (e.g., linear regression, logistic regression, survival model), but using the IPW estimator instead of the usual ordinary least squares or maximum likelihood estimator. As long as assumption (e) holds, an estimate of the indirect effect may be obtained by subtracting the CDE from the TE.

For controlled effects, we will also examine the modified G-estimator for the rank preserving model (RPM) described in Ten Have et al. [34]. This estimator does not require assumption (c); however, it does require that individuals are randomized to the intervention (i.e., assumptions (a) and (b)). It also assumes that there are no interaction effects between baseline covariates, X_0, and the mediator and between baseline covariates, X_0, and intervention assignment on the potential outcomes. However, there should be strong interaction effects between the baseline covariates, X_0, and intervention assignment on the mediator. Essentially, this estimator is using the interactions between baseline covariates, X_0, and intervention assignment as instrumental variables. The G-estimator also requires assumption (e). Thus, in summary, the G-estimator exchanges assumption (c) for an assumption of strong interaction effects between the baseline covariates and intervention assignment on the mediator. Although not a stated assumption of the G-estimator (see [34, 35]), assumption (d) is also required. The assumptions for each estimator are summarized in Table 14.1.

5 Simulation Study: Method

5.1 Simulation Study Conditions

The simulation study crosses four assumption violation conditions with four confounding conditions. The first confounding scenario (A) does not involve any confounders. The second confounding scenario (B) involves a pre-treatment confounder, X_0, of M and Y that has not been influenced by T. The third confounding scenario (C) involves a post-treatment but pre-mediator confounder, X_1, of M and Y that has been influenced by T. The fourth confounding scenario (D) involves a pre-treatment confounder of T, M, and Y, such that there is not random assignment to T. These confounding conditions are crossed with two sample size conditions, $N = 100$ and $N = 500$, and three other conditions that systematically violate the assumptions of the different approaches; specifically, monotonicity, the exclusion restriction, and the no-interaction between T and M assumption. A fourth condition in which none of these assumptions are violated is also included. To summarize, for each sample size, there are 16 simulation conditions as follows: no confounders/no violations, no confounders/exclusion restriction violated, no confounders/monotonicity violated, no confounders/no-interaction violated, unmeasured pre-T confounder of M and Y/no violations, unmeasured pre-T confounder of M and Y/exclusion restriction violated, unmeasured pre-T confounder of M and Y/monotonicity violated, unmeasured pre-T confounder of M and Y/no-interaction violated, post-T confounder of M and Y/no violations, post-T confounder of M and Y/exclusion restriction violated, post-T confounder of M and Y/monotonicity violated, post-T confounder of M and Y/no-interaction violated, unmeasured pre-T confounder of T, M, and Y/no violations, unmeasured pre-T confounder of T, M, and Y/exclusion restriction violated, unmeasured pre-T confounder of T, M, and Y/monotonicity violated, unmeasured pre-T confounder of T, M, and Y/no-interaction violated.

In each of the simulation conditions, we generated 1000 data sets and estimated the following causal effects: principal strata effects with TSLS IV estimator, principal strata effects with Bayesian estimator, controlled effects using the IPW estimator, controlled effects using the RPM G-estimator, and natural effects using the Imai et al. [4] estimator.

5.2 Data Generation

The goal is for the data generation to be general enough that it does not favor one approach over another. However, we also need to know the population values for each of the effects. Therefore, we generated all of the potential outcomes for each individual, including the ones that would never be observed for any individual—$Y(1, M(0))$ and $Y(0, M(1))$—so that the causal effects defined previously may be directly computed for each individual. By generating data for all potential outcomes, the true values in all conditions are known.

Each of the simulation study conditions described above dictates the specific values of population parameters (given in Table 14.2), but here we describe the data generation generally. M is binary so that the comparison between principal stratification and the other approaches is more straightforward. However, note that a binary M is not necessary for estimating the controlled or natural effects. T is binary and is generated from a binomial distribution with probability of 0.5 in confounding scenarios A, B, and C. In confounding scenario D, T was generated from a binomial distribution with a probability dependent on X_0. In other words, T is randomized in confounding scenarios A, B, and C but not in D. Y is a continuous, normally distributed variable.

The potential outcomes for M were generated according to a multinomial distribution,

$$[M(0), M(1)] = \begin{Bmatrix} 1,1 \\ 0,0 \\ 1,0 \\ 0,1 \end{Bmatrix} \begin{matrix} p_{11} \\ p_{00} \\ p_{10} \\ p_{01} \end{matrix},$$

where, for confounding scenarios B, C, and D,

$$p_{ij} = \frac{e^{\gamma_0^{ij} + \gamma_1^{ij} X_0 + \gamma_2^{ij} X_1}}{\sum_{i=0}^{1} \sum_{j=0}^{1} e^{\gamma_0^{ij} + \gamma_1^{ij} X_0 + \gamma_2^{ij} X_1}}.$$

Thus, $p_{00} = P[M(0) = 0, M(1) = 0]$, $p_{11} = P[M(0) = 1, M(1) = 1]$, $p_{10} = P[M(0) = 1, M(1) = 0]$, and $p_{01} = P[M(0) = 0, M(1) = 1]$. For confounding scenario A, the multinomial probabilities were set to particular values depending on whether or not the monotonicity assumption was violated.

The potential outcomes for Y were generated according to a multivariate normal distribution with mean,

$$E\left[Y\left(t, M(t)\right)\right] = \beta_0 + \beta_1 t + \beta_2 M(t) + \beta_3 t M(t) + \beta_4 X_0 + \beta_5 X_1,$$

Table 14.2 Population parameter values for simulation study

Conf. scenario	Violation	Population parameters																	
		β_0	β_1	β_2	β_3	β_4	β_5	p_{00}			p_{01}			p_{10}			p_{11}		
No conf.	No violation	0.2	0	0.39	0	0	0	0.33			0.33			0			0.33		
	Exclusion rest.	0.2	0.39	0.39	0	0	0	0.33			0.33			0			0.33		
	Monotonicity 1	0.2	0	0.39	0	0	0	0.25			0.25			0.25			0.25		
	Monotonicity 2	0.2	0	0.39	0	0	0	0.2			0.5			0.1			0.2		
	No T-M interact.	0.2	0.39	0.39	0.39	0	0	0.33			0.33			0			0.33		
								γ_0	γ_1	γ_2	γ_0	γ_1	γ_2	γ_0	γ_1	γ_2	γ_0	γ_1	γ_2
Pre-T unmeasured conf. of M and Y	No violation	0.2	0	0.39	0	0.2	0	0.3	0.3	0	0.3	1	0	0^{a}			0.3	0.7	0
	Exclusion rest.	0.2	0.39	0.39	0	0.2	0	0.3	0.3	0	0.3	1	0	0^{a}			0.3	0.7	0
	Monotonicity	0.2	0	0.39	0	0.2	0	0.3	0.3	0	0.3	1	0	0.3	0.3	0	0.3	0.7	0
	No T-M interact.	0.2	0.39	0.39	0.39	0.2	0	0.3	0.3	0	0.3	1	0	0^{a}			0.3	0.7	0
Post-T conf. of M and Y	No violation	0.2	0	0.39	0	0.2	0.2	0.3	0.3	0.3	0.3	1	1	0^{a}			0.3	0.7	0.7
	Exclusion rest.	0.2	0.39	0.39	0	0.2	0.2	0.3	0.3	0.3	0.3	1	1	0^{a}			0.3	0.7	0.7
	Monotonicity	0.2	0	0.39	0	0.2	0.2	0.3	0.3	0.3	0.3	1	1	0.3	0.3	0.3	0.3	0.7	0.7
	No T-M interact.	0.2	0.39	0.39	0.39	0.2	0.2	0.3	0.3	0.3	0.3	1	1	0^{a}			0.3	0.7	0.7

Note. There is a fourth confounding scenario: Pre-T confounder of T, M, and Y. In this scenario, all parameters are the same as the pre-T confounder of M and Y scenario, with the addition that the Pre-T confounder influences the probability of T ($\beta = .2$). T = treatment, M = mediator, Y = outcome, Conf. = confounder, Interact. = interaction, Rest. = restriction

[a] The value of p_{10} was set to 0 in these conditions

where X_0 is a pre-treatment confounder and X_1 is a post-treatment/pre-mediator confounder. The correlations among the four potential outcomes were set to 0.3 and the error variance was set to 1.0. For the confounders, X_0 was generated from an $N(0,1)$ distribution and X_1 was generated from $T + N(0,1)$, such that the intervention had an effect on X_1.

5.3 Population Values

The population values for each condition of the simulation study are given in Table 14.2. For confounding scenario A in the conditions in which the monotonicity assumption holds, $p_{10} = 0$ and $p_{00} = p_{11} = p_{01} = 1/3$. The proportions for confounding scenario A when monotonicity was violated were set to $p_{00} = 0.2$, $p_{11} = 0.2$, $p_{10} = 0.1$, and $p_{01} = 0.5$. For confounding scenario A only, we also studied a condition in which the monotonicity assumption was violated and all proportions were set to 0.25. The purpose of this condition was to examine what happens as the proportion of defiers increases. In addition, because the proportions are equal, the indirect effect is zero because for 25 % of the sample the indirect effect is positive and for another 25 % of the sample, the indirect effect is equally negative. Thus, the effects cancel out. Although it is unlikely that the stratum proportions would ever be exactly equal or that the proportion of defiers would ever be as large as 0.25, this condition provides some idea of how extreme the bias may become. In the mediation context, the proportion of defiers represents the proportion of individuals for whom the intervention has an iatrogenic effect on the mediator.

For confounding scenario D, the parameter settings were the same as confounding scenario B. However, in confounding scenario D, T was generated from a Bernoulli distribution with $p = 1/(1 + \exp(-0.2 * X_0))$ so that the pre-T confounder had an effect on intervention assignment. For only the $N = 500$ sample size condition, we examined a large effect size condition in which we replaced 0.39 with 0.59 for β_1, β_2, and β_3 in Table 14.2.

6 Simulation Study: Results

The true values for each of the effects were computed according to the definitions presented previously using the potential outcomes. For estimation of the effects, we used only the data that would be available to an investigator (e.g., $M_i(1)$, $Y_i(1,M_i(1))$ if $T_i = 1$). We computed the Monte Carlo (MC) mean and standard deviation (SD) across the 1000 replications. We computed the bias as the difference between the MC mean and the true value, the mean squared error (MSE) as the squared bias plus the squared MC SD, and the 95 % coverage as the number of times the confidence interval (CI) included the true value divided by 1000 and multiplied by 100. The results of the simulations, along with the true values, are given in Tables 14.3,

Table 14.3 Confounding scenario A (no unmeasured confounders) results ($N = 500$) for medium effect size

	NIE	NDE	TE	IPWCDE	IPWθ_M	RPMθ_M	RPMCDE	TSLSIV	Bayesian
	No violations								
TRUE	0.13	0	0.13	0	0.39	0.39	0	0.39	0.39
MEAN	0.131	−0.002	0.129	−0.002	0.391	0.391	−0.002	0.383	0.390
BIAS	0.001	−0.002	−0.001	−0.002	0.001	0.001	−0.002	−0.007	0.000
SD	0.037	0.094	0.090	0.095	0.095	0.095	0.095	0.273	0.146
MSE	0.001	0.009	0.008	0.009	0.009	0.009	0.009	0.072	0.021
Coverage	94.4 %	94.9 %	94.8 %	94.0 %	93.9 %	93.9 %	94.0 %	95.9 %	99.8 %
	Exclusion restriction violated								
TRUE	0.13	0.39	0.52	0.39	0.39	0.39	0.39	0.78	0.78
MEAN	0.130	0.391	0.521	0.391	0.392	0.392	0.391	1.591	0.783
BIAS	−0.000	0.001	0.001	0.001	0.002	0.002	0.001	0.811	0.003
SD	0.034	0.096	0.091	0.095	0.095	0.095	0.095	0.318	0.151
MSE	0.001	0.009	0.0083	0.009	0.008	0.008	0.009	0.771	0.023
Coverage	96.2 %	95.3 %	95.2 %	95.6 %	95.7 %	95.7 %	95.6 %	23.2 %	99.9 %
	Monotonicity violated (all proportions equal)								
TRUE	0	0	0	0	0.39	0.39	0	0.39	0.39
MEAN	−0.000	0.003	0.003	0.003	0.388	0.388	0.003	1.444	0.365
BIAS	−0.000	0.003	0.003	0.003	−0.002	−0.002	0.003	1.054	−0.025
SD	0.018	0.094	0.097	0.090	0.090	0.090	0.090	43448.8	0.285
MSE	0.000	0.009	0.009	0.009	0.008	0.008	0.009	2236.32	0.082
Coverage	95.1 %	92.9 %	92.7 %	93.4 %	95.7 %	95.7 %	93.4 %	99.7 %	99.3 %
	Monotonicity violated ($p_{10} = 0.1$, $p_{01} = 0.5$, $p_{00} = p_{11} = 0.2$)								
TRUE	0.156	0	0.156	0	0.39	0.39	0	0.39	0.39
MEAN	0.154	−0.001	0.154	−0.001	0.387	0.387	−0.001	0.387	0.390
BIAS	−0.002	−0.001	−0.002	−0.001	−0.003	−0.003	−0.001	−0.003	−0.001
SD	0.042	0.100	0.093	0.098	0.098	0.098	0.098	0.227	0.159
MSE	0.002	0.010	0.009	0.010	0.010	0.010	0.010	0.054	0.025
Coverage	96.3 %	94.7 %	94.7 %	94.8 %	95.3 %	95.3 %	94.8 %	94.8 %	97.8 %

14.4, 14.5, 14.6, and 14.7 for the $N = 500$ sample size condition. The results for $N = 100$ were similar; therefore, they are not presented here but are available as supplementary online materials. Likewise, the results for the large effect size condition were similar and are not presented here but are available as supplementary online materials.

6.1 No Confounders

The results for the no confounders/no violations, no confounders/exclusion restriction violated, and no confounders/monotonicity violated conditions are reported

Table 14.4 Results for violation of no-interaction between T and M assumption for all confounding scenarios

	NIE_0	NIE_1	$NDE_{M(0)}$	$NDE_{M(1)}$	TE	CDE_0	CDE_1	$IPW\theta_{M\|t=0}$	$IPW\theta_{M\|t=1}$	TSLSIV	Bayesian
	Scenario A (no confounders)										
TRUE	0.13	0.26	0.52	0.65	0.78	0.39	0.78	0.39	0.78	1.17	1.17
MEAN	0.129	0.260	0.513	0.645	0.773	0.380	0.778	0.386	0.785	2.356	1.166
BIAS	−0.001	0.000	−0.007	−0.005	−0.007	−0.010	−0.002	−0.004	0.005	**1.186**	−0.004
SD	0.049	0.056	0.101	0.098	0.091	0.134	0.136	0.135	0.135	**0.363**	0.147
MSE	0.002	0.003	0.010	0.010	0.008	0.019	0.018	0.018	0.019	1.543	0.021
Coverage	94.7 %	96.1 %	94.6 %	95.8 %	95.9 %	95.4 %	95.4 %	95.0 %	95.3 %	**2.2 %**	99.8 %
	Scenario B (unmeasured pre-T confounder of M and Y)										
TRUE	0.132	0.263	0.515	0.647	0.779	0.39	0.78	0.39	0.78	1.17	1.17
MEAN	0.135	0.295	0.480	0.640	0.775	0.327	0.804	0.405	0.882	2.339	1.245
BIAS	0.004	*0.031*	*−0.035*	−0.007	−0.004	*−0.063*	0.024	0.015	*0.102*	**1.169**	*0.075*
SD	0.051	0.059	0.103	0.106	0.097	0.135	0.140	0.138	0.137	**0.357**	0.149
MSE	0.003	0.004	0.012	0.011	0.009	0.022	0.021	0.021	0.029	1.494	0.028
Coverage	94.8 %	92.6 %	94.1 %	93.9 %	94.4 %	92.5 %	93.2 %	94.6 %	*89.5 %*	**3.2 %**	99.7 %
	Scenario C (post-T confounder of M and Y)										
TRUE	0.154	0.307	0.711	0.864	1.018	0.59	0.98	0.39	0.78	1.37	1.37
MEAN	0.176	0.345	0.509	0.677	0.854	0.475	1.048	0.316	0.889	2.102	1.165
BIAS	0.023	0.038	*−0.202*	*−0.187*	*−0.164*	*−0.115*	0.068	−0.074	*0.109*	**0.732**	*−0.205*
SD	0.061	0.076	0.119	0.120	0.105	0.159	0.144	0.146	0.157	**0.318**	0.234
MSE	0.004	0.007	0.055	0.049	0.038	0.039	0.025	0.025	0.037	0.652	0.097
Coverage	95.3 %	93.6 %	*58.7 %*	*63.8 %*	*65.2 %*	*88.0 %*	92.1 %	93.4 %	*89.2 %*	**33.5 %**	98.1 %
	Scenario D (unmeasured pre-T confounder of T, M, and Y)										
TRUE	0.132	0.264	0.515	0.647	0.779	0.39	0.78	0.39	0.78	1.17	1.17
MEAN	0.140	0.306	0.522	0.688	0.828	0.367	0.846	0.402	0.881	2.403	1.24
BIAS	0.008	0.042	0.007	0.041	0.049	−0.023	0.066	0.012	*0.101*	**1.233**	0.07
SD	0.052	0.057	0.107	0.108	0.099	0.137	0.139	0.138	0.138	**0.347**	0.15
MSE	0.003	0.005	0.012	0.013	0.012	0.020	0.025	0.020	0.028	1.644	0.027
Coverage	93.5 %	90.5 %	94.2 %	92.3 %	90.6 %	95.0 %	91.5 %	94.0 %	89.6 %	**1.4 %**	99.8 %

Note: Boldface type indicates more severe bias and very poor coverage rates. Italics indicates moderate bias and poor coverage rates. Underlining indicates slight bias

Table 14.5 Confounding scenario B (unmeasured pre-T confounder of M and Y) results ($N = 500$) for medium effect size

	NIE	NDE	TE	IPWCDE	IPWθ_M	RPMθ_M	RPMCDE	TSLSIV	Bayesian
	No violations								
TRUE	0.132	0	0.132	0	0.39	0.39	0	0.39	0.39
MEAN	0.153	−0.021	0.132	−0.021	0.450	0.450	−0.021	0.388	0.462
BIAS	0.021	−0.021	−0.000	−0.021	0.060	0.060	−0.021	−0.002	0.072
SD	0.038	0.097	0.094	0.097	0.097	0.097	0.097	0.276	0.158
MSE	0.002	0.010	0.009	0.010	0.012	0.012	0.010	0.077	0.030
Coverage	93.3 %	93.5 %	94.5 %	94.0 %	91.7 %	91.7 %	94.0 %	95.6 %	99.8 %
	Exclusion restriction violated								
TRUE	0.132	0.390	0.522	0.39	0.39	0.39	0.39	0.78	0.78
MEAN	0.150	0.369	0.519	0.369	0.449	0.449	0.369	1.567	0.815
BIAS	0.019	−0.021	−0.002	−0.021	0.059	0.059	−0.021	0.787	0.035
SD	0.038	0.099	0.095	0.097	0.097	0.097	0.097	0.315	0.165
MSE	0.002	0.010	0.009	0.010	0.013	0.013	0.010	0.719	0.028
Coverage	93.0 %	94.5 %	95.0 %	94.6 %	90.1 %	90.1 %	94.6 %	25.7 %	99.7 %
	Monotonicity violated								
TRUE	0.005	0	0.005	0	0.39	0.39	0	0.39	0.39
MEAN	0.005	−0.001	0.005	−0.001	0.427	0.427	−0.001	0.300	0.426
BIAS	0.000	−0.001	−0.000	−0.001	0.037	0.037	−0.001	−0.090	0.036
SD	0.020	0.092	0.094	0.091	0.091	0.091	0.091	277.276	0.233
MSE	0.000	0.009	0.009	0.009	0.010	0.010	0.009	795.214	0.056
Coverage	94.5 %	94.9 %	95.0 %	95.1 %	91.9 %	91.9 %	95.1 %	98.5 %	99.7 %

in Table 14.3. For these conditions, there is no interaction between T and M. Therefore, $\text{NIE}_1 = \text{NIE}_0$ and only one value, NIE, is reported; likewise for NDE and CDE. In the no confounders/no-interaction violated condition, there are two NIEs, NDEs, and CDEs, and results are reported in the top panel of Table 14.4. For the principal strata effects, the results reported in the tables are for the estimand, $E[Y(1) - Y(0)|M(1) = 1, M(0) = 0]$.

For the condition in which all assumptions hold (i.e., exclusion restriction, monotonicity, no interaction between T and M), all approaches give the same unbiased results for all effects. For the condition in which the exclusion restriction is violated, the TSLS IV results are biased with 24 % coverage but the Bayesian principal strata effects are unbiased. Natural and controlled effect estimates are all unbiased.

For the no confounders/monotonicity violated condition, we examined different values (0.1 and 0.25) for the proportion of defiers. Thus, Table 14.3 reports two sets of results for this condition. When monotonicity is violated, the TSLS IV results are biased, with the bias increasing as the proportion of defiers increases from 0.1 to 0.25. The TSLS IV estimates are only slightly biased when the proportion of defiers is small (0.1). When the proportion of defiers is larger (0.25), the MC SD and therefore the MSE became very large. A proportion of defiers of 0.25 is an

Table 14.6 Confounding scenario C (post-T confounder of M and Y) results ($N = 500$) for medium effect size

	NIE	NDE	TE	IPWCDE	IPWθ_M	RPMθ_M	RPMCDE	TSLSIV	Bayesian
	No violations								
TRUE	0.153	0.2	0.353	0.2	0.39	0.39	0.2	0.59	0.59
MEAN	0.173	0.003	0.176	0.200	0.392	0.401	0.000	0.398	0.389
BIAS	0.020	−0.197	−0.177	0.000	0.002	0.011	−0.200	−0.192	−0.201
SD	0.047	0.109	0.103	0.107	0.107	0.966	0.397	0.260	0.235
MSE	0.003	0.051	0.042	0.012	0.010	0.932	0.197	0.107	0.096
Coverage	94.4 %	55.4 %	59.6 %	94.9 %	96.3 %	96.5 %	94.9 %	88.5 %	97.8 %
	Exclusion restriction violated								
TRUE	0.153	0.590	0.743	0.59	0.39	0.39	0.59	0.98	0.98
MEAN	0.171	0.388	0.559	0.583	0.388	0.387	0.389	1.390	0.772
BIAS	0.018	−0.203	−0.185	−0.007	−0.002	−0.003	−0.201	0.410	−0.208
SD	0.050	0.111	0.104	0.107	0.107	0.925	0.379	0.287	0.228
MSE	0.003	0.053	0.045	0.012	0.011	0.854	0.184	0.258	0.095
Coverage	94.4 %	52.2 %	55.8 %	93.2 %	95.8 %	95.9 %	93.0 %	74.1 %	99.6 %
	Monotonicity violated								
TRUE	0.040	0.2	0.240	0.2	0.39	0.39	0.2	0.59	0.59
MEAN	0.055	0.004	0.058	0.199	0.389	0.412	0.001	−1.278	0.362
BIAS	0.014	−0.196	−0.182	−0.001	−0.001	0.022	−0.199	−1.868	−0.228
SD	0.023	0.100	0.101	0.095	0.096	0.438	0.110	3539.23	0.344
MSE	0.001	0.048	0.043	0.009	0.009	0.193	0.359	5859.25	0.170
Coverage	96.0 %	51.9 %	58.0 %	95.4 %	94.9 %	94.2 %	56.0 %	99.3 %	95.6 %

extreme case, as it is unlikely that the proportions in each stratum would be equal or that the intervention would have an iatrogenic effect on this many individuals. Also note that in this case, because the proportions for all strata were equal, the NIE true value is zero because there is an equal proportion with a positive indirect effect and a negative indirect effect and they cancel out. Natural and controlled effect estimates are all unbiased regardless of the proportion of defiers.

For the condition in which the no-interaction between T and M assumption is violated, TSLS IV estimates are biased with 2 % coverage. As mentioned previously when discussing principal strata effects, this condition is also a violation of the exclusion restriction. All other effect estimates were unbiased (see Table 14.4) including the Bayesian principal strata estimate.

6.2 Pre-*T* Confounder of *M* and *Y*

The models fitted to the simulated data in this confounding scenario did not adjust for the pre-T confounder. Thus, this set of conditions represents a violation of the no unmeasured confounding assumption. Results are reported in Table 14.5

Table 14.7 Confounding scenario D (unmeasured pre-T confounder of T, M, and Y) results ($N = 500$) for medium effect size

	NIE	NDE	TE	IPWCDE	IPWθ_M	RPMθ_M	RPMCDE	TSLSIV	Bayesian
	No violations								
TRUE	0.132	0	0.132	0	0.39	0.39	0	0.39	0.39
MEAN	0.157	0.014	0.171	0.014	0.452	0.452	0.014	0.492	0.468
BIAS	0.026	0.014	0.040	0.014	0.062	0.062	0.014	0.102	0.078
SD	0.039	0.097	0.093	0.098	0.098	0.098	0.098	0.267	0.152
MSE	0.002	0.010	0.010	0.010	0.013	0.013	0.010	0.082	0.029
Coverage	91.3 %	94.6 %	92.3 %	94.5 %	90.5 %	90.5 %	94.5 %	93.8 %	99.5 %
	Exclusion restriction violated								
TRUE	0.132	0.390	0.522	0.39	0.39	0.39	0.39	0.78	0.78
MEAN	0.158	0.404	0.561	0.404	0.452	0.452	0.404	1.624	0.851
BIAS	0.026	0.014	0.040	0.014	0.062	0.062	0.014	0.844	0.071
SD	0.039	0.097	0.093	0.098	0.098	0.098	0.098	0.304	0.150
MSE	0.002	0.010	0.010	0.010	0.013	0.013	0.010	0.803	0.027
Coverage	91.7 %	94.4 %	92.4 %	94.5 %	90.5 %	90.5 %	94.5 %	13.9 %	99.3 %
	Monotonicity violated								
TRUE	0.005	0	0.005	0	0.39	0.39	0	0.39	0.39
MEAN	0.008	0.041	0.049	0.041	0.426	0.426	0.041	0.152	0.452
BIAS	0.004	0.041	0.045	0.041	0.036	0.036	0.041	−0.238	0.062
SD	0.019	0.091	0.092	0.091	0.091	0.091	0.091	134.72	0.237
MSE	0.000	0.010	0.011	0.010	0.010	0.010	0.010	212.72	0.060
Coverage	95.2 %	92.2 %	92.4 %	92.5 %	93.4 %	93.4 %	92.5 %	51.3 %	99.6 %

for the unmeasured pre-T confounder of M and Y/no violations, unmeasured pre-T confounder of M and Y/exclusion restriction violated, and unmeasured pre-T confounder of M and Y/monotonicity violated conditions. Results for the unmeasured pre-T confounder of M and Y/no-interaction violation condition are reported in the second panel of Table 14.4.

For the unmeasured pre-T confounder of M and Y/no violations condition, the TSLS IV estimates and the natural direct and indirect, and controlled direct effects are unbiased. However, the Bayesian principal strata effects, and the estimates of θ_M using either IPW or the RPM are slightly biased.

For the condition in which the exclusion restriction is violated, the TSLS IV estimates are more severely biased, and the IPW and RPM estimates of θ_M are moderately biased. The remaining effects are unbiased. Coverage for the TSLS IV estimate is 26 % but the NIE, NDE, and CDE have adequate coverage (approx. 93–95 %).

For the condition in which monotonicity is violated, the NIE, NDE, and CDE are unbiased. However, the TSLS IV and Bayesian principal strata estimates and the estimates of θ_M using either IPW or RPM are biased. The TSLS IV estimate is more biased than the RPM or IPW estimates of θ_M and the Bayesian estimate. Again, the MC SD and therefore MSE for the TSLS IV estimate are extremely large.

For the condition in which the no-interaction between T and M assumption is violated, the NIE_1, $\text{NDE}_{M(0)}$, CDE_0, $\theta_{M|t=1}$, and the Bayesian principal strata estimates were slightly biased, the NIE_0, $\text{NDE}_{M(1)}$, CDE_1, and $\theta_{M|t=0}$ estimates were unbiased, and the TSLS IV estimate was severely biased with unacceptable 95 % coverage (3 %, see Table 14.4).

6.3 Post-T Confounder of M and Y

In this confounding scenario, the exclusion restriction is violated in all the conditions due to the post-T confounder. For this confounding scenario, the models fit to the simulated data included both the pre- and post-T confounders. Thus, the no-unmeasured confounding assumptions are *not* violated. Results are reported in Table 14.6 for the post-T confounder of M and T/no violations, post-T confounder of M and T/exclusion restriction violated, and post-T confounder of M and T/monotonicity violated conditions. Results for the post-T confounder of M and T/no-interaction violated condition are reported in the third panel of Table 14.4.

For the post-T confounder of M and T/no violations condition (however, the exclusion restriction is violated due to the post-T confounder although there is not otherwise a direct effect of T on Y), the TSLS IV and Bayesian principal strata estimates, the NDE, and the CDE estimated via the RPM are biased to approximately the same degree. The CDE estimated via IPW, the NIE, and θ_M estimated via either IPW or the RPM are unbiased although the MSE of θ_M for the RPM is much larger than the MSE for the IPW estimates. The 95 % coverage for the NDE and TSLS IV estimates is unacceptable.

For the condition in which the exclusion restriction is violated (i.e., there is a direct effect of T on Y in addition to the effect through the post-intervention confounder), the TSLS IV and Bayesian principal strata estimates, the NDE, and the CDE estimated via the RPM are biased. The CDE estimated via IPW, the NIE, and θ_M estimated via either IPW or the RPM are unbiased although the MSE of θ_M for the RPM is much larger than the MSE for the IPW estimates. The 95 % coverage for the NDE and TSLS IV estimates is unacceptable.

For the condition in which monotonicity is violated, the TSLS IV and Bayesian principal strata estimates, the NDE, and the CDE estimated via the RPM are biased. The CDE estimated via IPW, the NIE, and θ_M estimated via either IPW or the RPM are unbiased although the MSE of θ_M for the RPM is much larger than the MSE for the IPW estimate. The 95 % coverage for the NDE and TSLS IV estimates is unacceptable.

For the condition in which the no-interaction between T and M assumption is violated, all effects are biased to some degree. The TSLS IV estimate is the most severely biased with unacceptable 95 % coverage (33.5 %, see Table 14.4). The $\text{NDE}_{M(0)}$, $\text{NDE}_{M(1)}$, CDE_0, the IPW $\theta_{M|t=1}$, and the Bayesian principal strata effect estimates were moderately biased. Coverage for these effects was also unacceptable. The NIE_1, NIE_0, CDE_1, and IPW $\theta_{M|t=0}$ estimates were slightly biased.

6.4 Pre-*T* Confounder of *T*, *M*, and *Y*

The models fitted to the simulated data in this confounding scenario did not adjust for the pre-T confounder. Thus, these conditions represent a violation of the no-unmeasured-confounders of T and Y, T and M, and M and Y assumptions. Results are reported in Table 14.7 for the unmeasured pre-T confounder of T, M, and Y/no violations, unmeasured pre-T confounder of T, M, and Y/exclusion restriction violated, and unmeasured pre-T confounder of T, M, and Y/monotonicity violated conditions. Results for the unmeasured pre-T confounder of T, M, and Y/no-interaction violated condition are reported in the fourth panel of Table 14.4. There is no post-T confounder in any of the conditions for this confounding scenario (Table 14.8).

For the unmeasured pre-T confounder of T, M, and Y/no violations condition, the TSLS IV estimate is biased. In this confounding scenario, the use of T as an IV is not justified for the TSLS IV estimator. All other estimates are slightly biased. The bias is most notable when compared to the corresponding bias in Tables 14.3 and 14.5. For example, in Table 14.3, the no-unmeasured-confounding assumption holds and there is no bias. In Table 14.5, the no-unmeasured-confounding assumption holds with regard to T but not M. Bias for the NIE and θ_M estimates using either IPW or the RPM are essentially the same between Tables 14.5 and 14.7. However, the bias for the NDE and CDE estimated using either IPW or the RPM are larger in Table 14.7 than in Table 14.5 because the no-unmeasured-confounding assumption for T is also violated in Table 14.7. Finally, the bias for the Bayesian principal strata estimates increased in Table 14.7 compared to Table 14.5.

For the condition in which the exclusion restriction is violated, the results follow the exact same pattern except that now the TSLS IV estimate is more severely biased due to violation of the exclusion restriction. In addition, the 95 % coverage for the TSLS IV estimate is unacceptably low (13.9 %). For the condition in which monotonicity is violated, the results again follow the same pattern except that, in addition, the MC SD for the TSLS IV estimate, and therefore the MSE, is extremely large (Table 14.9).

For the condition in which the no-interaction between T and M assumption is violated, the NIE_0, $\text{NDE}_{M(0)}$, and $\theta_{M|t=0}$ estimates are unbiased. The TSLS IV estimates were again severely biased with unacceptable coverage (1.4 %, see

Table 14.8 Results of empirical data analysis for natural effects

	Without interaction			With interaction		
	Estimate	95 % CI		Estimate	95 % CI	
NIE_0	1.384	0.215	2.825	2.251	0.348	4.859
NIE_1	1.384	0.215	2.825	0.898	−0.494	2.588
$\text{NDE}_{M(0)}$	1.313	−1.249	3.680	1.595	−0.885	4.334
$\text{NDE}_{M(1)}$	1.313	−1.249	3.680	0.242	−2.631	3.034
TE	2.697	0.321	4.917	2.493	0.300	4.870

Table 14.9 Results of empirical data analysis for controlled effects using inverse propensity weighted estimator

	Without interaction				With interaction			
	Estimate	SE	95 % CI		Estimate	SE	95 % CI	
CDE_0	2.879	1.063	0.796	4.963	3.824	1.076	1.714	5.933
CDE_1	2.879	1.063	0.796	4.963	0.972	2.228	−3.394	5.338
$\theta_{M\|t=0}$	2.194	1.183	−0.125	4.513	3.847	2.073	−0.216	7.911
$\theta_{M\|t=1}$	2.194	1.183	−0.125	4.513	0.996	1.350	−1.650	3.641

Table 14.10 Results of empirical data analysis for Bayesian principal strata effects

	Estimate	SE	95 % CI	
CACE	5.1318	2.6897	−0.5840	10.4499
AACE	−1.3870	4.8571	−9.7951	10.0298
NACE	2.4501	2.9214	−3.2872	8.5401
DACE	0.3493	6.1052	−11.2820	13.2813

Table 14.4). The $\theta_{M|t=1}$ estimate was moderately biased. The NIE_1, $NDE_{M(1)}$, CDE_0, CDE_1, and Bayesian principal strata estimates were all slightly biased (Table 14.10).

6.5 *Additional Overall Observations Regarding Results of Simulation Study*

The MC SD for the Bayesian estimates was generally larger than the MC SD for the other methods. The MC SD for the TSLS IV estimates were much larger than that for the other methods when there was an interaction between T and M. Coverage for the Bayesian principal strata estimates was over 99 % in almost all simulation conditions. The results were similar for the $N = 100$ sample size condition, which are included in supplementary online materials. We also examined a large effect size, 0.59 (see [42]), and obtained similar results. That is, all 0.39 values in Table 14.2 were replaced with 0.59. These results are included in supplementary online materials.

The IPW CDE and θ_M estimates were unbiased when the no-interaction between T and M assumption is violated (see top panel of Table 14.4). These estimates were also unbiased when there was a post-T confounder of M and Y (see Table 14.6). However, when both of these assumptions were violated, these estimates were biased (see third panel of Table 14.4). We examined this situation further by generating 1000 replications for a sample size $N = 10{,}000$ and estimating the IPW CDE and θ_M effects. Although the MC SD decreased as would be expected due to the increased sample size, the bias did not decrease. In fact, it remained consistent with the bias reported in the third panel of Table 14.4. Thus, IPW CDE and θ_M estimates are not robust for the post-T confounder of M and Y/no-interaction violation condition.

7 Discussion

The simulation study results illustrate that if the identifying assumptions used by an estimator hold, then the estimator performs well in terms of bias, and if they do not hold, then the estimator does not perform well in terms of bias. In addition, some estimators seem to be more robust than others when assumptions are violated. Specifically, the simulation study illustrates that the TSLS IV estimator of the principal strata effects and the RPM G-estimator, which relies on interaction terms that act as instrumental variables, require that the instrumental variable assumptions hold and if they do not, these methods are just as biased as those that rely on sequential ignorability. This problem has been known for quite some time when attempting to estimate the causal effect of an endogenous variable on an outcome [43] and it carries over to mediation analysis as well. Unfortunately, many of the assumptions cannot be verified in empirical data, leaving the researcher to attempt to justify the assumptions based on rational argument. However, we suggest that researchers who use instrumental variable methods, such as the RPM, report the strength of the interaction term on the mediator, as well as the strength of the interaction term on the outcome. Note that the lack of an effect of the interaction term on the outcome does not guarantee that the exclusion restriction holds and that violation of the exclusion restriction cannot be verified or refuted from the observed data [44]. Furthermore, weak instruments may actually amplify bias in comparison with an unadjusted estimate (see e.g., [43–45]). In other words, using no instrument can be better than using a weak instrument. We propose that researchers take the following steps: define the causal estimand, justify the identification assumptions, and try several estimators.

7.1 Comparison of Approaches in Terms of Definitions

The definitions of the various approaches coincide in very limited situations in which all assumptions of the various approaches hold. Specifically, when there are no confounders and all assumptions hold (i.e., exclusion restriction, monotonicity, no interaction between T and M), then $\text{NIE}_0 = \text{NIE}_1 = \text{CACE}$ and $\text{NDE}_{M(0)} = \text{NDE}_{M(1)} = \text{CDE}_m$. Thus, one consideration in choosing an approach is clearly articulating the scientific question of interest. For example, if the researcher is interested in the causal effect of the intervention on the outcome among those individuals for whom the intervention had an effect on the mediator in the intended direction, then the principal strata effects are of interest. If the researcher is interested in the effect of the mediator on the outcome, then the controlled effects are of interest, because the natural effects do not define this effect separately from the indirect effect. If the researcher is interested in the causal effect of the intervention on the outcome that is due to the mediator, then the NIEs are of interest.

7.2 Comparison of Approaches in Terms of Identification

For different empirical data sets, certain assumptions are more likely to hold than others. For example, in some studies the exclusion restriction may be plausible, and in other studies no post-intervention confounders may be more plausible. Thus, one consideration in choosing an approach is the plausibility of the various assumptions for a particular data set. For an extensively studied research area, scientists may have knowledge about the validity of model assumptions but this knowledge is unlikely in relative new research areas.

The assumption of no post-treatment confounders (assumption (d)), in which there might be multiple mediators or confounders of the mediator and outcome that have been influenced by the intervention, is likely to be violated in many studies. Suppose a researcher is interested in the NIE, but assumption (d) is not plausible. If instead assumption (e), no $T \times M$ interaction, is plausible, then an estimate of the CDE can be obtained and subtracted from the TE to obtain an estimate of the indirect effect. Another alternative is to include measures of the additional mediators of the intervention in the statistical analysis, known as a multiple mediator model. Accurate estimation of causal effects in this model is an active research area in the field of causal inference (e.g., [46, 47]).

If a researcher is not able to justify any of the identifying assumptions, or is particularly interested in a specific estimand and cannot justify the identifying assumptions for that estimand, then it is important to find ways to assess the sensitivity of the estimates to violations of the assumptions. In some cases, sensitivity analysis has been developed. For instance, Imai et al. [4] proposed sensitivity analysis to the no-unmeasured-confounding assumptions used in identifying natural effects and implemented it in the R *mediation* package. VanderWeele [28] has proposed a sensitivity analysis for the no-unmeasured-confounding assumptions used in identifying controlled effects. Sensitivity analysis for the presence of a post-treatment confounder for natural effect estimates has recently been developed [46]. However, one type of sensitivity analysis that researchers could try is using several different estimators that rely on different identifying assumptions for the particular definition of interest. If the results generally agree, it seems safe to conclude that either the assumptions are not violated or that the estimates are not sensitive to violations of them. Of course, if the results do not agree, the researcher does not know which are correct. In any case, identifying causal effects will require assumptions; thus, it seems development of sensitivity analysis is an important direction for future research. Another alternative is to design future research studies in order to reduce or eliminate the violation of assumptions.

7.3 Comparison of Approaches in Terms of Estimation

If one particular definition of mediation is of scientific interest and the identifying assumptions of a particular estimator are not plausible, then a different estimator using different identifying assumptions may be used. For example, if controlled effects are of interest and the no-unmeasured-confounders assumption is not plausible, then the RPM using intervention-by-baseline-covariate interaction effects on the mediator as IVs may be more plausible. New estimators for each of the approaches that use different identifying assumptions are rapidly being developed in the statistical literature (see e.g., [38, 39, 48]). However, none of these estimators relaxes the no post-treatment confounders assumption for estimation of the natural effects. If natural effects are of interest, and the no post-treatment confounders assumption is unlikely to hold, then investigators may be able to define and estimate the NDEs and NIEs on the treated as described in Vansteelandt and VanderWeele [32].

7.4 Limitations and Future Directions

In this study, we only considered estimation—we did not consider hypothesis testing and power. This and sensitivity analysis are directions for future work. We also did not vary the strength of the confounding because the size of the simulation study was already large. We would expect that as the effect of the unmeasured pre-T confounder of M and Y (confounding scenario B), or of T, M, and Y (confounding scenario D) increases, the bias resulting from not accounting for the confounder would also increase. We also did not vary the strength of post-T confounder of M and Y or of the interaction between T and M; rather we examined only the presence or absence of violations of these assumptions.

There are other estimators for each approach that we did not consider here. For the principal stratification approach, we did not implement the Jo et al. [49] estimator, which uses reference stratification and propensity scores. Elliott et al. [37] proposed a Bayesian estimator for principal strata effects, although this estimator is only applicable when there is a binary mediator and a binary outcome. For estimating the natural effects, Hogan [39] proposed an imputation-based estimator, Daniels et al. [38] proposed a Bayesian estimator, Vansteelandt et al. [50] proposed an imputation-based estimator, and VanderWeele and Vansteelandt [51] and Valeri and VanderWeele [52] proposed an estimator for dichotomous outcomes based on the mediation formula [6, 53]. Several other estimators for the CDEs have been proposed, including a sequential G-estimation approach proposed by Vansteelandt [54] and an estimator proposed by Emsley et al. [55] that is very similar to the RPM G-estimator. Albert [56] proposed a TSLS estimator that is similar to those proposed by Dunn and Bentall [57] and Joffe and Greene [58].

8 Conclusions

We examined three different definitions of causal effects in the mediation context. For each of these definitions, we presented commonly used identifying assumptions along with estimation methods using different sets of these identifying assumptions. Specifically, we examined the TSLS IV and a Bayesian estimator for principal strata effects, the Imai et al. [4] estimator for natural effects, and IPW and the RPM G-estimator for controlled effects. In conclusion, we demonstrated that effect estimates may be biased when the identifying assumptions underlying each method are violated. We recommend that researchers specify which definition (i.e., causal effect) they wish to estimate along with consideration of the plausibility of assumptions made. For the mediation case with randomized T, two critical assumptions are the extent to which there is confounding of the M to Y relation and the extent to which there are effects of the treatment (i.e., post-treatment confounders) that confound the M to Y relation. Mediation analysis from a potential outcomes framework provides a more detailed approach to understanding mediating processes by specifying the definitions and assumptions necessary for causal inferences. Finally, we suggest that whenever possible researchers conduct sensitivity analyses.

References

1. MacKinnon, D.P.: Introduction to Statistical Mediation Analysis. LEA, New York (2008)
2. Coffman, D.L.: Estimating causal effects in mediation analysis using propensity scores. Struct. Equ. Model. **18**, 357–369 (2011)
3. Coffman, D.L., Zhong, W.: Assessing mediation using marginal structural models in the presence of confounding and moderation. Psychol. Methods (2012). doi:10.1037/a0029311
4. Imai, K., Keele, L., Tingley, D.: A general approach to causal mediation analysis. Psychol. Methods **15**, 309–334 (2010)
5. Jo, B.: Causal inference in randomized experiments with mediational processes. Psychol. Methods **13**, 314–336 (2008)
6. Pearl, J.: The causal mediation formula – a guide to the assessment of pathways and mechanisms. Prev. Sci. **13**, 426–436 (2012)
7. Holland, P.W.: Causal inference, path analysis, and recursive structural equations models. Sociol. Methodol. **18**, 449–484 (1988)
8. Holland, P.W.: Statistics and causal inference. J. Am. Stat. Assoc. **81**, 945–970 (1986)
9. Rubin, D.B.: Estimating causal effects of treatments in randomized and nonrandomized studies. J. Educ. Psychol. **66**, 688–701 (1974)
10. Rubin, D.B.: Causal inference using potential outcomes: design, modeling, decisions. J. Am. Stat. Assoc. **100**, 322–331 (2005)
11. Little, R.J.A., Rubin, D.B.: Causal effects in clinical and epidemiological studies via potential outcomes: concepts and analytical approaches. Annu. Rev. Public Health **21**, 121–145 (2000)
12. Schafer, J.L., Kang, J.D.Y.: Average causal effects from non-randomized studies: a practical guide and simulated example. Psychol. Methods **13**, 279–313 (2008)
13. Winship, C., Morgan, S.L.: The estimation of causal effects from observational data. Annu. Rev. Sociol. **25**, 659–706 (1999)
14. VanderWeele, T.J.: Concerning the consistency assumption in causal inference. Epidemiology **20**(6), 880–883 (2009)

15. Westreich, D., Cole, S.R.: Invited commentary: positivity in practice. Am. J. Epidemiol. **171**, 674–677 (2010)
16. Frangakis, C.E.: Principal stratification. In: Gelman, A., Meng, X.L. (eds.) Applied Bayesian Modeling and Causal Inference from Incomplete Data Perspectives, pp. 97–108. Wiley, New York (2004)
17. Frangakis, C.E., Rubin, D.B.: Principal stratification in causal inference. Biometrics **58**, 21–29 (2002)
18. Rubin, D.B.: Direct and indirect causal effects via potential outcomes. Scand. J. Stat. **31**, 161–170 (2004)
19. Pearl, J.: Direct and indirect effects. In: Besnard, P., Hanks, S. (eds.) Proceedings of the Seventeenth Conference on Uncertainty in Artificial Intelligence. Morgan Kaufman, San Francisco (2001)
20. Robins, J.M., Greenland, S.: Identifiability and exchangeability for direct and indirect effects. Epidemiology **3**, 143–155 (1992)
21. VanderWeele, T.J., Vansteelandt, S.: Conceptual issues concerning mediation, interventions and composition. Stat. Interface **2**, 457–468 (2009)
22. Imai, K., Keele, L., Yamamoto, T.: Identification, inference, and sensitivity analysis for causal mediation effects. Stat. Med. **25**, 51–71 (2010)
23. VanderWeele, T.J.: Simple relations between principal stratification and direct and indirect effects. Stat. Probab. Lett. **78**, 2957–2962 (2008)
24. Sobel, M.E.: Identification of causal parameters in randomized studies with mediating variables. J. Educ. Behav. Stat. **33**, 230–251 (2008)
25. Gallop, R., Small, D.S., Lin, J.Y., Elliott, M.R., Joffe, M.M., Ten Have, T.R.: Mediation analysis with principal stratification. Stat. Med. **28**, 1108–1130 (2009)
26. VanderWeele, T.J.: Marginal structural models for the estimation of direct and indirect effects. Epidemiology **20**, 18–26 (2009)
27. Pearl, J.: Interpretation and identification of Causal Mediation. Psychol. Meth. **19**(4), 459–481 (2014)
28. VanderWeele, T.J.: Bias formulas for sensitivity analysis for direct and indirect effects. Epidemiology **21**, 1–12 (2010)
29. Baron, R.M., Kenny, D.A.: The moderator–mediator variable distinction in social psychological research: conceptual, strategic and statistical considerations. J. Person. Soc. Psychol. **51**, 1173–1182 (1986)
30. Avin, C., Shipster, I., Pearl, J.: Identifiability of path-specific effects. In: Proceedings of the International Joint Conferences on Artificial Intelligence, pp. 357–363. Department of Statistics, UCLA, Los Angeles (2005)
31. Hafeman, D.M., VanderWeele, T.J.: Alternative assumptions for identification of direct and indirect effects. Epidemiology **22**, 753–764 (2011). doi:10.1097/EDE.0b013e3181c311b2
32. Vansteelandt, S., VanderWeele, T.J.: Natural direct and indirect effects on the exposed: effect decomposition under weaker assumptions. Biometrics **68**(4), 1019–1027 (2012)
33. Ten Have, T.R., Joffe, M.M.: A review of causal estimation of effects in mediation analysis. Stat. Meth. Med. Res. **21**, 77–107 (2012)
34. Ten Have, T.R., Joffe, M.M., Lynch, K.G., Brown, G.K., Maisto, S.A., Beck, A.T.: Causal mediation analyses with rank preserving models. Biometrics **36**, 926–934 (2007)
35. Lynch, K.G., Kerry, M., Gallop, R., Ten Have, T.R.: Causal mediation analyses for randomized trials. Health Serv. Outcome Res. Methodol. **8**, 57–76 (2008)
36. Angrist, J.D., Imbens, G.W., Rubin, D.B.: Identification of causal effects using instrumental variables. J. Am. Stat. Assoc. **91**, 444–472 (1996)
37. Elliott, M.R., Raghunathan, T.E., Li, Y.: Bayesian inference for causal mediation effects using principal stratification with dichotomous mediators and outcomes. Biostatistics **11**, 353–372 (2010)
38. Daniels, M.J., Roy, J., Kim, C., Hogan, J.W., Perri, M.: Bayesian inference for the causal effect of mediation. Biometrics **68**(4), 1028–1036 (2012)

39. Hogan, J.W.: Imputation-based inference for natural direct and indirect effects. Presented at the Workshop on Causal Inference in Health Research, Montreal, Canada, May 2011
40. Keele, L., Tingley, D., Yamamoto, T., Imai, K.: Mediation: R package for causal mediation analysis [Computer software manual] (2009). Available from http://CRAN.R-project.org/package=mediation (R package version 2.1)
41. Robins, J.M., Hernan, M.A., Brumback, B.A.: Marginal structural models and causal inference in epidemiology. Epidemiology **11**, 550–560 (2000)
42. MacKinnon, D.P., Lockwood, C.M., Hoffman, J.M., West, S.G., Sheets, V.: A comparison of methods to test mediation and other intervening variable effects. Psychol. Methods **7**, 83–104 (2002)
43. Bound, J., Jaeger, D.A., Baker, R.M.: Problems with instrumental variables estimation when the correlation between the instruments and the endogenous explanatory variable is weak. J. Am. Stat. Assoc. **90**, 443–450 (1995)
44. Hernan, M.A., Robins, J.M.: Instruments for causal inference: an epidemiologist's dream? Epidemiology **17**(4), 360–371 (2006)
45. Pearl, J.: On a class of bias-amplifying covariates that endanger effect estimates. UCLA Cognitive Systems Laboratory, Technical Report (R-356). In: Grunwald, P., Spirtes, P. (eds.) Proceedings of the Twenty-Sixth Conference on Uncertainty in Artificial Intelligence, pp. 417–424. Corvallis, OR (2010)
46. Imai, K., Yamamoto, T.: Identification and sensitivity analysis for multiple causal mechanisms: revisiting evidence from framing experiments. Polit. Anal. **1**, 1–31 (2013). doi:10.1093/pan/mps040
47. Wang, W., Nelson, S., Albert, J.M.: Estimation of causal mediation effects for a dichotomous outcome in multiple-mediator models using the mediation formula. Stat. Med. **32**(24), 4211–4228 (2013)
48. Lange, T., Vansteelandt, S., Bekaert, M.: A simple unified approach for estimating natural direct and indirect effects. Am. J. Epidemiol. **176**, 190–195 (2012)
49. Jo, B., Stuart, E.A., MacKinnon, D.P., Vinokur, A.D.: The use of propensity scores in mediation analysis. Multivar. Behav. Res. **46**, 1–28 (2011). doi:10.1080/00273171.2011.576624
50. Vansteelandt, S., Bekaert, M., Lange, T.: Imputation strategies for the estimation of natural direct and indirect effects. Epidemiol. Methods **1**, 131–158 (2012)
51. VanderWeele, T.J., Vansteelandt, S.: Odds ratios for mediation analysis for a dichotomous outcome. Am. J. Epidemiol. **172**, 1339–1348 (2010)
52. Valeri, L., VanderWeele, T.J.: Mediation analysis allowing for exposure-mediator interactions and causal interpretation: theoretical assumptions and implementation with SAS and SPSS macros. Psychol. Methods (2013)
53. Pearl, J.: Interpretable conditions for identifying direct and indirect effects. UCLA Cognitive Systems Laboratory Technical Report (R-389) (2012)
54. Vansteelandt, S.: Estimating direct effects in cohort and case-control studies. Epidemiology **20**(6), 851–860 (2009)
55. Emsley, R., Dunn, G., White, I.R.: Mediation and moderation of treatment effects in randomised controlled trials of complex treatments. Stat. Methods Med. Res. **19**(3), 237–270 (2010)
56. Albert, J.M.: Mediation analysis via potential outcomes models. Stat. Med. **27**, 1282–1304 (2008)
57. Dunn, G., Bentall, R.: Modelling treatment-effect heterogeneity in randomized controlled trials of complex interventions (psychological treatments). Stat. Med. **26**, 4719–4745 (2007)
58. Joffe, M.M., Greene, T.: Related causal frameworks for surrogate outcomes. Biometrics **65**, 530–538 (2009)
59. Gallop, R.: Principal stratification for assessing mediation with a continuous mediator. Paper presented at the Eastern North American Region of the International Biometric Society, Washington, April 2012
60. Cole, S.R., Frangakis, C.: The consistency statement in causal inference: a definition or an assumption. Epidemiology **20**(1), 3–5 (2009)

61. MacCallum, R.C., Zhang, S., Preacher, K.J., Rucker, D.D.: On the practice of dichotomization of quantitative variables. Psychol. Methods **7**(1), 19–40 (2002). doi:10.1037/1082-989X.7.1.19
62. Rosenbaum, P.R.: The consequences of adjustment for a concomitant variable that has been affected by the treatment. J. R. Stat. Soc. Ser. A (General) **147**, 656–666 (1984)
63. West, S.G., Biesanz, J.C., Pitts, S.C.: Causal inference and generalization in field settings: experimental and quasi-experimental designs. In: Reis, H.T.J., Judd, C. (eds.) Handbook of Research Methods in Social and Personality Psychology, pp. 40–84. Cambridge University Press, New York (2000)
64. Cox, M.G., Kisbu-Sakarya, Y., Miočević, M., MacKinnon, D.P.: Sensitivity plots for confounder bias in the single mediator model. Eval. Rev. **37**(5), 405–431 (2014)

Chapter 15
Causal Mediation Analysis Using Structure Equation Models

Douglas Gunzler, Nathan Morris, and Xin M. Tu

Abstract Structural equation modeling (SEM) is an extremely flexible, powerful technique for causal mediation analysis. In this chapter we discuss advantages to using the SEM framework in the context of causal mediation analysis. SEM is designed, in part, to test these more complicated mediation models in a single analysis. Thus the approach allows for ease of interpretation and estimation, while simplifying testing of mediation hypotheses. SEM can be used when extending a mediation process to multiple independent variables, mediators or outcomes, including latent constructs and performing longitudinal data analyses. In this chapter we also discuss SEM model fit information about the consistency of the hypothesized mediational model to the data. Standard causal inference assumptions used when deriving causal indirect effects can be applied in the SEM framework for inference with non-continuous outcomes and mediators.

Current SEM methods impose various restrictions on the study designs and data distributions, limiting the utility of the information they provide in real study applications. In particular, in longitudinal studies missing data is commonly addressed under the assumption of missing at random (MAR), where current methods are unable to handle such missing data if parametric assumptions are violated. We also discuss in this chapter a robust approach to address the limitations of current SEM within the context of longitudinal mediation analysis by utilizing a class of functional response models (FRM). Being distribution-free, the FRM-based approach does not impose any parametric assumption on data distributions and can handle different types of outcomes (i.e., continuous, count outcomes). In addition, by extending the inverse probability weighted (IPW) estimates to the

D. Gunzler (✉)
Center for Health Care Research & Policy, MetroHealth Medical Center, Case Western Reserve University, 2500 MetroHealth Drive, Cleveland, OH 44109, USA
e-mail: dgunzler@metrohealth.org

N. Morris
Department of Epidemiology and Biostatistics, Case Western Reserve University, 10900 Euclid Ave, Cleveland, OH 44106, USA

X.M. Tu
Department of Biostatistics and Computational Biology, University of Rochester, Rochester, NY 14642, USA

H. He et al. (eds.), *Statistical Causal Inferences and Their Applications in Public Health Research*, ICSA Book Series in Statistics, DOI 10.1007/978-3-319-41259-7_15

current context, the FRM-based SEM provides valid inference for longitudinal mediation analysis under the two most popular missing data mechanisms; missing completely at random (MCAR) and missing at random (MAR). We illustrate the SEM approaches discussed in this chapter with real data.

1 Basic Advantages of Using Structural Equation Modeling for Causal Mediation Analysis

There are many advantages for using the structural equation modeling (SEM) framework in the context of mediation analysis [1–3].

- SEM allows for the inclusion of latent variables such as happiness and quality of life.
- SEM allows for the joint estimation of all parameters of a mediation model in a single analysis.
- SEM allows for the extension of the mediation process to include multiple independent variables, mediators, or outcomes in a single model.
- Many techniques are available (i.e., full information maximum likelihood) for handling missing data under various assumptions for a structural equation model in a single analysis.
- SEM approach provides model fit information about consistency of the hypothesized mediational model to the data.
- SEM implies a functional relationship among variables via a conceptual model, path diagram, and mathematical equations thus giving a rich, natural language for expressing causal relationships.

Causal inference methods can be directly applied in the SEM framework for causal mediation [4–6]. Thus, these approaches address the issues of potential confounders of the mediator–outcome relationship, potential interaction between the mediator and treatment, as well as provide definitions for deriving effects for analyses involving mediators and outcomes that are not on an interval scale (i.e., count data, categorical data) all within the SEM framework. These approaches can be readily implemented in MPlus (Múthen, 2011) [5]. MPlus is more generally a program for latent variable modeling of which classical SEM is a special case [7].

SEM allows for ease of extension to longitudinal data within a single framework, corresponding with a study's conceptual framework for clear hypothesis articulation [8]. Latent growth modeling (LGM) is an SEM extension for longitudinal data, and shows great flexibility in evaluating mediating relationships between multiple time-varying measures [8]. For example, the parallel process LGM framework can be used to evaluate how growth in the mediator influences growth in the outcome [9]. This LGM framework assumes no strong temporal relationship between the mediator and outcome, only in the growth of the mediator and growth of the outcome. Autoregressive and latent difference scores have also been used for longitudinal mediation analyses with SEM given a temporal relationship between the mediator and outcome. For more information on the topic of SEM extensions for longitudinal data in the context of mediation, see MacKinnon [10].

2 An Overview of Structural Equation Modeling

SEM is an extremely powerful and flexible multivariate modeling framework which addresses two central issues of real-world importance: measurement error (i.e., latent variables) and causal networks such as mediation [1, 2]. Classic approaches to SEM model the relationship between the covariance and the parameters. Suppose we aim to estimate v unknown model parameters from a total of w observed and w^* unobserved (latent) variables. Then using the classic SEM we model the covariance matrix as $\boldsymbol{\Sigma} = \boldsymbol{\Sigma}(\boldsymbol{\theta})$ for a vector of unknown parameters $\boldsymbol{\theta}$ of dimension $v \times 1$ and the variance–covariance matrix of our observed variables $\boldsymbol{\Sigma}(\boldsymbol{\theta})$ of dimension $w \times w$.

In more modern approaches to SEM, we often have to look outside of the covariance between variables [11]. For example, with categorical variables the covariance between variables alone is not a sufficient statistic for determining the likelihood. In such cases we may require information from individual level data or information from the fourth-moment instead of the covariance matrix. Numerous traditional statistical techniques such as ANOVA, linear regression and factor analysis can all be expressed in the SEM framework. Many more specialized techniques such as modeling feedback loops, latent constructs, and path analytic models can also be handled with SEM.

More specific to the nomenclature of SEM, in the context of mediation, we can view SEM as a conceptual model, path diagram, and system of linked regression-style equations to capture the mediating relationships among a web of observed and unobserved variables. The conceptual model is a general idea of the mediating relationships under study. We explain the concepts of the path diagram and the general LISREL form of structural equations in Sects. 3 and 4.

Conducting SEM analyses in the context of mediation analysis may involve four steps (1) specifying the model (2) assessing model fit (3) making any model modifications (4) testing mediation hypotheses of interest. David A. Kenny defines model specification as the "translation of theory, previous research, design, and common sense into a structural model" [12]. In this process, a researcher indicates causal paths and directionality between variables (latent or observed) under study based on a hypothesized mediation model. In the context of mediation analysis, we may be interested in evaluating a specific mediation effect adjusted for other model components, which in SEM is all done in a single analysis. We discuss model fit in Sect. 5. Model modifications based on empirical criteria (such as comparing between competing models or examining modification indices) may be useful in more complex structural equation models with multiple independent variables, mediators, and outcomes. However, since we mostly focus on simpler mediation models in this chapter to showcase the advantages of SEM in this context, we refer an interested reader to other resources that discuss model modification in SEM such as Kline [1] or Gunzler and Morris [11].

3 Path Diagrams

A path diagram for a mediation model will consist of nodes representing the variables, and arrows showing relations among them. In a path diagram, latent variables (e.g., *depression* or *stress*) are distinguished from their observed counterparts in convention by using a circle or ellipse rather than the rectangular or square box used for the observed variables. Error terms are generally denoted by a letter or symbol (i.e., e or ε) not enclosed in a shape. Arrows are generally used to represent relationships among the variables. A single straight arrow indicates a causal relation from the base of the arrow to the head of the arrow. Two straight single-headed arrows in opposing directions connecting two variables may be used to indicate a feedback loop. A curved two-headed arrow indicates there may be some association between the two variables.

Path diagrams can be understood as implying certain conditional independence relations among variables. Such conditional independence relations can be extracted from the path diagram using the "d-separation" rule. D-separation is a criterion for determining, from a given diagram, whether a set X of variables is independent of another set Y, given a third set Z [4]. If the particular variables x and y are not d-separated by z (i.e., z does not block the causal path between them), then they are said to be d-connected by z. Note that d-connected is another way of describing mediation [13]. See Bollen [2] and Pearl [4] for a more complete explanation of these rules and for details about modeling complex relationships involving latent constructs using path diagrams and SEM.

As an example of a path diagram for a hypothesized mediation, Fig. 15.1 represents the path diagram for the causal path from *time since symptom onset*

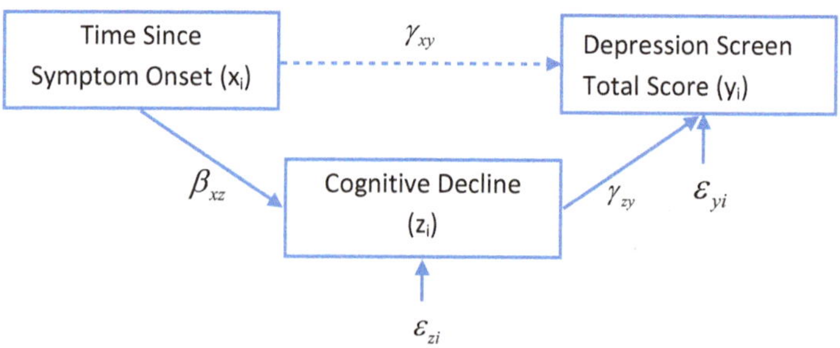

Fig. 15.1 Path diagram for the hypothesized mediation model for the causal path from *time since symptom onset* to *depression*

in multiple sclerosis (MS) patients to *depression*. Here, also indirectly, *symptom onset* effects *depression* through *cognitive decline*. Since this path between *symptom onset* and *depression* is not blocked by *cognitive decline*, these two measures are d-connected by *cognitive decline*. All variables in this path diagram are observed. However, a concept like *depression* can potentially be a latent variable constructed from multiple observed items that are indicators of *depression* instead of a sum total score as depicted on this diagram.

In SEM, there are two classes of variables: *exogenous* and *endogenous*. Endogenous variables act as an outcome in at least one of the structural equations, while exogenous variables are always independent variables. Thus, endogenous variables are those nodes with no arrows pointing into them, and, thus, their variation is not explained by other factors in the model. From Fig. 15.1, *depression* and *cognitive decline* are endogenous variables, while *symptom onset* is an exogenous variable.

4 The LISREL Formulation of SEM

LISREL is a popular way to express structural equation models in a general matrix form [2, 14]. In this approach, two sets of equations are formed: the structural equations and the measurement equations [2]. The *measurement* equations explicitly model measurement error and latent variables. The *structural* equations show potential causal links between *endogenous* and *exogenous* variables [1, 2].

The measurement equations can be expressed in the following form:

$$\begin{aligned} \mathbf{y} &= \boldsymbol{\mu}_{\mathbf{y}} + \boldsymbol{\Lambda}_y \boldsymbol{\eta} + \boldsymbol{\varepsilon} \\ \mathbf{x} &= \boldsymbol{\mu}_{\mathbf{x}} + \boldsymbol{\Lambda}_x \boldsymbol{\xi} + \boldsymbol{\delta} \end{aligned} \tag{15.1}$$

Here, $\boldsymbol{\eta}$ is a vector of m unobserved latent *endogenous* variables which are measured by the p observed variables $\mathbf{y}$. Similarly, $\boldsymbol{\xi}$ represents a vector of r unobserved latent *exogenous* variables which are measured by the q observed variables $\boldsymbol{x}$. The equations for $\mathbf{y}$ and $\mathbf{x}$ include vectors of intercepts, $\boldsymbol{\mu}_y$ and $\boldsymbol{\mu}_x$, matrices of slopes, $\boldsymbol{\Lambda}_y$ and $\boldsymbol{\Lambda}_x$, respectively, and vectors of corresponding random error terms, $\boldsymbol{\varepsilon}$ of dimension $p \times 1$ and $\boldsymbol{\delta}$ of dimension $q \times 1$, respectively. $\boldsymbol{\mu}_y$ is of dimension $p \times 1$ and $\boldsymbol{\mu}_x$ is of dimension $q \times 1$, and $\boldsymbol{\Lambda}_y$ is of dimension $p \times m$ and $\boldsymbol{\Lambda}_x$ is of dimension $q \times r$. $\boldsymbol{\Lambda}_y$ and $\boldsymbol{\Lambda}_x$ are often referred to as loading matrices.

The structural model, which relates the unobserved latent variables to each other, can be expressed in the following form:

$$\boldsymbol{\eta} = \boldsymbol{\mu}_\eta + \mathbf{B}\boldsymbol{\eta} + \boldsymbol{\Gamma}\boldsymbol{\xi} + \boldsymbol{\zeta} \tag{15.2}$$

$\boldsymbol{\mu}_\eta$ is an $m \times 1$ matrix of intercepts for the unobserved endogenous latent variables, $\mathbf{B}$ is an $m \times m$ matrix of slopes relating the unobserved endogenous latent variables to each other, $\boldsymbol{\Gamma}$ is $m \times n$ matrix of slopes for the unobserved exogenous latent variables, and $\boldsymbol{\zeta}$ is an $m \times 1$ vector of random error terms for the unobserved endogenous latent variables.

In the special case of no latent variables, such as in the mediation model corresponding to Fig. 15.1, there is no measurement model because all variables are measured without error (i.e., $\mathbf{y} = \boldsymbol{\eta}$ and $\mathbf{x} = \boldsymbol{\xi}$). Thus the form of the structural model in (15.2) can be simplified to observed variables only:

$$\mathbf{y} = \boldsymbol{\mu}_y + \mathbf{B}\mathbf{y} + \boldsymbol{\Gamma}\mathbf{x} + \boldsymbol{\zeta} \quad (15.3)$$

Here, $\boldsymbol{\mu}_y$ is a $p \times 1$ matrix of intercepts, $\mathbf{B}$ is a $p \times p$ matrix of slopes for the observed endogenous variables, $\boldsymbol{\Gamma}$ is $p \times q$ matrix of slopes for the observed exogenous variables, and $\boldsymbol{\zeta}$ is $p \times 1$ vector of random error terms for the observed endogenous variables.

Given a little algebra, under the assumption that $\mathbf{I} - \mathbf{B}$ is invertible, from (15.2) all endogenous variables can be moved onto the left side of the equation, while all exogenous variables remain on the right side:

$$\begin{aligned}
\boldsymbol{\eta} &= \boldsymbol{\mu}_\eta + \mathbf{B}\boldsymbol{\eta} + \boldsymbol{\Gamma}\boldsymbol{\xi} + \boldsymbol{\zeta} \\
(\mathbf{I} - \mathbf{B})\,\boldsymbol{\eta} &= \boldsymbol{\mu}_\eta + \boldsymbol{\Gamma}\boldsymbol{\xi} + \boldsymbol{\zeta} \\
\boldsymbol{\eta} &= (\mathbf{I} - \mathbf{B})^{-1}\boldsymbol{\mu}_\eta + (\mathbf{I} - \mathbf{B})^{-1}\boldsymbol{\Gamma}\boldsymbol{\xi} + (\mathbf{I} - \mathbf{B})^{-1}\boldsymbol{\zeta} \\
\boldsymbol{\eta} &= (\mathbf{I} - \mathbf{B})^{-1}\left(\boldsymbol{\mu}_\eta + \boldsymbol{\Gamma}\boldsymbol{\xi}\right) + \boldsymbol{\zeta}*
\end{aligned} \quad (15.4)$$

where $\mathbf{I}$ is the identity matrix of dimension $m \times m$ and $\boldsymbol{\zeta}* = (\mathbf{I} - \mathbf{B})^{-1}\boldsymbol{\zeta}$. As follows, (15.3) can be rewritten in a similar form to (15.4) given no latent variables:

$$\mathbf{y} = (\mathbf{I} - \mathbf{B})^{-1}\left(\boldsymbol{\mu}_y + \boldsymbol{\Gamma}\mathbf{x}\right) + \boldsymbol{\zeta}* \quad (15.5)$$

5 SEM for a Mediation Model

The SEM for the typical mediation process with a single independent variable, mediator, and outcome as depicted in Fig. 15.1 can be expressed by the following structural equations:

$$\begin{aligned}
y_i &= \gamma_0 + \gamma_{zy} z_i + \gamma_{xy} x_i + \varepsilon_{yi}, \\
z_i &= \beta_0 + \beta_{xz} x_i + \varepsilon_{zi}
\end{aligned} \quad (15.6)$$

Note that the two structural equations are linked together and inference about them is simultaneous, based on a joint distribution, unlike two, independent standard regression equations.

Here, we might assume, given that multivariate normality is an appropriate assumption

$$\begin{pmatrix} \varepsilon_{yi} \\ \varepsilon_{zi} \end{pmatrix} \sim N(0, \Psi), \quad \Psi = \begin{pmatrix} \sigma_y^2 & 0 \\ 0 & \sigma_z^2 \end{pmatrix}. \tag{15.7}$$

Note that we are assuming that $\text{Cov}\left(\varepsilon_{yi}, \varepsilon_{zi}\right) = 0$ in assuming no mediator–outcome confounding.

This mediation model is not linear (i.e., it is curvilinear) in terms of the parameters [15]. To see this, we can express these equations in the form of (15.3):

$$\begin{bmatrix} y_i \\ z_i \end{bmatrix} = \begin{bmatrix} \gamma_0 \\ \beta_0 \end{bmatrix} + \begin{bmatrix} 0 & \gamma_{zy} \\ 0 & 0 \end{bmatrix} \begin{bmatrix} y_i \\ z_i \end{bmatrix} + \begin{bmatrix} \gamma_{xy} \\ \beta_{xz} \end{bmatrix} x_i + \begin{bmatrix} \varepsilon_{yi} \\ \varepsilon_{zi} \end{bmatrix} \tag{15.8}$$

Likewise, we can express them in a form similar to (15.5):

$$\begin{aligned} \begin{bmatrix} y_i \\ z_i \end{bmatrix} &= \begin{bmatrix} 1 & -\gamma_{zy} \\ 0 & 1 \end{bmatrix}^{-1} \left(\begin{bmatrix} \gamma_0 \\ \beta_0 \end{bmatrix} + \begin{bmatrix} \gamma_{xy} \\ \beta_{xz} \end{bmatrix} x_i \right) + \begin{bmatrix} 1 & -\gamma_{zy} \\ 0 & 1 \end{bmatrix}^{-1} \begin{bmatrix} \varepsilon_{yi} \\ \varepsilon_{zi} \end{bmatrix} \\ &= \begin{bmatrix} \gamma_0 + \gamma_{zy}\beta_0 + \left(\gamma_{xy} + \gamma_{zy}\beta_{xz}\right) x_i + \varepsilon_{yi}^* \\ \beta_0 + \beta_{xz} x_i + \varepsilon_{zi}^* \end{bmatrix} \end{aligned} \tag{15.9}$$

The above SEM is clearly not linear in the parameters because of the terms $\gamma_{zy}\beta_0$ and $\gamma_{zy}\beta_{xz}$ in the first row of the matrix in (15.9).

The *direct effect* is the pathway from the exogenous variable to the outcome while controlling for the mediator. Therefore, in our path diagram in Fig. 15.1 γ_{xy} is the direct effect. The *indirect effect* describes the pathway from the exogenous variable to the outcome through the mediator. This path is represented through the product of β_{xz} and γ_{zy}. Finally, the *total effect* is the sum of the direct and indirect effects of the exogenous variable on the outcome, $\gamma_{xy} + \beta_{xz}\gamma_{zy}$.

The primary hypothesis of interest in a mediation analysis is to see whether the effect of the independent variable or intervention on the outcome can be mediated by a change in the mediating variable. In a *full mediation* process, the effect is 100 % mediated by the mediator that is, in the presence of the mediator, the pathway connecting the intervention to the outcome is completely broken so that the intervention has no direct effect on the outcome. In most applications, however, *partial mediation* is more common, in which case the mediator only mediates part of the effect of the intervention on the outcome, that is, the intervention has some residual direct effect even after the mediator is introduced into the model.

Inference (standard errors and p-values) for testing mediation effects in the SEM framework is easily performed using the Delta method (e.g., Sobel [16]; Clogg et al. [17]). Currently a popular approach to assessing mediation is to bootstrap confidence intervals (percentile, bias-corrected, and bias-corrected, and accelerated) for total and specific indirect effects [18].

The sample size or power for mediation analysis might be derived using simulation techniques under full mediation, where the direct effect is equal to zero, vs. a suitable alternative effect size to be considered for the direct effect.

Significant advances have been made over the past few decades in the theory and applications as well as software development for fitting SEM models that can be used in the context of mediation analysis. For example, in addition to specialized packages such as LISREL [14], MPlus [19], EQS [20], and Amos [21], procedures for fitting SEM are also available from general-purposes statistical packages such as R, SAS, STATA, and Statistica. These packages provide inference based on maximum likelihood, generalized least squares, and weighted least squares.

Typically, robust maximum likelihood approaches are used for SEM analysis. For example, in MPlus, the MLR approach uses ML to estimate the parameters, but uses a robust sandwich type estimator (Huber–White sandwich estimator) to calculate standard errors that are robust to model assumptions such as multivariate normality [22]. Bootstrapping is a similar but more computationally intensive approach to creating robust standard errors [23].

Both ML and MLR provide a method for dealing with missing data under the missing at random (MAR) assumption. For example, a slight modification of ML, full information ML (FIML) is one such approach to handle missing data under MAR assumption as implemented in MPLUS [19]. In this approach, all parameters and standard errors are derived from the joint distribution of the endogenous and exogenous variables, given assumptions such as multivariate normality and conditional independence. Under these assumptions, the marginal likelihood after integrating out the missing values can be maximized. Individual level data is needed for FIML.

6 Model Fit

Model fit indices are measures of the discrepancy between the model and data. In SEM analyses we evaluate a collective group of model fit measures which each represents different aspects of model fit. We provide a brief introduction here of some of the statistics and indices that will be useful for mediation analyses.

For starters, an asymptotically chi-squared distributed test statistic (or robust corrected statistic) provides a basis for assessing model fit, and in itself tests overall model fit. The null hypothesis is that there is no difference between the proposed model and the data structure, while the alternative hypothesis is that there is a difference between the proposed model and the data structure. *Thus, a large chi-squared test with a corresponding small p-value indicates that the model does* ***not*** *fit the data.* However, commonly, studies will reject the null as the chi-squared statistic is affected by nonnormality, correlation size, low power, and sample size (both too small or too large).

A commonly used index, Root Mean Square Error of Approximation (RMSEA) [24], is a point estimate that builds on this chi-squared statistic but is parsimony and sample size corrected. Confidence intervals can be constructed around the point estimate. A close fit hypothesis can be tested for the model using RMSEA. There are several limitations to the fit index, namely, RMSEA may not exactly follow an assumed non-central chi-square distribution, may be sensitive to nonnormality, and may favor larger models.

Another commonly reported fit index, the Comparative Fit Index (CFI) [25], is an incremental fit measure comparing the fit of the model to a baseline model (typically the model for the data of interest with no covariance) on a zero to one continuous scale. The closer the CFI index is to one, the better the model fit.

The Tucker–Lewis Index (TLI) [26] is another commonly reported incremental fit measure with a higher penalty for adding parameters than CFI, and without the zero to one range restriction. A commonly used absolute fit index, based on standardized difference between the observed correlation and the predicted correlation, is the Standardized Root Mean Square Residual (SRMR) [27].

Some general rule of thumb guidelines in SEM literature are that RMSEA ≤ 0.05 indicates an excellent fit while <0.08 is acceptable; CFI and TLI < 0.90 are acceptable and <0.95 are excellent fit. In addition, all three indices should reach acceptable (preferably excellent) levels before designating a model as good fitting. SRMR value ≤ 0.08 represent a good fit with the model.

7 Limitations of SEM for Mediation Analysis

The specified mediation model in an application of SEM must be plausible to obtain meaningful results. Causal assumptions, such as sequential ignorability or temporal order of variables, should be based on strong scientific theory and prior evidence. SEM often requires a large sample size, where the number of parameters vs. sample size is an important consideration.

Model identifiability is an issue that often arises when performing SEM analyses. Essentially, models are not identifiable if different values for the parameters (e.g., multicollinearity) can lead to the same distribution. Similarly, there may be multiple equivalent models that fit the data equally well. Again, there is no statistical inference, only scientific theory and prior evidence, that will allow a researcher to choose between these equivalent models. Further, there are some similar limitations with SEM as with traditional methods, as covariance and correlation matrices analyzed may be influenced by missing data and outliers.

8 Applications to a Multiple Sclerosis-Depression Study Using MPlus

Cleveland Clinic's Knowledge Program (KP) links patient-reported PHQ-9 data to its EPIC electronic health record (EHR) [28]. The Mellen Center [29] for Multiple Sclerosis manages more than 20,000 visits and 1000 new patients every year for MS treatment. The KP tracks illness severity and treatment efficacy over time across the Mellen Center population.

The PHQ-9 is a nine-item self-reported depression screening tool [30]. Patients specify frequency in the past 2 weeks (0 = not at all to 3 = every day) of nine symptoms, yielding a total score (range: 0–27). The KP collects eight Performance Scales and three assimilated scales (PS) [31–33] which are single item patient-reported disability measures. These include MS-related fatigue, cognitive, hand function and mobility domains with six ordinal responses.

We discuss here a retrospective study with observational data from 3507 MS patients from 2008 to 2011. All patients had a PHQ-9 score, and approximately 90 % of patients had no missing data on the measures assessed in this section. All missing data could be handled using FIML approach in MPlus.

Considering the context of the study and prior theory about the relationship between MS and depression, mediation analysis was used to evaluate the hypothesis that a longer *time since symptom onset* leads to increased *cognitive decline* which leads to higher levels of *depression* (see Fig. 15.1 for path diagram).

Mediation analysis with SEM was performed in MPlus using a maximum likelihood estimator while bootstrapping 95 % bias-corrected (BC) confidence intervals using the percentile method. Age, gender, and race were controlled for in the structural equations for each endogenous variable in the structural model. *Standardized estimates* were reported rather than raw estimates, so that estimates from different structural equations are on the same scale and are straightforward to assess in terms of magnitude (between -1 and 1).

In the mediation model, both estimated paths for the indirect effect were significant ($p < 0.001$) with $\widehat{\beta}_{xz} = 0.08$ and $\widehat{\gamma}_{zy} = 0.60$. The direct effect was also significant ($p = 0.001$) with $\widehat{\gamma}_{xy} = -0.06$. The estimate of the indirect (mediated) effect was thus $\widehat{\beta}_{xz}\widehat{\gamma}_{zy} = 0.08 \times 0.60 = 0.05$.

The 95 % bootstrapped BC confidence intervals around this point estimate of the indirect effect were (0.023, 0.075) which does not contain zero. Thus, we could conclude from this analyses that while a longer time since symptom onset lead to a decreased PHQ-9 total score (see direct effect), cognitive decline was a mechanism of change (partial mediator). In line with our prior hypothesis a longer time since symptom onset leads to increased cognitive decline which leads to a higher PHQ-9 total score. Since this model is just identified (fit the data exactly) tests of model fit cannot yield useful results.

We discuss an application of a more complex mediation model to showcase advantages of the SEM framework for mediation and to assess model fit. The path diagram in Fig. 15.2 corresponds to an extension of the previous mediation model. Instead of just using the PHQ-9 total score as an outcome, we form a latent variable for *depression* using the nine individual PHQ-9 items [11, 34]. This model includes additional mediators (hand function and mobility) along with another outcome (fatigue) describing MS symptoms. The relationships specified for our analyses were derived from a priori theory from MS specialists and prior studies [34–37]. However, there does not seem to be clear prior evidence for making causal assumptions about the relationship between cognitive decline and fatigue. Thus, we omit this relationship.

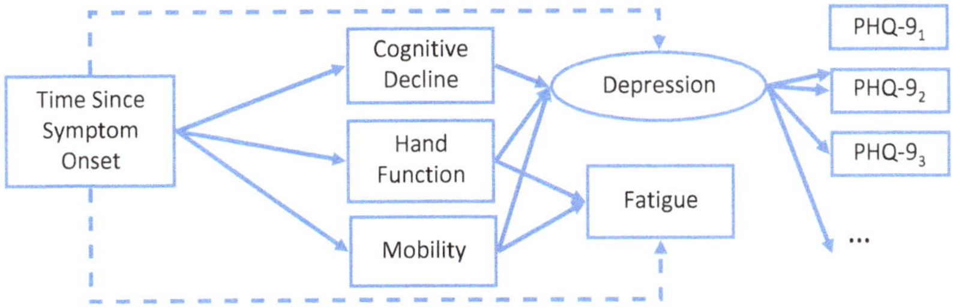

Fig. 15.2 Path diagram for the hypothesized multiple mediator multiple outcome model for the causal path from time since *symptom onset* to *depression* and *fatigue*. For visual ease we leave out of this diagram all error terms for the endogenous variables and correlations (all are significant $p < 0.001$ and of a positive magnitude) among the three mediators (cognitive decline, hand function, and mobility) and among the two outcomes (depression and fatigue)

Given all the additional causal paths, the model did not show a good model fit according to multiple SEM fit statistics and indices in comparison with our rule of thumb guidelines: $\chi^2(92) = 3722.673$, $p \leq 0.001$; RMSEA (90 % Confidence Interval) = 0.106 (0.103, 0.109); CFI = 0.845; TLI = 0.781; SRMR = 0.055. Therefore, potentially a researcher may want to modify the model (based on both clinical theory and empirical criteria) before reporting these findings. For more information on how to perform model modification using modification indices, see Kline [1] or Gunzler and Morris [11].

In Table 15.1 we show all the model derived specific and total indirect effects. We can assess these specific and total indirect effects while accounting for other model relationships. For example, given this more complex model, controlling for other relationships, and a latent construct for *depression*, the mediated effect is still significant, but of a lower magnitude, from *symptom onset* → *cognitive decline* → *depression* (see Table 15.3) compared to the simpler model corresponding to Fig. 15.1. The total indirect effect from *symptom onset* to *depression*, while adjusting for all other model relationships, is the sum of the three individual indirect effects from *symptom onset* to *depression* ($0.025 + 0.015 + 0.035 = 0.075$).

9 FRM-Based Distribution-Free SEM Approach for Mediation

In mediation analysis for a longitudinal study, missing data is commonly addressed under the assumption of MAR using a maximum likelihood-based estimator (i.e., ML or MLR). However, current methods are unable to handle such missing data if parametric assumptions are violated. For an example of this, we simulated a mediation model with missing responses over MAR at three repeated measures at different sample sizes ($n = 50$, 100, 2000) [38]. This model included central t-distributed random error terms with 3 degrees of freedom and no missing data at baseline ($t = 1$) and about 15 % (30 %) missing data at time $t = 2$ (3). Since the

Table 15.1 Assessment of potential mediation effects for multiple mediator multiple outcome MS-depression example using the bias-corrected bootstrapping method

Pathway	Estimate	95 % lower CI	95 % upper CI
Symptom onset → *Depression*			
Total effect	−0.013	−0.058	0.032
Direct effect	−0.088	−0.125	−0.051
Total indirect effect	0.075	0.048	0.102
Specific indirect effects via			
Cognitive decline	0.025	0.012	0.039
Hand function	0.015	0.003	0.028
Mobility	0.035	0.024	0.045
Symptom onset → *Fatigue*			
Total effect	0.056	0.014	0.098
Direct effect	−0.026	−0.063	0.010
Total indirect effect	0.082	0.058	0.107
Specific indirect effects via			
Hand function	0.018	0.003	0.033
Mobility	0.064	0.049	0.079

error terms are t-distributed, the joint normal distribution assumption is not met in the presence of missing data following MAR. For more technical details about the simulated model, see Gunzler et al. [38].

As shown in Fig. 15.3 ML-based methods will show bias in estimating the primary parameters of interest of the mediation model at all sample sizes (small to large) while the robust Functional Response Modeling (FRM)-based approach [15] will exhibit little bias that decreases as the sample size increases [38].

We now provide details about the FRM-based approach. Consider the mediation model in Eq. (15.6). We can replace our outcome, mediator, and independent variable (y_i, z_i, x_i) with appropriate time-varying versions (y_{i3}, z_{i2}, x_{i1}) given data collected at three repeated measures at $t = 1, 2, 3$ and temporality among the measures for assessing longitudinal mediation $(x_{i1} \rightarrow z_{i2} \rightarrow y_{i3})$.

$$\begin{aligned} y_{i3} &= \gamma_0 + \gamma_{zy} z_{i2} + \gamma_{xy} x_{i1} + \varepsilon_{yi3}, \\ z_{i2} &= \beta_0 + \beta_{xz} x_{i1} + \varepsilon_{zi2} \end{aligned} \qquad (15.10)$$

In performing robust inference, we can relax our assumption of multivariate normality and instead just assume that the distribution of the error terms is continuous:

$$\begin{pmatrix} \varepsilon_{yi3} \\ \varepsilon_{zi2} \end{pmatrix} \sim (0,\ \Psi), \quad \Psi = \begin{pmatrix} \sigma_y^2 & 0 \\ 0 & \sigma_z^2 \end{pmatrix}, \quad x_{i1} \perp \varepsilon_{zi2}, \quad x_{i1}, z_{i2} \perp \varepsilon_{yi3} \qquad (15.11)$$

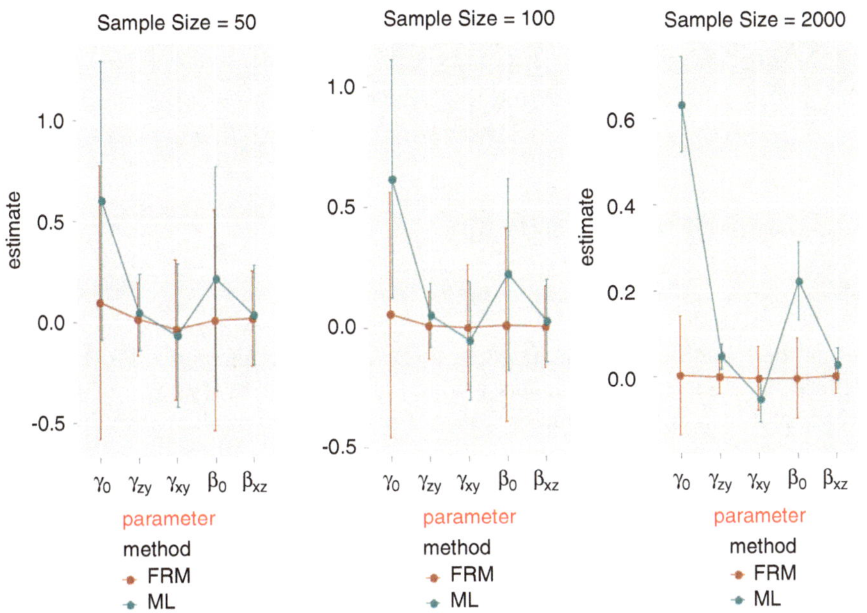

Fig. 15.3 Simulation results: mean estimates – population estimates (± standard errors) show the bias in ML while FRM performs well with missing data. Adapted from Gunzler D Lu N Tang W Wu P Tu XM A Class of Distribution-free Models for Longitudinal Mediation Analysis. *Psychometrika* 2014, 17(4), 543–568

In (15.11) we apply a stronger independence assumption for no correlation between the mediator and outcome (termed *pseudo-isolation*) than in (15.6). It is then readily checked that:

$$\mathrm{Cov}\left(\varepsilon_{yi3}, \varepsilon_{zi2}\right) = \mathrm{Cov}\left(\varepsilon_{yi3}, z_{i2}\right) = 0 \tag{15.12}$$

To apply FRM in our setting to the revised mediation model for (y_{i3}, z_{i2}, x_{i1}) to estimate a set of parameters $\boldsymbol{\theta} = \left(\gamma_0, \gamma_{zy}, \gamma_{xy}, \beta_0, \beta_{xz}, \sigma_z^2, \sigma_y^2\right)$, then let

$$\begin{aligned}
&\mathbf{f}_i = \left(\mathbf{f}_{1i}^T, \mathbf{f}_{2i}^T\right)^T, \ \mathbf{h}_i(\boldsymbol{\theta}) = \mathbf{h}(\mathbf{x}_i, \boldsymbol{\theta}) = \left(\mathbf{h}_{1i}^T(\boldsymbol{\theta}), \mathbf{h}_{2i}^T(\boldsymbol{\theta})\right)^T, \\
&i = 1, 2, \ldots, n, \\
&\mathbf{f}_{1i} = \left(y_{i3}, z_{i2}\right)^T, \ \mathbf{f}_{2i} = \left(y_{i3}^2, y_{i3}z_{i2}, z_{i2}^2\right)^T, \ \mathbf{x}_i = x_{i1}, \\
&\mathbf{h}_{1i}(\boldsymbol{\theta}) = \left(\left(\gamma_0 + \gamma_{zy}\beta_0\right) + \left(\gamma_{xy} + \gamma_{zy}\beta_{xz}\right)x_{i1}, \ \beta_0 + \beta_{xz}x_{i1}\right)^T, \\
&\mathbf{h}_{2i}(\boldsymbol{\theta}) = E\left(\mathbf{f}_{2i} \mid \mathbf{x}_i\right) = \left(E\left(y_{i3}^2 \mid \mathbf{x}_i\right), E\left(y_{i3}z_{i2} \mid \mathbf{x}_i\right), E\left(z_{i2} \mid \mathbf{x}_i\right)\right)^T,
\end{aligned} \tag{15.13}$$

where

$$
\begin{aligned}
E\left(z_{i2}^2 \mid \mathbf{x}_i\right) &= \sigma_{\varepsilon z}^2 + \left(\beta_0 + \beta_{xz} x_{i1}\right)^2, \\
E\left(y_{i3} z_{i2} \mid \mathbf{x}_i\right) &= \gamma_{zy}\left(\beta_0 + \beta_{xz} x_{i1}\right)\left(\beta_0 + \beta_{xz} x_{i1}\right) + \gamma_{zy}\sigma_{\varepsilon z}^2 \\
&\quad + \left(\beta_0 + \beta_{xz} x_{i1}\right)\left[\left(\gamma_0 + \gamma_{zy}\beta_0\right) + \left(\gamma_{xy} + \gamma_{zy}\beta_{xz}\right) x_{i1}\right], \\
E\left(y_{i3}^2 \mid \mathbf{x}_i\right) &= \gamma_{zy}^2\sigma_{\varepsilon z}^2 + \sigma_{\varepsilon y}^2 + \left[\left(\gamma_0 + \gamma_{zy}\beta_0\right) + \left(\gamma_{xy} + \gamma_{zy}\beta_{xz}\right) x_{i1}\right]^2.
\end{aligned} \tag{15.14}
$$

Then, the FRM for the SEM in (15.10) is

$$
E\left(\mathbf{f}_i \mid x_i\right) = \mathbf{h}_i\left(\boldsymbol{\theta}\right), \ i = 1, 2, \ldots, n. \tag{15.15}
$$

Given the pseudo-isolation assumption as in (15.11), an alternative FRM can be defined to estimate the parameters of primary interest $\boldsymbol{\theta} = \left(\gamma_0, \gamma_{zy}, \gamma_{xy}, \beta_0, \beta_{xz}\right)$ without the help of higher-order moments. For details on this alternative FRM, see Gunzler et al. [38].

Let

$$
S_i = \mathbf{f}_i - \mathbf{h}_i\left(\boldsymbol{\theta}\right), \ \ Di = \frac{\partial}{\partial \boldsymbol{\theta}} \mathbf{h}_i\left(\boldsymbol{\theta}\right) \tag{15.16}
$$

The following estimating equations are well defined and readily evaluated in closed form:

$$
\mathbf{w}_n\left(\boldsymbol{\theta}\right) = \frac{1}{n}\sum_{i=1}^{n} \mathbf{w}_{ni} = \frac{1}{n}\sum_{i=1}^{n} D_i V_i^{-1} S_i = \mathbf{0} \tag{15.17}
$$

V_i is the working variance matrix. A necessary condition is to select V_i to ensure that $E\left(\mathbf{w}_n\right) = \mathbf{0}$. A sufficient condition is that $E\left(V_i^{-1} S_i \middle| \mathbf{x}_i\right) = V_i^{-1} E\left(S_i \middle| \mathbf{x}_i\right)$. One trivial solution for V_i is the identity matrix. The estimating equations in (15.17) can be solved using, for example, the Newton Raphson algorithm.

Under mild regularity conditions, regardless of data distributions:

$$
\begin{aligned}
&\sqrt{n}\left(\widehat{\boldsymbol{\theta}} - \boldsymbol{\theta}\right) \xrightarrow{d} N\left(0, \Sigma_{\boldsymbol{\theta}}\right), \\
&\Sigma_{\boldsymbol{\theta}} = B^{-1} E\left(D_i V_i^{-1} S_i S_i^T V_i^{-1} D_i^T\right) B^{-T}, \\
&B = E\left(D_i^T V_i^{-1} D_i\right)
\end{aligned} \tag{15.18}
$$

Both Wald and Score Tests have been developed to test the true value of parameters of a mediation model based on the sample estimates using the FRM-based approach [38].

While these estimating equations provide valid inference under complete data and the MCAR assumption, weighted estimating equations are necessary for valid inference when the missing data follows the MAR assumption. Using Inverse Probability Weighting (IPW) we can develop a set of weighted estimating equations for inference about θ. We provide a sketch here. Assume no missing data at baseline ($t = 1$) and monotone missing data for $t = 2$ and 3. Then, let

$$r_{it} = \begin{cases} 1 & \text{if } z_{it} \text{ and } y_{it} \text{ are observed} \\ 0 & \text{if } z_{it} \text{ and } y_{it} \text{ are missing} \end{cases}, \quad \mathbf{r_i} = (r_{i1},\ r_{i2},\ r_{i3})^T \tag{15.19}$$

$$\pi_{it} = \Pr\left(r_{it} = 1 \middle| \mathbf{x}_i, \mathbf{z}_i, \mathbf{y}_i\right), \quad \Delta_{it} = \frac{r_{it}}{\pi_{it}}, \quad 2 \le t \le 3. \tag{15.20}$$

Now let

$$\Delta_i = \begin{pmatrix} \frac{r_{i3}}{\pi_{i3}} & 0 & 0 & 0 & 0 \\ 0 & \frac{r_{i2}}{\pi_{i2}} & 0 & 0 & 0 \\ 0 & 0 & \frac{r_{i3}}{\pi_{i3}} & 0 & 0 \\ 0 & 0 & 0 & \frac{r_{i3}}{\pi_{i3}} & 0 \\ 0 & 0 & 0 & 0 & \frac{r_{i2}}{\pi_{i2}} \end{pmatrix} \tag{15.21}$$

Then, the weighted estimating equations are

$$\mathbf{w}_n(\boldsymbol{\theta}) = \frac{1}{n}\sum_{i=1}^{n} \mathbf{w}_{ni} = \frac{1}{n}\sum_{i=1}^{n} D_i V_i^{-1} \Delta_i S_i = \mathbf{0} \tag{15.22}$$

For details about solving these weighted estimating equations and the asymptotic properties, see Gunzler et al. [38].

The distribution-free FRM-based approach is straightforward to extend to non-continuous mediators and outcomes (i.e., count, categorical). For example, if y_{it} is a binary outcome, the revised model

$$\begin{aligned} &z_{i2} = \beta_0 + \beta_{xz} x_{i1} + \varepsilon_{zi}, \quad y_{i3} \middle| x_{i1}, z_{i2} \sim \text{Binomial}\,(\mu_i, 1)\,, \\ &\mu_i = E\left(y_{i3} \middle| x_{i1}, z_{i2}\right), \quad \text{logit}\,(\mu_i) = \gamma_0 + \gamma_{xy} x_{i1} + \gamma_{zy} z_{i2}, \\ &\varepsilon_{zi} \sim N\left(0, \sigma_z^2\right), \quad x_{i1} \perp \varepsilon_{zi} \end{aligned} \tag{15.23}$$

Binomial(μ_i, 1) denotes a Binomial distribution with mean μ_i and size 1, i.e., a Bernoulli with mean μ_i. We can now use the same definitions and formulas (15.13) through (15.22) to apply the FRM-based approach for inference for this binary outcome model.

9.1 Illustration of FRM-Based Approach with Child Resilience Example

To illustrate the approach to real study data, we applied the FRM to a longitudinal study known as the Child Resilience Project [39]. Data was collected for this study from 2006 to 2011. This analysis included 401 students from first up to third grade in five Rochester City School District elementary schools. The study examines how children with a higher risk of developing behavioral problems with a mentor socially improve compared to the control and lower risk children over periods of 6 and 18 months.

We examined what role a potential mediator, self-reported verbal, declarative knowledge of the skills the child is learning in the Resilience Project at 6 months, plays in a cause and effect relationship between the treatment at baseline and the child's self-initiated demonstration of skills at 18 months (Fig. 15.4). Thus we have longitudinal data with three assessment times, baseline, 6 months, and 18 months and temporally the mediator is hypothesized to occur before the outcome.

The treatment is a binary indicator as children either had a mentor or no mentor. In the hypothesis of interest, the treatment would be expected to predict a higher demonstration of skills, which would indicate that the children receiving a mentor improved their social skills over time. The distributions of both the mediator and outcome were skewed as shown in Fig. 15.5.

We had full information on whether each child received the treatment at baseline. However, there were a high percentage of missing observations for both the mediator (37 %) and outcome (59 %). We modeled this missing data using logistic regression:

$$\begin{aligned} &\operatorname{logit}(p_{i2}) = \eta_{02} + \eta_{x1}x_{i1}, \ \operatorname{logit}(p_{i3}) = \eta_{03} + \eta_{z2}z_{i2}, \\ &p_{it} = \Pr\left(r_{it} = 1 \mid r_{i(t-1)} = 1\right). \end{aligned} \quad (15.24)$$

This is a simplified special case of a missing data model for applying IPW in which we are building our missing data models with only observed data at the previous time point (without using any other information). We estimated the parameters in R program using the *glm* function. Since we modeled our missing data at $t = 2$ based on the treatment information at baseline, we used all 401 observations.

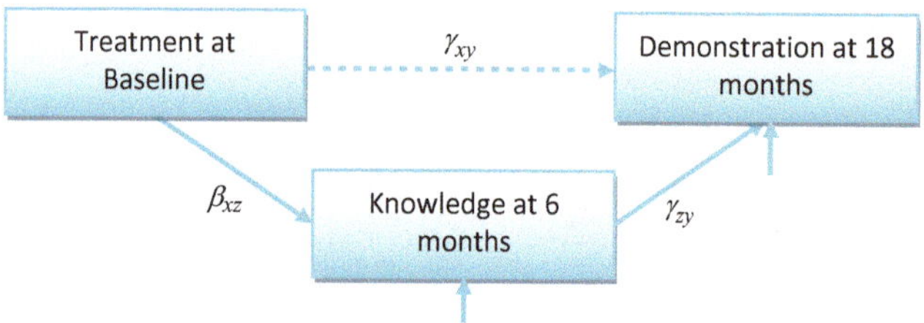

Fig. 15.4 Path diagram for the mediation model for the Child Resilience Study with MAR Data

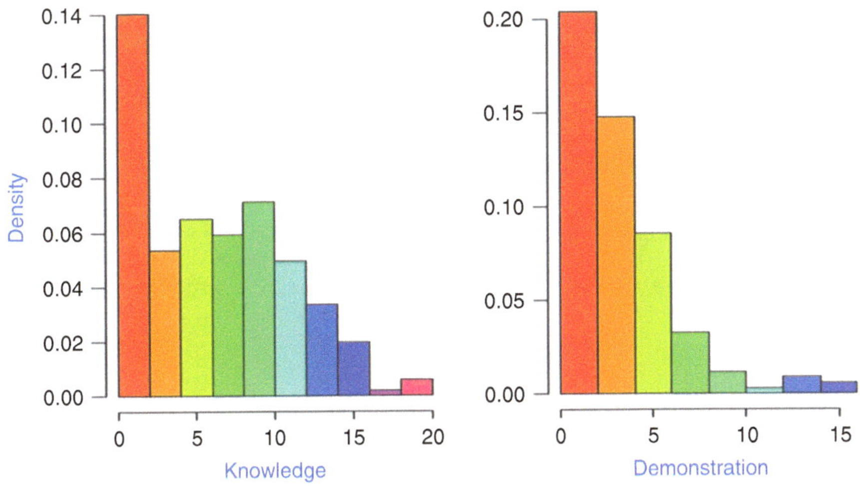

Fig. 15.5 Histograms of verbal, declarative knowledge of skills and demonstration of skills for the Child Resilience Study

Table 15.2 Parameter estimates, standard errors, and p-values for the missing data model for the Child Resilience Study

Estimates, standard errors, and p-value Child Resilience Example under missing data			
η	Estimate	Standard error asymptotic	p-value
Sample size = 401			
η_{02}	0.546	0.147	<0.001
η_{x1}	−0.019	0.207	0.926
η_{03}	0.250	0.201	0.214
η_{z2}	0.067	0.029	0.022

Shown in Table 15.2 are the estimates for the missing data model in (15.24). The p-value for η_{z2} was significant, indicating a MAR mechanism for the missing data at time 3. Since the p-value for η_{x1} was not significant, missing data at time 2 was MCAR and we would expect no bias for the estimates of time 2 parameters $\boldsymbol{\beta} = (\beta_0, \beta_{xz})^T$ in ML. However, we expect to see a bias for the estimates of the time 3 parameters $\boldsymbol{\gamma} = (\gamma_0, \gamma_{xy}, \gamma_{zy})^T$ in ML. In the hypothesis of interest, the treatment would be expected to predict a higher demonstration of skills at 18 months, which would indicate that the children receiving a mentor improved their social skills over time.

Shown in Table 15.3 are the estimates of the main parameters of $\boldsymbol{\theta} = (\gamma_0, \gamma_{zy}, \gamma_{xy}, \beta_0, \beta_{xz})$ and associated standard errors and type I errors for this mediation model obtained from the alternative FRM and ML. From the table, we see that the estimates for FRM and ML were practically the same for the $\boldsymbol{\beta}$ parameters, but different for $\boldsymbol{\gamma}$ parameters. In the $\boldsymbol{\beta}$ parameter estimates, FRM had a smaller standard error than ML. We saw from the simulation for longitudinal missing data in Fig. 15.3 that ML would produce a value of γ_{xy} biased less in magnitude than

Table 15.3 Parameter estimates, standard errors, and type I error rates for the mediation model for the Child Resilience Study with missing data

Estimates, standard errors, and type I errors Child Resilience Study example under missing data (37 %/59 %)				
	Estimate method		Standard error method	
θ	FRM	ML	FRM	ML
Sample size = 401				
γ_0	1.812	1.810	0.278	0.352
γ_{zy}	−0.042	−0.039	0.053	0.050
γ_{xy}	2.330	2.283	0.503	0.480
β_0	3.429	3.429	0.370	0.374
β_{xz}	4.390	4.390	0.528	0.529
	Type I α for H_0: $\gamma_{xy} = 0$		Wald < 0.001	<0.001
			Score < 0.001	

the true estimate. This appeared true again as the FRM estimate was higher in magnitude, confirming that the treatment predicted a higher demonstration of skills at 18 months. The parameter γ_{zy} was not significant for either FRM or ML in this model ($p > 0.421$ for Wald Test in both FRM and ML), implying a non-significant indirect effect in this mediation analysis.

10 Chapter Conclusion

Structural equation modeling provides a very general, powerful framework for performing causal mediation analysis. By taking advantage of the functional response models (FRM), we have developed a robust approach to systematically address the limitations of SEM as it applies to mediation analysis. This class of FRM-based SEM requires no parametric models for the data distribution and provides valid inference for longitudinal mediation hypotheses under the two most popular missing data mechanisms, missing completely at random (MCAR) and missing at random (MAR). The approach can be extended for noncontinuous mediators and outcomes.

References

1. Kline, R.B.: Principles and Practice of Structural Equation Modeling. Guilford Press (2011)
2. Bollen, K.: Structural Equations with Latent Variables. Wiley, New York (1989)
3. Gunzler, D., et al.: Introduction to mediation analysis with structural equation modeling. Shanghai Arch. Psychiatry **25**(6), 390–394 (2013)
4. Pearl, J.: Causality: Models, Reasoning and Inference, 2nd edn. Cambridge Univ Press (2009)

5. Muthén, B.: Applications of causally defined direct and indirect effects in mediation analysis using SEM in Mplus. Download at www.statmodel.com/download/causalmediation.pdf (2011)
6. Imai, K., Keele, L., Tingley, D.: A general approach to causal mediation analysis. Psychol. Methods **15**(4), 309 (2010)
7. Muthén, B.O.: Beyond SEM: general latent variable modeling. Behaviormetrika **29**(1; ISSU 51), 81–118 (2002)
8. Preacher, K.J.: Latent Growth Curve Modeling. Sage (2008)
9. Cheong, J., MacKinnon, D.P., Khoo, S.T.: Investigation of mediational processes using parallel process latent growth curve modeling. Struct. Equ. Model. **10**(2), 238–262 (2003)
10. MacKinnon, D.P.: Introduction to Statistical Mediation Analysis. Routledge (2008)
11. Gunzler, D.D., Morris, N.: A tutorial on structural equation modeling for analysis of overlapping symptoms in co-occurring conditions using MPlus. Stat. Med. **34**(24), 3246–3280 (2015)
12. Kenny, D.A.: Terminology and Basics of SEM. (2011). Available from: http://davidakenny.net/cm/basics.htm
13. Baron, R.M., Kenny, D.A.: The moderator–mediator variable distinction in social psychological research: conceptual, strategic, and statistical considerations. J. Pers. Soc. Psychol. **51**(6), 1173 (1986)
14. Joreskog, K., Sorbom, D.: LISREL 8 User's Reference Guide. Scientific Software Chicago (1996)
15. Kowalski, J., Tu, X.M.: Modern Applied U-Statistics, vol. 714. Wiley (2008)
16. Sobel, M.E.: Asymptotic confidence intervals for indirect effects in structural equation models. Sociol. Methodol. **13**(1982), 290–312 (1982)
17. Clogg, C.C., Petkova, E., Shihadeh, E.S.: Statistical methods for analyzing collapsibility in regression models. J. Educ. Behav. Stat. **17**(1), 51–74 (1992)
18. Preacher, K.J., Hayes, A.F.: Asymptotic and resampling strategies for assessing and comparing indirect effects in multiple mediator models. Behav. Res. Methods **40**(3), 879–891 (2008)
19. Muthén, L.K., Muthén, B.O.: *Mplus*. The Comprehensive Modelling Program for Applied Researchers: User's Guide, vol. 5 (2012)
20. Bentler, P.M.: EQS Structural Equations Program Manual, p. 254. BMDP Statistical Software (1989)
21. Arbuckle, J.: Amos 6.0 User's Guide. Marketing Department, SPSS Incorporated (2005)
22. Huber, P.J.: The behavior of maximum likelihood estimates under nonstandard conditions. In: Proceedings of the Fifth Berkeley Symposium on Mathematical Statistics and Probability (1967)
23. Nevitt, J., Hancock, G.R.: Performance of bootstrapping approaches to model test statistics and parameter standard error estimation in structural equation modeling. Struct. Equ. Model. **8**(3), 353–377 (2001)
24. Browne, M.W., et al.: Alternative ways of assessing model fit. Sage Focus Editions **154**, 136 (1993)
25. Bentler, P.M.: Comparative fit indexes in structural models. Psychol. Bull. **107**(2), 238 (1990)
26. Tucker, L.R., Lewis, C.: A reliability coefficient for maximum likelihood factor analysis. Psychometrika **38**(1), 1–10 (1973)
27. Hu, L.-t., Bentler, P.M.: Fit indices in covariance structure modeling: sensitivity to underparameterized model misspecification. Psychol. Methods **3**(4), 424 (1998)
28. Katzan, I., et al.: The Knowledge Program: an innovative, comprehensive electronic data capture system and warehouse. In: AMIA Annual Symposium Proceedings, pp. 683–692 (2011)
29. Mellen Center for Multiple Sclerosis Treatment and Research, C.C., Neurological Institute (2013). Available from: Retrieved from http://my.clevelandclinic.org/neurological_institute/mellen-center-multiple-sclerosis/default.aspx
30. Blacker, D.: Psychiatric rating scales. In: Sadock, B.J., Sadock, V.A. (eds.) Kaplan and Sadock's Comprehensive Textbook of Psychiatry, 8th edn, pp. 929–955. Lippincott Williams & Wilkins, Philadelphia (2005)

31. Schwartz, C.E., Vollmer, T., Lee, H.: Reliability and validity of two self-report measures of impairment and disability for MS. Neurology **52**(1), 63–70 (1999)
32. Chamot, E., Kister, I., Cutter, G.R.: Item response theory-based measure of global disability in multiple sclerosis derived from the Performance Scales and related items. BMC Neurol. **14**(1), 192 (2014)
33. Marrie, R.A., Goldman, M.: Validity of performance scales for disability assessment in multiple sclerosis. Mult. Scler. **13**(9), 1176–1182 (2007)
34. Gunzler, D., et al.: Disentangling multiple sclerosis & depression: an adjusted depression screening score for patient-centered care. J. Behav. Med. **38**(2), 237–250 (2015)
35. Beal, C.C., Stuifbergen, A.K., Brown, A.: Depression in multiple sclerosis: a longitudinal analysis. Arch. Psychiatr. Nurs. **21**(4), 181–191 (2007)
36. Brown, R., et al.: Longitudinal assessment of anxiety, depression, and fatigue in people with multiple sclerosis. Psychol. Psychother. Theory Res. Pract. **82**(1), 41–56 (2009)
37. Krupp, L.B.: Fatigue in Multiple Sclerosis: A Guide to Diagnosis and Management. Demos Medical Publishing (2004)
38. Gunzler, D., et al.: A class of distribution-free models for longitudinal mediation analysis. Psychometrika **79**(4), 543–568 (2014)
39. Wyman, P.A., et al.: Intervention to strengthen emotional self-regulation in children with emerging mental health problems: proximal impact on school behavior. J. Abnorm. Child Psychol. **38**(5), 707–720 (2010)

Index

H. He et al. (eds.), *Statistical Causal Inferences and Their Applications in Public Health Research*, ICSA Book Series in Statistics, DOI 10.1007/978-3-319-41259-7

D

E

T

W

Kolibri 360 GmbH
Ferdinand-Jühlke-Straße 7
99095 Erfurt, Deutschland
produktsicherheit@kolibri360.de